Discovering Life Science

By Keith Wilhelmi

Based on Globe Life Science by Mary C. Hicks and Bryan Bunch

Special thanks to:

- **Bryan Bunch**, for allowing me to extend the life of his unique book (which I used for many years), and for being a patient and supportive mentor.

- **Mary C. Hicks**, for her efforts to create a textbook that would inspire young people to believe in themselves - and to love biology.

- **Gayle Wilhelmi**, always my first sounding board, for actively supporting me in committing the first years of my retirement to this endeavor.

- the authors of the To Do Yourself activities in the Globe Life Science textbook:
  - Thomas Covotsos   - Burton Goldfeld   - Jack Rothman   - Jean M. Squires

Keith Wilhelmi taught life science to middle school students in Kentucky for 42.5 years, retiring in May, 2020. Details may be found at *lifesciencetextbook.com*

Copyright 2022

Self published by Keith Wilhelmi, Louisville, KY. 40220. All rights reserved. No part of this book may be reproduced or transmitted in any form or by any means, electrical or mechanical, including photocopying, recording, or by any information storage and retrieval system, without permission in writing from Keith Wilhelmi.

ISBN: 979-8-9859572-1-1

Table of Contents

Unit 1 -   LIVING THINGS IN THEIR ENVIRONMENT

Lesson			
	1	What Are Living Things?	1
	2	How Do Scientists Study Living Things?	4
	3	Ready for More Scientific Thinking?	9
	4	What is an Ecosystem?	12
	5	How Are Materials Cycled in Nature?	14
	6	What Is a Food Chain?	16
	7	What Is a Food Web?	18
	8	What Are Energy Pyramids?	20
	9	What Is Succession?	22
	10	What Is a Biome?	24
	11	What Are Habitats and Niches?	28
	12	Can an Ecosystem Get Out-Of-Balance?	30
		Review What You Know	33
		Careers in Life Science	37

Unit 2 -   HOW LIVING THINGS ARE ORGANIZED

Lesson			
	1	What Units Make Up Living Things?	39
	2	What Are the Jobs of a Cell's Parts?	42
	3	How Do Cells Work Together?	44
	4	How Are Living Things Grouped?	46
	5	What Are the Simplest Living Things?	49
	6	What Are the Main Groups of Plants?	52
	7	What Are the Main Groups of Animals?	54
		Review What You Know	60
		Careers in Life Science	63

Unit 3 -   FOOD FOR LIFE

Lesson			
	1	What Chemicals Make Up Living Things?	65
	2	How Do Green Cells Make Food?	68
	3	How Do Living Things Get Energy?	71
	4	What Are the Parts of Plants?	74
	5	What Foods Give Us Energy?	78
	6	Why Do We Need Vitamins and Minerals?	82
	7	What Are Calories?	87
	8	How Can You Balance Your Diet?	90
		Review What You Know	93
		Summing Up	95

Unit 4 - DIGESTION AND TRANSPORT

Lesson			
	1	What is the Job of Your Digestive System?	97
	2	How Does Digestion Begin?	100
	3	How Does Digestion End?	102
	4	What Happens After Food Is Digested?	104
	5	What Is the Work of the Blood?	106
	6	What Is the Job of Your Transport System?	108
	7	What is the Path of Your Blood?	112
		Review What You Know	116
		Careers in Life Science	119

Unit 5 - BREATHING AND MOVEMENT

Lesson			
	1	What Is the Job of Your Respiratory System?	121
	2	What Happens When You Breathe?	124
	3	How Does Your Body Get Rid of Wastes?	126
	4	What Is Your Body's Framework?	130
	5	How Do Your Muscles Work?	134
		Review What You Know	138
		Summing Up	141

Unit 6 - BEHAVIOR AND CONTROL

Lesson			
	1	Why Do Living Things Act as They Do?	143
	2	How Does Your Body Receive Messages?	146
	3	What Is the Job of Your Nervous System?	149
	4	What Are the Jobs of Your Brain?	153
	5	How Do You Learn New Behavior?	155
	6	What Are Chemical Messengers?	159
		Review What You Know	164
		Careers in Life Science	167

Unit 7 - REPRODUCTION IN SIMPLE ORGANISMS AND PLANTS

Lesson			
	1	How Do Living Things Reproduce?	169
	2	How Do Molds Reproduce?	172
	3	How Do Animals Grow Back Lost Parts?	174
	4	How Do Plants Reproduce Asexually?	176
	5	What Is the Work of a Flower?	180
	6	What Are Seeds and Fruits?	184
		Review What You Know	188
		Summing Up	191

Unit 8 - REPRODUCTION IN HIGHER ANIMALS

Lesson			
	1	How Do Animals Reproduce Sexually?	193
	2	How Do Fish Reproduce?	196
	3	How Do Amphibians and Reptiles Reproduce?	200
	4	How Do Birds Reproduce?	204
	5	How Do Mammals Reproduce?	207
	6	How Do Insects Reproduce?	210
		Review What You Know	214
		Careers in Life Science	217

Unit 9 - HEREDITY AND CHANGE THROUGH TIME

Lesson			
	1	How Do Parents Pass Traits to Offspring?	219
	2	What Happens When Cells Divide?	223
	3	How Does DNA Make Proteins?	228
	4	What Are Dominant and Recessive Genes?	234
	5	Can Heredity Be Predicted?	238
	6	How Is Sex Inherited?	244
	7	What Are Genetic Disorders?	249
	8	What Is Evolution by Natural Selection?	253
	9	Are Fossils Evidence of Evolution?	256
	10	What Other Evidence is There for Evolution?	259
		Review What You Know	262
		Summing Up	265

Unit 10 - PROTECTING HEALTH

Lesson			
	1	What Are Infectious Diseases?	267
	2	How Does the Body Fight Infection?	272
	3	What Is Heart Disease?	277
	4	What Is Cancer?	281
	5	What Is Drug Abuse?	285
		Review What You Know	288
		Summing Up	290
		Careers in Life Science	292
		Glossary	293
		Credits	302
		Index	303

--- To The Student ----

Welcome to **Discovering Life Science**!

What is life? What type of living things are there? What keeps life going? How do our bodies work? Are you ready to discover some of the answers?

First, notice that each lesson of <u>Discovering Life Science</u> is short. Most lessons can be read in much less than a class period, or as one homework assignment.

Pause now. Take a few seconds to flip through the pages of the first lesson. All lessons follow the same pattern.

- The **title** is always in the form of a question.

- Next, there is a short story called ***Exploring Science / Historical Steps.***
    It helps you see why this lesson's topic is important.

- Now, view the main part of the lesson, with the key terms in bold print.

- Most lessons also have a brief, hands-on activity called ***To Do Yourself***.

- Once you've read a lesson thoughtfully, the **review** questions remind you of the main points. These questions will also help the information "stick" in your mind.

  Not sure of an answer? Give your brain a second chance!
    Take a minute to find the answer in the reading.
    Each time that you do so, you actually save yourself study time!

As you saw in the Table of Contents, there are ten units.
    Each unit is followed by review questions.

Every other unit ends with a **careers** page.
    Some of you will discover your life's career this year.
        All of you will better understand life itself!

Before you begin, here are three tips:

   1) Read at *your* pace. The goal is to learn, not just "finish."

   2) Let the illustrations help you.

   3) Break any unfamiliar word apart. Check what the parts mean.
       The word will quickly seem easy to understand - and to remember.
           For example - for the word biology ......
                                    bio = "life"
                                    ology = "study of"
       So, <u>biology</u> is -- the study of life!

Enjoy **<u>Discovering Life Science</u>**!

UNIT ONE

LIVING THINGS IN THEIR ENVIRONMENT

What are Living Things?

U-1 L-1

Exploring Science

Are Robots Alive? The meaning of the word robot has changed over time. For years a robot was expected to look like a human. Today, looks don't matter. The word *robot* now means a machine that does work for us - with the help of a computer. Some experts argue that true robots are also able to make decisions.

What about a drone? Is it a robot? One type of drone is able to sense objects and avoid them. This drone is certainly a robot.

In fact, in many ways this drone seems to be alive! Compare it to us. We have a brain; its computer functions as a brain. We take in food for energy; it takes in electrical energy. We can see nearby objects; it can "see" - and even share its views with people. We can move; it can move. In fact, it can fly!

Drones have been designed to deliver packages, check on farm crops, and locate forest fires or lost hikers.

You have likely seen other robots, like those that cut grass or vacuum carpets. These can "feel" and avoid things in their path.

Other robots are able to "smell" a small indoor fire and put it out. And there are robots that can "hear" and obey certain commands.

What do you think? Are robots alive?

➤ In some schools students learn to build robots? Can you guess where this happens in a school?

A drone on the lookout!

What Living Things Do

Look around you. What living things do you see? Anything that is alive is an **organism** (OR-guh-niz-um). Any animal, plant, or other living thing is an organism. You, too, are an organism. Let's look at the things that organisms are able to do. Scientists call these **life functions**.

▸ <u>MOVE</u>. Living things can **move**, even if they stay in one place. The hydra (HY-druh) is an animal that usually stays in place by attaching its base to a twig or stone. It waves its arms to catch food. Some plants can move from place to place.

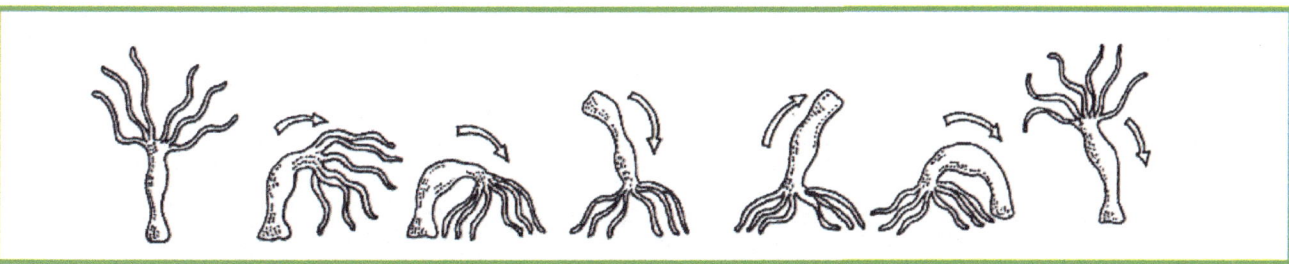

A hydra usually stays in one place, in the position shown at the left. If it needs to move, it somersaults.

1

Mimosa leaves are open during the day. They close at night, or when they are touched.

▸ GROW. Living things become larger in size, or **grow**. At birth, a red kangaroo is less than two centimeters long and weighs one gram. When fully grown, it may be 210 centimeters long and weigh 90 kilograms.

▸ RESPOND. Living things react, or **respond**, to changes in the environment. An organism's **environment** (en-VY-run-munt) is everything that is around it. Touch the leaves of a mimosa tree, and they respond by closing up. Living things also respond to changes from within themselves. What do you do when you feel hungry?

➤ To Do Yourself   How Do Brine Shrimp React to Changes in Their Environment?

You will need:

Brine shrimp eggs, large wide-mouth jar, aged tap water, 6 teaspoons of table salt, hand lens, flashlight, timer (stopwatch)

1. In the jar, mix 6 teaspoons of salt in a liter of dechlorinated water.
2. Add a teaspoon of brine shrimp eggs.
3. Let the jar with eggs sit for a day.
4. After 24 hours, observe the jar with a hand lens. Record your observations.
5. Shine a flashlight on the water for five minutes. Examine the shrimp with your hand lens. Describe what happens to the shrimp. Do they move toward or away from the light?

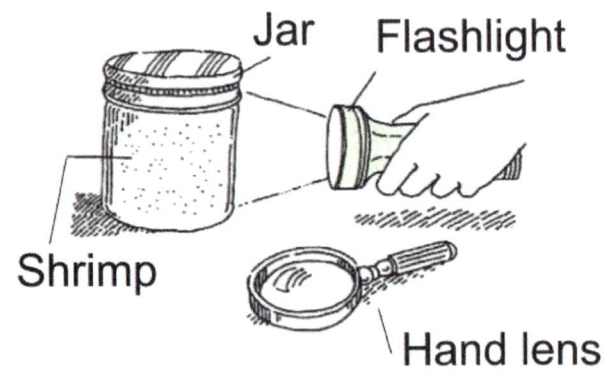

Questions:

1. How long did it take for the shrimp eggs to hatch?_____

2. Which way did the brine shrimp move when you shone the light on them?_____

_____

• REPRODUCE. living things can make more of their own kind, or **reproduce**. A pig can produce more pigs. The seeds of a corn plant become new corn plants. Yogurt gets its tangy taste from the work of tiny bacteria (bak-TEER-ee-uh) that reproduce by splitting in half.

• USE FOOD AND OXYGEN. Living things need energy to stay alive. **Food** supplies energy. Animals and other organisms take in food. Plants and plant-like living things make their own food.

Most living things also need and use oxygen (OK-si-jun) to help get energy from food. **Oxygen** is a part of the air that you breathe. Other basic needs of living things are water and the right temperature. The environment supplies all of these needs: food, oxygen, water, and the right temperature.

• GET RID OF WASTES. Living things get rid of wastes. When you breathe, your body gives off carbon dioxide as a waste product. Carbon dioxide is given off when energy is released from food.

All of the things that organisms do to stay alive are its **life functions**. Moving, growing, and responding are life functions. So are reproducing, using food and oxygen, and getting rid of wastes. How then, can you tell a living thing from a nonliving thing?

In some ways, nonliving things seem to carry out some life functions. A hang glider moves in the wind. Blow air into a beach ball and it grows larger. Stretch and let go of a rubber band, and it reacts. A robot can even make another robot. You cannot use any *single* life function to tell if something is alive. Only living things can carry on all of the life functions.

---------- **REVIEW** ---------- U-1 L-1

I. In each blank write the word that fits best. Choose from the words below.

| energy | environment | wastes | respond | functions | grow |
| organism | oxygen | move | reproduce | temperature | robot |

Any living thing is an _____. Four things that all living things do

are _____, _____, _____, and

_____. Living things get _____ from air. They

get rid of _____. The things that an organism does to stay alive

are its life _____. Everything around a living thing is

its _____.

II. Write the word that matches each statement.

    grow   respond   reproduce

A. _____ An earthworm moves away from light.
B. _____ A dog gives birth to puppies.
C. _____ A tree becomes taller.

III. You are an astronaut on another planet. You find an object that moves. How might you decide if it is an organism?

_____

How Do Scientists Study Living Things? U-1 L-2

Exploring Science / Historical Steps

Live Comes From Life. "Where did I come from?" asks the young child. "From us," say her parents. Today, all scientists agree that living things come from other living things. Like a young human, a young fly, frog, or mouse has parents like itself.

Long ago, people thought that some living things came from nonliving things. This belief was known as **spontaneous generation**. Believers in spontaneous generation felt that rotting meat produced maggots (young flies). They thought that mud could turn into frogs. Placing wheat on dirty shirts was believed to produce mice.

Some people doubted such beliefs. They wanted proof. In the 1660s, **Francesco Redi**, an Italian doctor, did an experiment to test spontaneous generation of maggots. Redi put some rotting meat in an open jar and watched it closely. He soon saw some adult flies near the meat. Three days later, maggots covered the meat.

After 19 days, the maggots stopped moving. Then they turned into small hard "footballs". Redi placed some of the footballs into another jar that was empty. After eight more days, the footballs broke open. Out came flies!

The flies had laid eggs on the meat. The eggs then hatched into **maggots**. Finally, the maggots developed into flies.

Redi set up more jars of rotting meat. He sealed some; others he covered with netting. He left more jars open. No maggots appeared in the sealed jars or the jars covered with netting. As before, maggots soon appeared in the open jars. Redi had shown that baby flies (maggots) come only from eggs laid by the parent flies.

Redi's work helped to change beliefs about where living things come from. Other scientists showed that frogs come from eggs laid by other frogs. They showed that all baby mice come from other mice. It just happens that mice like to eat wheat and to nest in dirty shirts.

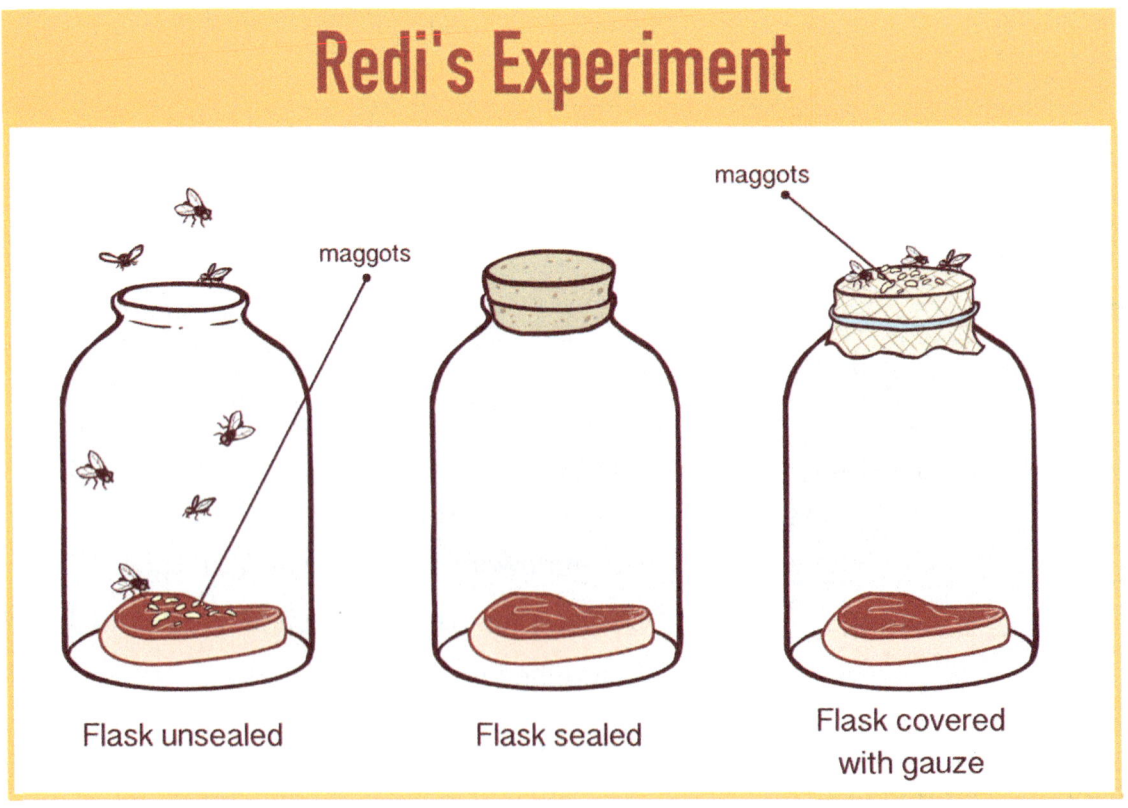

4

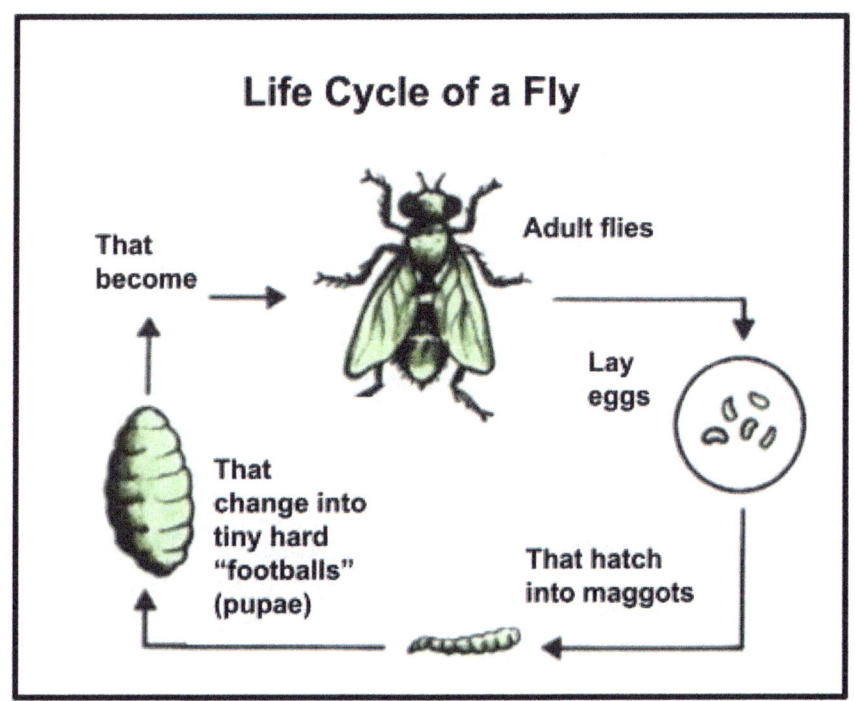

Louis Pasteur with a child

After microscopes were invented, tiny living things called **microbes** were found. By this time, scientists no longer believed that *animals* could come from nonliving matter. Yet some thought that *microbes* could!

Late in the 1800s, spontaneous generation was also shown to be wrong by one of history's most famous scientists, **Louis Pasteur** (Loo-ee pah-STUR), of France. Pasteur showed that microbes also have parents.

No one has shown that *any* living thing comes from nonliving matter. All organisms have parents.

➤ Scientists long wondered where a certain kind of eel, which lives in rivers, comes from. The statement more likely to be true is:

　A. River mud can turn into eels.
　B. Eels lay eggs in the ocean. The eggs hatch, and the baby eels swim into a river.

➤ Want more? Research this phrase: "Pasteur's swan neck flask."

The Ways of the Scientist

The study of living things is life science - also called **biology**. Scientists who study biology are **biologists** (by-OL-uh-jists). Some biologists are **ecologists** (ih-KOL-uh-jists). They study **ecology** (ih-KOL-uh-jee), the relationship of living things to their environment.

Like other scientists, biologists may work in laboratories. Or they may work outside - in forests, in fields, or in fresh or saltwater. They may work anywhere in the **biosphere** (BY-uh-sfeer). The biosphere is the part of the earth in which life exists. A layer of soil, water, and air makes up the biosphere. Some biologists even work in space, aboard spacecraft or in space suits.

How Scientists Often Seek Answers

1 State a problem   2 Observe   3 Collect facts   4 Make a hypothesis   5 Test the hypothesis   6 Conclude

Scientists have many ways of working. Often, however, their work follows a general pattern, called the **scientific method**. The parts of this method are described below:

- **STATING A PROBLEM.** Scientists are curious. They ask many questions. For example, Redi wondered where maggots come from. His problem could simply be stated, "Where do maggots come from?"

- **OBSERVING.** Using one's senses to gain information is called observing. Observing can involve seeing, hearing, touching, smelling, or tasting. All scientists use **observation**. When Redi saw maggots appear on rotting meat, he was observing.

In science, **measurement** is a way of observing. Counting is a kind of measuring. Redi was measuring when he counted the days between changes in his jars. Scientists also measure with tools. Their tools include balances, thermometers, and meter sticks. Using a microscope helped Pasteur observe and study microbes.

- **COLLECTING FACTS.** Scientists collect facts, or data, by observing. They also read reports of other scientists. This helps them build on the work of other scientists. Of course, they are very careful not to simply trust everything that they read, especially on the internet. The sources of all information must be checked and compared with trusted sources.

- **MAKING HYPOTHESES.** Scientists use the information that they collect to make hypotheses (hy-POTH-ih-seez). A hypothesis is a guess or possible answer. Redi's hypothesis was that maggots come from flies.

- **TESTING HYPOTHESES.** One way to test a hypothesis is to make new observations. Scientists also do experiments to test hypotheses. This is what Redi did.

- **CONCLUSION.** From the results of tested hypotheses, scientists make **conclusions**, (explanations). Redi concluded that maggots come from flies, not from rotting meat. Pasteur concluded that microbes could grow only from other microbes.

Scientists now accept that living things come only from other living things. Before this idea was tested, it was a hypothesis. After being tested many times, the idea became a **theory** (THEE-uh-ree). Even a solid theory may later be shown to be incorrect. If new observations do not fit the theory, the theory must be changed.

You have likely heard someone say, "That idea is just a theory!" Many people think that the word theory is simply another word for "guess." Obviously, this is not the way that scientists use the term theory.

➤ To Do Yourself Does Mold Grow on Any Kind of Bread?

You will need:
2 slices of packaged white bread (with preservatives), 2 slices of home-made or preservative-free white bread, 4 sealable plastic sandwich bags

1. Place the four slices of bread inside of the plastic bags, one slice per bread.
2. Dampen each slice of bread with a few drops of water. Seal the bags.
3. Label each bag's bread type and the date.
4. Keep the bread in a dark, warm place.
5. Make a hypothesis about which bread will become moldy first.
6. Observe your bread slices every day (through the plastic).
7. Record your data in a notebook.
8. Dispose of the bags properly.

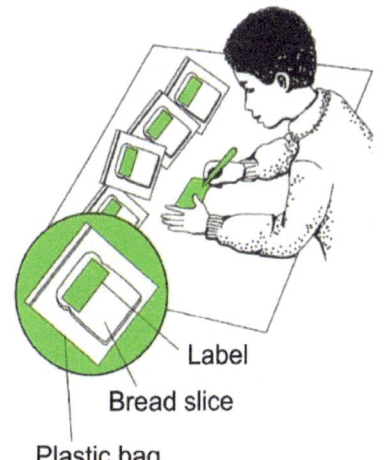
Label
Bread slice
Plastic bag

Questions

1. Which kind of bread became moldy first? _____

2. Make a hypothesis about where the bread mold came from. _____

_____

3. What can you conclude about the two types of bread? _____

_____

REVIEW — U-1 L-2

I. Fill each blank with the word that fits best. Choose from the words below.

biosphere	measuring	theory	biology	observing
hypothesis	ecology	guessing	biologists	

Life science is also called _____. Scientists who study biology are _____. The _____ is the part of the earth in which life exists. Using one's senses is _____. Counting or using a tool like a meter stick is _____. A possible explanation is a _____. A hypothesis that is repeatedly tested and fits all observations may become a _____.

II. Show the order of events in developing a scientific theory. Place 1, 2, 3, or 4 in front of each item.

A. ___ testing hypotheses
B. ___ making hypothesis
C. ___ conclusion
D. ___ stating a problem

III. Write the word that matches each statement.

observation   measurement   hypothesis   facts

A. _____ A written report or Redi's experiment.

B. _____ A guess that maggots come from flies.

C. _____ Looking at microbes with a microscope.

D. _____ Counting the days it takes for maggots to appear

Ready for More Scientific Thinking?

U-1 L-3

Exploring Science / Historical Steps

Scientific thinking and vaccines. In the last lesson you learned the basic steps of the scientific method. To learn a few more parts of scientific thinking, let's look at the work of another famous scientist from the past.

Edward Jenner - famous for his smallpox vaccine

In the late <u>1700</u>s, no one knew that many diseases were caused by things too small to be seen. Sadly, the virus causing **smallpox** was killing millions of people. This disease caused pus-filled blisters to form on its victims.

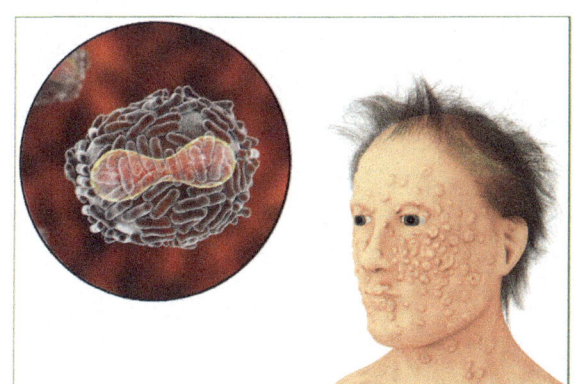

The variola virus that causes smallpox.
A man with smallpox blisters.

#1 - <u>RECOGNIZING PATTERNS</u>.
A pattern is a repeated observation - something that is seen, heard, smelled, touched or tasted more than once.

Milkmaids, young women who milked cows, often suffered from *cowpox*. This disease seemed much like smallpox. It caused illness and pus-filled blisters. But cowpox did not kill people.

Edward Jenner, a surgeon in England, noticed a pattern. Even when smallpox was killing people all around them, milkmaids were not dying.

#2 - <u>INFERRING</u>.
A guess, judgment, or conclusion based on observations or data.

In <u>1796</u>, Jenner made an inference. Maybe something in the pus of cowpox blisters protects milkmaids from whatever causes smallpox. He decided to try an experiment.

#3 - <u>EXPERIMENTING</u>.
Use of the scientific method. (See the previous lesson).

From the blister of a milkmaid with *cowpox*, Jenner removed a small amount of pus. He put this pus under the skin of a healthy child! The child was ill for nine days, and then recovered. Jenner then put pus from a <u>smallpox</u> blister under the same child's skin. No disease occurred!

#4 - <u>COMPARING,</u>
<u>OR CONTROLLING</u>.
Some experiments are simply comparisons. Others involve <u>controls</u> - which are described next.

Jenner *compared* the child that he helped to children who were not given the cowpox pus. Sadly, many children who were not given the cowpox pus died.

Later, Jenner compared more people who were given the cowpox pus to people who did not receive it. Over and over, people who were given the cowpox pus did not die from smallpox. And over and over, many people who never received the cowpox pus died of smallpox.

Today's scientists would complete Jenner's experiment a bit differently. First, they would put the cowpox pus under the skin of a large number of volunteers. Then, under the skin of a second large group of very similar volunteers, they would put clean water (instead of pus). Both groups would then be observed.

The second group is called the **control** group. Using this method, called a **controlled experiment**, scientists would know for sure that any difference between the groups was due to the pus.

#5 - AVOIDING BIASES.

A bias is simply an opinion based on past experiences.

Jenner was criticized by many people. This often happens to those who try something new. People in Jenner's time simply could not imagine how Jenner's cowpox pus could work. Their biases kept them from believing his claim. Even the most famous group of scientists, the **Royal Society** (in England), refused to publish Jenner's results.

Fortunately, within a few years, Jenner's method was accepted and widely used. A disease that had killed millions of people was gradually defeated.

The fact is, we all have biases. Scientists must ignore their biases in order to design experiments. They must also ignore their biases as they draw conclusions.

You probably know that Jenner's method is known today as **vaccination**. Of course, today's vaccines use products that are much safer than pus. (Vaccines are covered in more detail in Unit 10, Lesson 2).

#6 - VERIFYING

To verify is to repeat work and then obtain the same outcome.

Many important discoveries have occurred because scientists noticed a *pattern*, made an *inference*, and then designed a *controlled experiment* that was free of *bias*.

To be widely accepted, the results of an experiment must be able to be verified. That is, another scientist must be able to repeat the experiment - and obtain the same results. Of course, for verification to occur, all of the steps of an experiment must be very clearly described. Good scientists must also be good writers!

Six parts of scientific thinking...

- **Recognizing Patterns**
- **Inferring**
- **Experimenting**
- **Comparing, or Controlling**
- **Avoiding Biases**
- **Verifying**

➤ To Do Yourself   Are You Ready to Use the Six Parts of Scientific Thinking?

You will need: A notebook and a pencil. This activity is simply a way to practice *thinking* scientifically. You are not expected to do more than record your *thoughts*.

1. Walk down your street and use your observation skills to "wonder"!
2. Jot down what you see, hear, smell, and feel. (You probably don't want to taste anything at this point).
3. Write down any patterns that you notice. Are there some repeated images, sounds, smells or textures?
4. Does anything occur that leads you to guess "Why"? By guessing, you are inferring!
5. How might you find out if your inference is correct? Would you be able to learn the answer from more observations? Or from a comparison? Or from a controlled experiment?

6. Pause to ask yourself: Have experiences from my past influenced what I've recorded so far? Have my biases affected my thinking?
7. If you gather enough information to determine an answer to your question in step 4, ask yourself: If a friend wanted to repeat my work, what should I be sure to share so that she or he may verify my outcome?

Question:
• Which of your observations surprised you? _____
_____

------- REVIEW ------- U-1 L-3

I. Fill each blank with the word that fits best. Choose from the words below.

| control | verify | biases | patterns | infer | compare |

We all have past experiences, so it is not surprising that we all have _____.

A careful observer is alert for things that repeat - called _____.

If you have observed someone's cheerful behavior carefully, and have noticed a pattern, you might _____ why they smile so often. Among the volunteers in an experiment, some do <u>not</u> receive a treatment. This group is the _____.

II. In 2020, to develop vaccines against the Covid virus, thousands of volunteers were divided into two groups. One group <u>did</u> receive a vaccine. The other group (the <u>control</u> group) received clean water. Scientists kept track of how many people from each group became sick with Covid.

• An experiment with a control group allows a scientist to feel more confident about her conclusion. Explain why this is the case.

11

What Is An Ecosystem? U-1 L-4

Exploring Science / Historical Steps

<u>City Wildlife</u>. As people built towns and cities in America, wildlife moved away. Sometimes, wildlife comes back. New York's Bronx borough has foxes. Coyotes roam suburbs in California. Falcons nest on top of hotels in Atlantic City. Canada geese are seen nationwide.

The term for the variety of living things in an area is **biodiversity** (bye-oh-di-VER-si-tee). What level of biodiversity do you have in your town or city? Years ago, ecologists in New York City decided to find out. They counted the kinds of reasonably large animals in Central Park. They found 269 kinds of birds, 3 kinds of turtles, and 2 kinds of frogs. There were 9 kinds of fish, 6 kinds of bats, and 3 kinds of woodchucks.

Why do you think some animals are coming back to towns and cities? Perhaps more people are welcoming them. Ecologists believe that as habitats for some wild animals have declined, more animals are moving to cities.

➤ Which seems more likely to be true?
A. The level of biodiversity in towns and cities will stay the same.
B. As environments change, the wildlife both in cities and in the wild will change.

➤ Want more? Search "The Half-Earth Project" or the <u>E.O. Wilson Biodiversity Foundation</u>. Read the graphic novel <u>Naturalist</u> (adapted from E.O. Wilson's autobiography).

Populations, Communities, and Ecosystems

We share the earth with many other kinds, or **species** (SPEE-sheez), of living things. All the members of one species that live in an area make up a **population**. Do you know the population of people in your town or city? What other populations live there?

In a desert, the giant cactuses are one population. Kit foxes, sage, and lizards make up others. There are many more.

A **community** is made up of more than one population. These populations live together and interact with each other. Some species give shelter to others. In the desert, elf owls nest in the giant cactuses. In the heat of the day, coyotes rest in their shade.

One population may also provide food for another. Cactuses are green plants that make their own food. Ground squirrels eat the seeds of cactuses. Coyotes eat the squirrels.

The desert is an example of an ecosystem. A living thing and its nonliving environment make up an **ecosystem**. A lake, a rotting log, or the soil in a flowerpot are also ecosystems.

An ecosystem can be small or large. A tiny tidal pool on the beach is an ecosystem. So is the entire earth!

A Desert Ecosystem

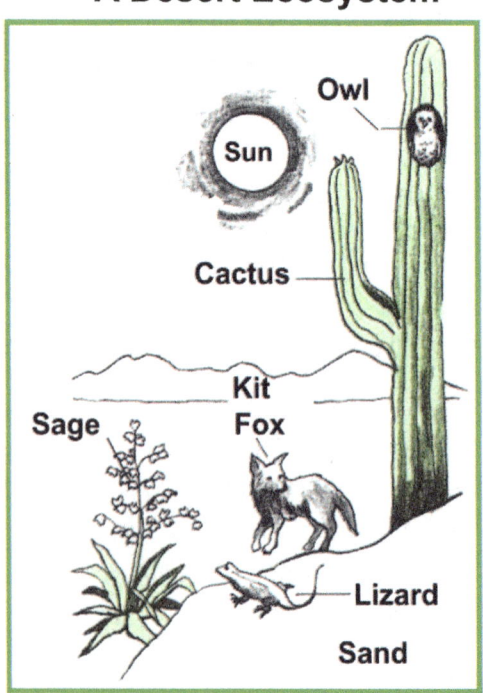

12

➤ To Do Yourself   Where Do Organisms Live Near You?

You will need:

Drawing paper, crayons or colored pencils

1. Make a map of your block or around your school building. Show the sidewalk, streets, buildings, playground or park, and fields.
2. Show where each of the animals on the list below can find shelter, food, and water. Use a different color or symbol for each animal.

dogs	squirrels	worms	roaches
cats	pigeons	rats	flies
mice	sparrows	ants	spiders

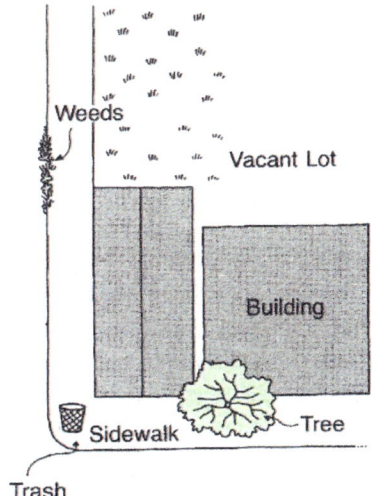

Questions

1. What other animals are there in your community? _____
2. Select a population of organisms on your block. Describe why it is important to your community.

---------- REVIEW ---------- U-1 L-4

I. In each blank, write the word that fits best. Choose from the words below.

| community | desert | living | populations |
| ecosystem | nonliving | species | town |

There are many different kinds (_____) of living things. A

_____ is made up of all the organisms of one species in an area.

Populations that live together and interact make up a _____.

All of the organism (the_____ things), plus their

_____ environment make up an _____.

II. Circle the word (between the brackets) that makes each statement correct.

A. All of the grasshoppers in a meadow make up a [population / ecosystem].

B. The living and nonliving things in a river make up [an ecosystem / a community].

C. In a forest, all of the organisms make up a [community / population].

III. What needs of living things are met in a space station? Is the space station an ecosystem? Explain.

13

25. Trees with needle-like leaves are the climax plants in a
    a. tundra   **b.** coniferous forest   **c.** desert   25. _____

26. A factor that limits the number of a species of animals is
    a. predators   **b.** sunlight   **c.** scavengers   26. _____

C. Apply What You Know

1. Professional problem-solvers are very good at *recognizing patterns*. If you overheard the following questions, who would you *infer* is asking them?
    1) "Is your stomach upset only after eating certain foods?"
    2) "Do you usually get nervous before recess, or before a specific class?"
    3) "Would this pie smell and taste better when cooked longer, or at a lower temperature?"

2. Besides the two that are underlined in the previous question, list at least two other parts of thinking scientifically. Which one involves a scientist describing her work in such detail that another scientist would be able to repeat it?

3. Study the drawing (on the next page) of a food chain in Africa.
Use the words below to fill in the blanks.

    carnivore   **energy**   **herbivore**   **producer**   **predator**
    third-level consumer   **decomposer**   **first-level consumer**
    omnivore   **second-level consumer**   **prey**   **scavenger**

    a. Three terms that describe the zebra are (1) _____
    (2) _____ (3) _____

    b. Three terms that describe the lion are (4) _____
    (5) _____ (6) _____

    c. The grass is a (7) _____ because it makes food, using
    (8) _____ from the sun.

    d. When the lion in the drawing dies, its body is consumed by the hyenas. Two terms that describe a hyena are (9) _____ and
    (10) _____.

4. Write the word (or words) that best completes each statement.
    a. The grass, zebras, lions, and hyenas are each members of different _____.
        (1) communities   (2) ecosystems   (3) biomes   (4) populations

b. If hunters are allowed to kill most of the lions in the area, the zebra population would probably _____.

    (1) get larger, then smaller;   (2) get smaller, then larger;

    (3) become extinct;   (4) stay the same. Explain your answer.

_____

c. Laws have been made to protect the lions because_____
_____

    (1) their food supply is almost gone;   (2) they are extinct;

    (3) they are endangered;   (4) nothing limits their numbers.

D. Find out more.

1. Visit a nearby ecosystem. It may be your school grounds, your backyard, a park, a vacant lot, a wooded area, or a field. Other possible ecosystems are a mud puddle, a pond, a stream, a river, a tidal pool, or a lake. Make a list of the populations that you observe. Describe the nonliving parts of the ecosystem. Draw one or more food chains that you observe.

2. Find out how the world's human population has changed since the year 1 AD. Draw a graph that shows its growth. Try to predict what the population will be in 2050.

Careers in Life Science

Research lab assistant. Do you like discovering new information and solving problems? Do you take pride in doing things well? Labs across the country hire assistants to help researchers carry out experiments. At some labs there is a good deal of "people work," such as testing patients who have a certain medical condition. At other labs, nearly all of an assistant's time is spent using science equipment - for example, to test blood. Some of the equipment is fairly simple - like microscopes. But some require a good deal of training. The key is a willingness to learn and a determination to be accurate. Most research assistants have at least two years of education beyond high school.

Researcher. Sometimes research assistants go on to complete a college degree, and then an advanced degree. Completing a PhD, which takes at least four years, could enable them to lead their own research team. There will always be questions to answer - how living things function, how diseases can be cured, how nature can be better protected. Many researchers share their advanced knowledge by also teaching college students.

Science teachers know that students enjoy hands-on learning - and working with others. Here, Evan and Anna prepare their class's annual turnip crop for a homeless shelter.

Science teacher. Knowing that you helped someone better understand a challenging concept - what could be more fulfilling? Equally rewarding is helping young people feel better about themselves! If you love biology and you communicate well - becoming a life science teacher might be for you! Science teachers are in demand for all ages of children - from elementary through high school.

A research assistant gathers data from a volunteer.

UNIT TWO

HOW LIVING THINGS ARE ORGANIZED

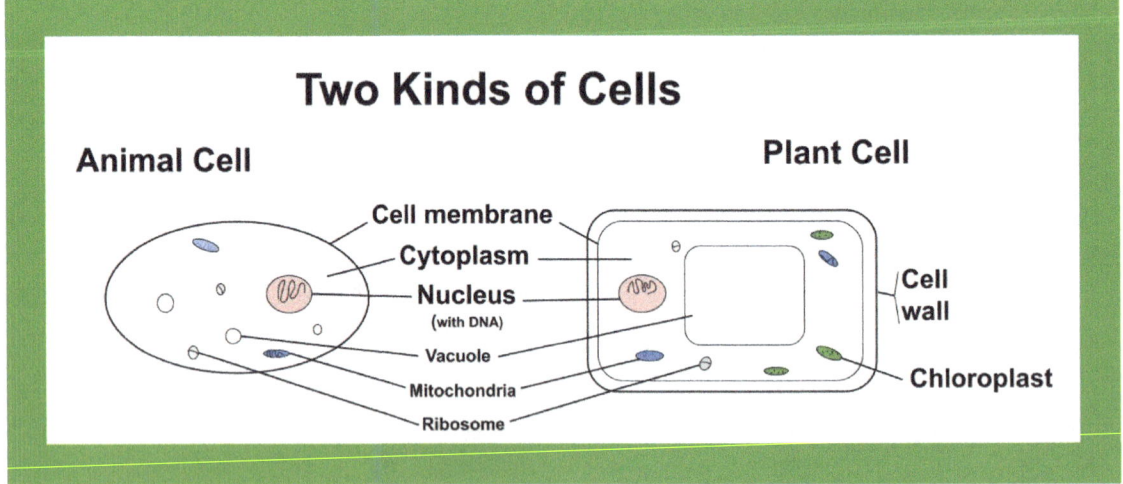

What Units Make Up Living Things? U-2 L-1

Exploring Science / Historical Steps

From 200 to 10,000,000 Times Life Size.

Anton van Leeuwenhoek (1632 - 1723)

It happened over 300 years ago, in Holland. **Anton van Leeuwenhoek** (An-tun van Lay-vun-hook) prepared a new microscope that he had made. He looked at a drop of lake water. What he saw surprised him!

The water was alive with what Leeuwenhoek called "wee beasties." The microscope made the tiny organisms look 200 times larger than life size. Leeuwenhoek was likely the first to see living things that small. His work was a giant step ahead for life science.

Today microscopes are much stronger. A special type of electron microscope can magnify up to 10 million times! Microscopes have come a long way in 300 years!

➤ Which statement is more likely to be true?
A. With the electron microscope, organisms smaller than "wee beasties" can be seen.
B. There are no organisms smaller than a "wee beastie."

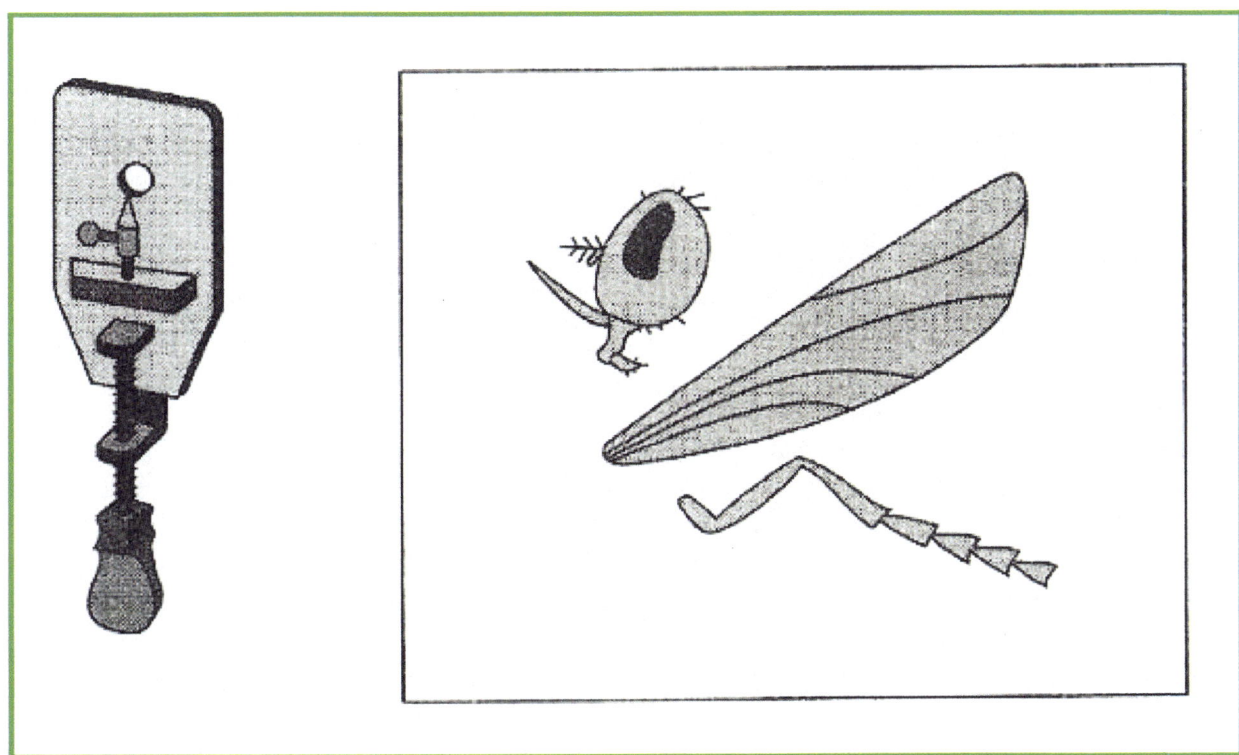

Leeuwenhoek's small hand-held microscope and his sketch of magnified insect parts

39

Microbes and Cells

Living things come in many sizes. Some, like whales and giant redwoods, are large. Others, like mites, are very small. A hand lens makes a mite look larger. The **lens** is made of a curved piece of clear material, such as glass or plastic. Some hand lenses make small things look eight times larger than they are. That means that their magnifying power is eight.

Microbes (MY-krobes) are living things too small to be seen with a hand lens. We use a microscope to see microbes. **Compound microscopes** have two or more lenses. Those used in middle and high schools often have a top magnifying power of 400 times.

The microbes that you see with a microscope may be plant-like **algae** or animal-like **protozoans** (proh-tuh-ZOH-uns).

A few years before Leeuwenhoek first saw microbes, the English scientist **Robert Hooke** had another "first." He also built microscopes. One day, while looking at a slice of cork, he noticed that it was made up of tiny boxes. He called these "**cells**." Later, it was found that *all* living things are made up of these tiny units. We still use Hooke's word for them!

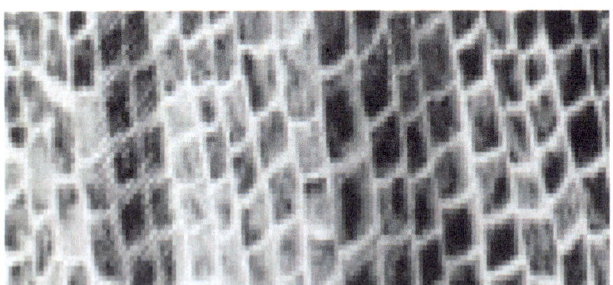

Magnified cork cells

Hooke's cork cells were dead and empty. Only their walls were left. Living cells, though, are anything but empty. They are filled with living matter. There are living cells in a frog's blood. A bit of muscle in your stomach or your heart is made up of living cells. So is a green leaf.

A cell has three main parts. These are somewhat like the parts of a peach. The **cell membrane** (MEM-brayn) covers the cell, like the skin of a peach. At the peach's center is the pit. A cell's center is its **nucleus** (NOO-klee-us).

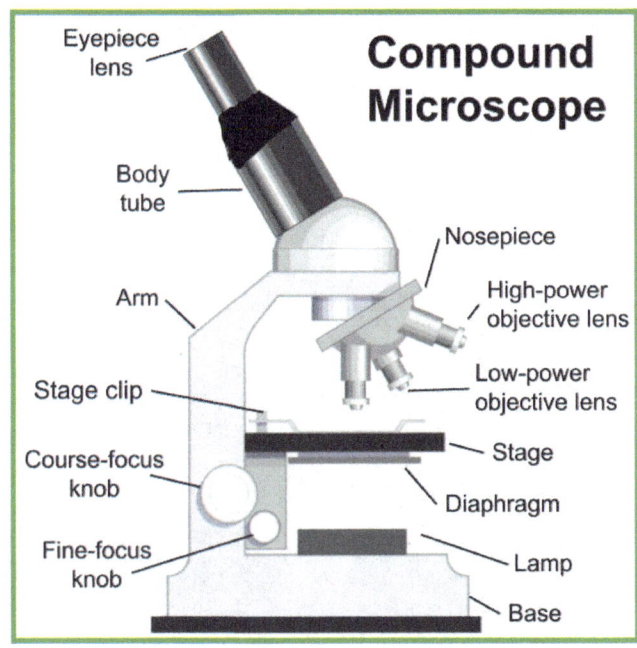

Between the peach's skin and pit is the "flesh." In a cell, the **cytoplasm** (SY-tuh-plaz-um) is between the cell membrane and the nucleus.

The number of cells in organisms ranges from only one, to trillions. Many microbes are made up of just one cell. Bacteria are one-celled living things. So are the protozoans and many algae. Larger organisms are made up of many cells. That includes a tree, an ant, and you.

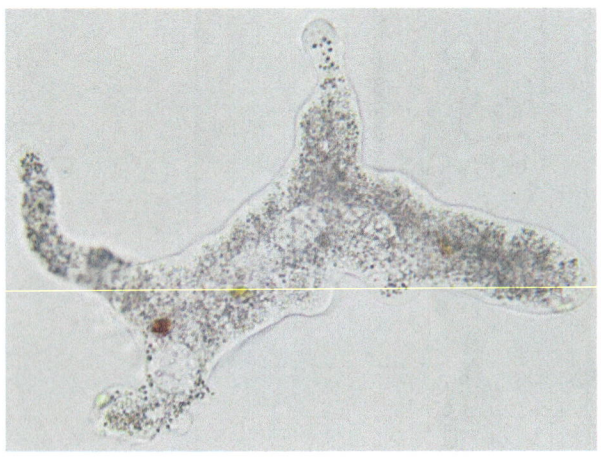

The nearly transparent (see through) ameba moves in a blob-like manner, extending thin portions of its single cell outward while the inner material flows into these 'false feet' (called pseudopodia). It eats by surrounding itself around its prey (which might be bacteria, algae, or other small single-celled organisms).

> **To Do Yourself**   **What Are the Parts of an Animal Cell?**

You will need:

Microscope, prepared slides of frog's blood, slide of human skin cells

1. Carefully place a prepared slide on the microscope stage. Use the low power objective lens to focus. Draw your view.
2. Calculate the magnification using low power. [Multiply the magnification of the eyepiece lens by that of the objective lens]. Record that number.
3. Switch to high power. Calculate the magnification now. Notice which animal cell parts you can now see more clearly.

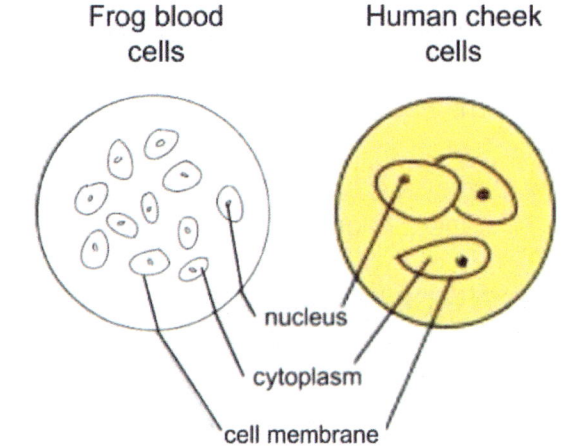

Questions

1. What cell parts could you see in both kinds of cells?
   _____
2. Why do you think that a stain was added to these cells?
   _____
3. Describe the shape of the cells that you saw.
   _____

--------------------------------- REVIEW --------------------------------- U-2 L-1

I. In each blank, write the word that fits best. Choose from the words below.

cells   **cytoplasm**   **nucleus**   **cell membrane**   **microbe**

All living things are made up of tiny units called _____. The

_____ is the cell's outside "skin." In the center of a

cell is the _____. Between the nucleus and the cell membrane is

the _____.

II. Circle the word (between the brackets) that makes each statement true.

   A. The cells in [cork / frog's blood] have no cytoplasm.

   B. A hand lens may have a magnifying power of [8 / 100] times.

   C. Examples of one-celled living things are bacteria and [ants / protozoans].

III. Show how you find the answer to the following problem. The top lens of a microscope magnifies 10 times, and the bottom lens 30 times. When you look through the microscope, how many times will your view be magnified? [Hint: See **To Do Yourself** above.]

41

What Are the Jobs of a Cell's Parts? U-2 L-2

Exploring Science

Animal, Vegetable, or Mineral? A detective at a crime scene collects a speck of dried red material. Is it blood? Paint? Fruit? What else could it be?

Back at the police laboratory, the speck is placed in a special liquid. With a microscope, a technician checks for cells. She finds cells, so she knows that the speck came from an organism. But was it a plant or an animal?

➤ Under a microscope, cells might be seen in a tiny piece of toothpick that is made of
   A. wood   **B.** silver   **C.** plastic

The Work of a Cell

How do you know if a cell is from an animal or a plant? Look for the special parts of the cell. A plant cell has a **cell wall**. So do plant-like microbes, such as algae. In addition, bacteria have cell walls. Animal cells, on the other hand, have no cell walls. Neither do animal-like microbes, such as the ameba.

In plants, the cell wall is made of nonliving matter called **cellulose** (SEL-yuh-lohs). This is what makes a plant's stem stiff.

Each part of a cell has a job to do. For a cell to do its work, it must take in food and oxygen. It must also give off wastes.

The cells of green plants contain oval parts called **chloroplasts** (KLOHR-uh-plasts). Chloroplasts contain a green chemical called **chlorophyll** (KLOHR-uh-fil) that is used to make food. Algae, the key food-maker in aquatic habitats, have chlorophyll in their cells, too.

In the cytoplasm of cells are bubble-like parts called **vacuoles** (VAK-yoo-ohls). Some vacuoles store water and food for later use. Other vacuoles store wastes until the cell gets rid of them.

Two other important cell parts are the **mitochondria** (my-tuh-KON-dree-uh) and the **ribosomes** (RY-buh-sohms). In mitochondria food is "burned" to make energy. Ribosomes are the sites of protein production.

The **nucleus** ("center") of the cell controls what goes on in the cell. In the nucleus is a special molecule called **DNA**. DNA contains a code or pattern. One job of DNA is to "tell" the ribosomes which proteins to build. This helps to define how organisms look and act. DNA is also key to an organism's ability to reproduce.

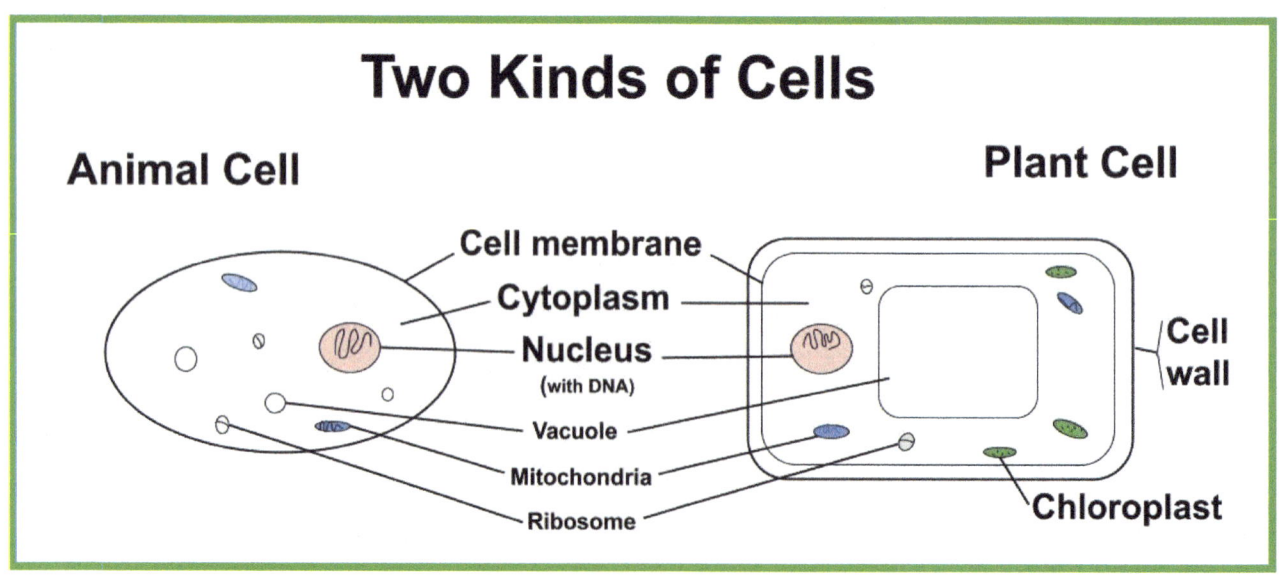

42

> **To Do Yourself**   How are Plant and Animal Cells Different?

You will need:

Microscope, slide, coverslip, medicine dropper, tweezers (forceps), iodine solution, an onion <u>cut into small pieces</u>, Elodea

1. An onion has flat, curved layers - as shown. But do not use the outside coating. Cut open an onion, select one layer, and cut it into small pieces (about the size of a pencil eraser).
2. Next, from the *<u>inner</u>* curve of one of these small pieces, peel off only the very thin "skin."
3. Place this thin "skin" flat (not folded) on a slide. Add a drop of iodine to the onion, and add a coverslip.
4. Observe the slide under low power, then the higher powers of the microscope.
5. Tear off the tip of an **elodea** leaf. Place it flat on a slide. Add a drop of water (not iodine) and a coverslip. If you scan all views of this slide, you might see some oval-shaped chloroplasts *moving* in a single file!

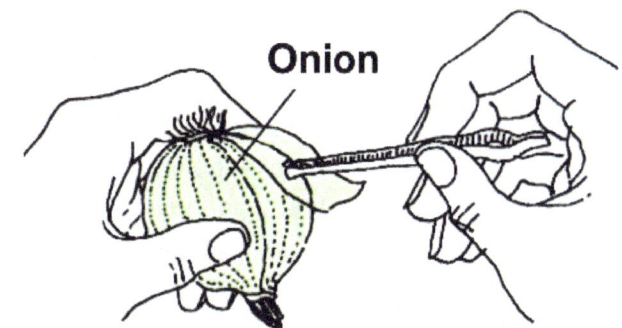

Onions have layers - but do <u>not</u> use this outer "skin."

Questions

1. How were the two plants' cells alike and different? _____
2. How were these plant cells similar to animal cells? _____
3. How were these plant cells different from animal cells? _____

---------------------------------- **REVIEW** ---------------------------------- U-2 L-2

I. In each blank, write the word that fits best. Choose from the words below.

DNA   **cell wall**   **chlorophyll**   **nucleus**   **vacuoles**

All plant cells have one part that animal cells do not have. That part is the

_____. Green plants also have _____ which

they use to make food. Water, food, or wastes may be stored in _____.

The _____ code defines how living things will look and act.

II. Write **A** for each part of an animal cell. Write **P** for each part of a plant cell. Write **A / P** if the part belongs to both plant and animal.

    A. ____ cell wall   **C.** ____ chlorophyll   **E.** ____ vacuole
    B. ____ chloroplast   **D.** ____ DNA   **F.** ____ cellulose

III. A human red blood cell loses its nucleus at one stage of its growth. A human brain cell always keeps its nucleus. Which cell do you think lives longer? Why?

How Do Cells Work Together? U-2 L-3

Exploring Science

The One and the Many. Here's a puzzle. When its life began, it had just one cell. Soon there were many cells. When it is full grown, it will have about four trillion cells. What is it? The answer: You.

You began life as one round cell. That one cell split into two. The two split again and again. By the time you were born, you already had billions of cells. When you are grown, you will have even more.

Here's a new puzzle. Now that you are almost grown, you have trillions of cells, but each one is different from the first one. Some are long and thin. Some are flat. Some have stripes across them. Some have no nucleus. How do you think that one round cell is able to turn into so many different shapes?

Scientists know part of the answer to this puzzle. Your first cell carries directions for new cells to become special kinds of cells. However, many of the details about how this happens is still a mystery.

➤ Which choice makes the statement true?

Cells of different shapes probably carry out functions that are [the same / different].

Cells Working Together

Cells do everything that living things do. In a one-celled organism, one cell does everything. An ameba, for example, has just one cell. It takes in food. It uses oxygen to get energy from food. It gets rid of waste. It responds to its environment. It grows and reproduces.

The bodies of many-celled organisms have several kinds of cells. Each kind is **specialized** (SPESH-uh-lyzed). That means each does a special job. For example, the **covering cells** in your skin are flat. Their job is to protect what they cover.

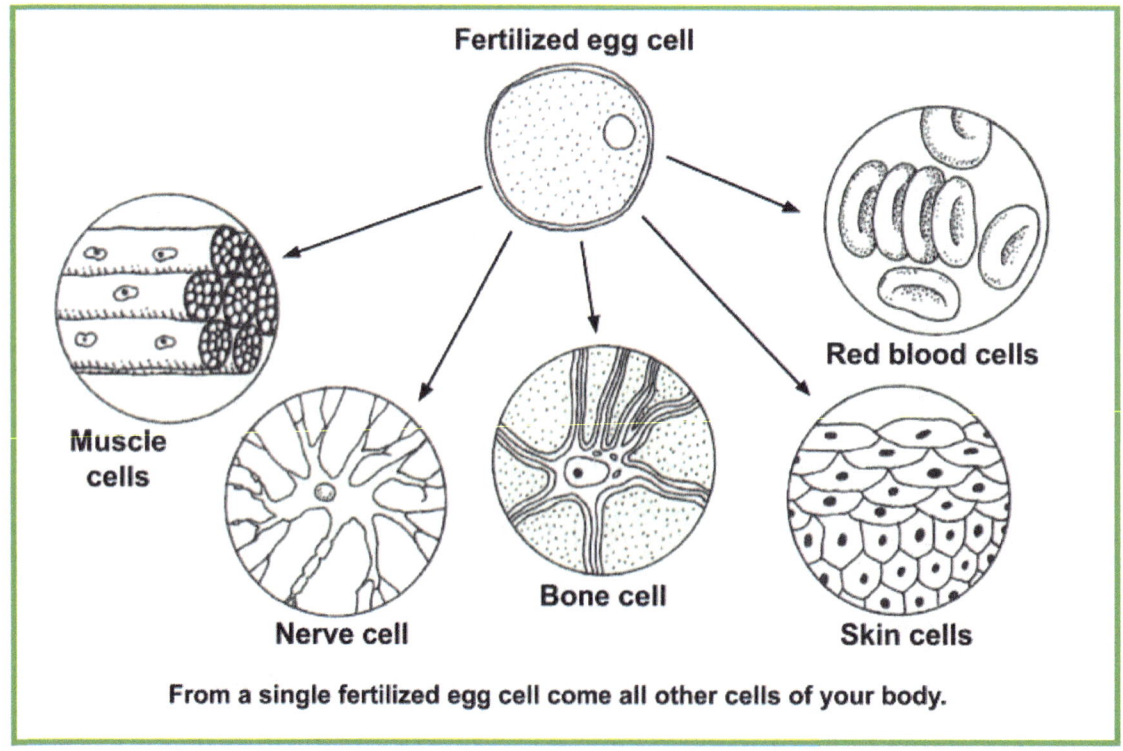

From a single fertilized egg cell come all other cells of your body.

44

Nerve cells (also called **neurons**) in your arm are long and thin. Their shape fits them for their job; that job is to carry messages.

A group of cells that does a special job makes up a **tissue** (TISH-yoo). The cells in the tissue all look similar and do the same work. Many nerve cells make up nerve tissue.

Groups of tissues that do special jobs make up **organs**. Your heart and your eyes are organs. The main job of your heart is to move blood through your body. This organ is made of a number of tissues, such as muscle tissue and nerve tissue.

A group of organs that work together make up a **system** (sometimes called an **organ system**). For example, your brain, nerves, eyes, and ears are some of the organs in your nervous system. This system controls your actions.

In a complex organism (like a human) several systems work together. Your nervous system works with other systems of your body, such as your skeletal system and your digestive system.

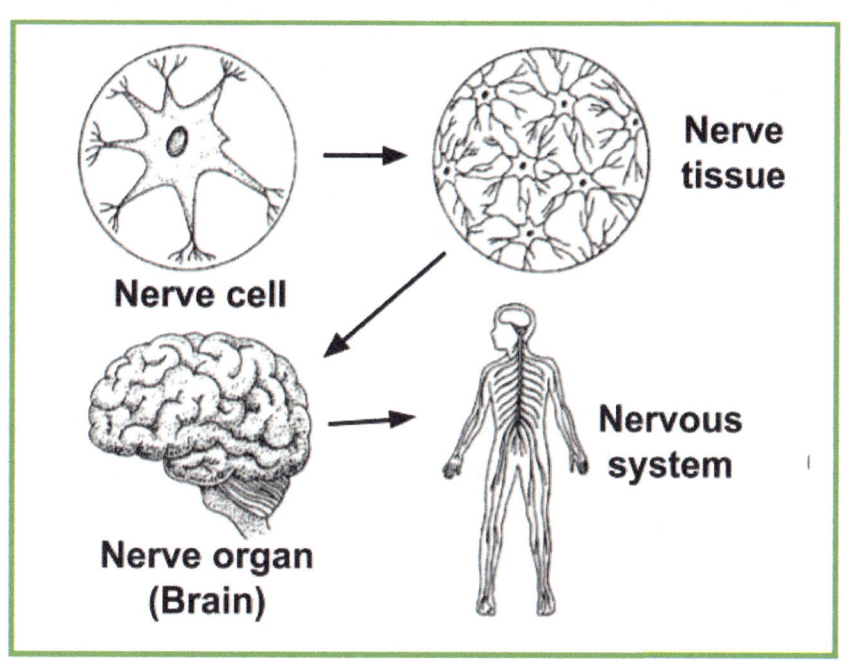

---------- REVIEW ---------- U-2 L-3

I. In each blank, write the word that fits best. Choose from the words below.

organ   specialized   system   ameba   organism   tissue   cell

Cells that do a special job are described as _____. A group of cells that are alike make up a _____ . In an _____ different tissues work together. Different organs make up a _____.
Several systems work together in a complex _____.

II. Fill in each blank in column **A** with the correct <u>word</u> from column **B**. (Write out the <u>words</u>).

	A	B
A. _____	brain, spinal cord, nerves	1. organ
B. _____	ear	2. organism
C. _____	whole human body	3. tissue
D. _____	group of nerve cells	4. system

III. What kind of organisms have no tissues? [Hint: Review the previous page.]

How Are Living Things Grouped? U-2 L-4

Exploring Science / Historical Steps

What's in a Name? Many animals and plants have more than one name. People in some places call a certain wild animal a mountain lion. People elsewhere call it a cougar .... or a puma .... or a panther.

Sometimes, *different* animals have the *same* name. In some places, a gopher has fur and digs burrows. In other places, a type of turtle is called a gopher. "Gopher" can also mean a kind of snake!

People may even use the same name for both an animal and a plant. Sea urchin is usually the name for a marine animal. However, the name sea urchin is also sometimes used for a type of shrub.

Carolus Linnaeus

In the 1700s, a scientist from Sweden, **Carolus Linnaeus,** helped to end the confusion. In his system, each kind of organism is given a **scientific name**. The system uses two Latin words for each name. The name for a mountain lion is *Felis concolor*. A bobcat is *Felis rufus*. A house cat is *Felis domesticus*. What do you think Felis means?

➤ Which name would a scientist probably give to the old cartoon character Mickey Mouse?
    A. *Mus musculus*   **B.** house mouse

➤ To Do Yourself   How Can Likenesses and Differences Help to Group Organisms?

You will need: A sheet of paper and a pencil

1. Look at the list of animal pairs below:

cat - lion	**chicken - duck**	**snake - worm**
gorilla - chimpanzee	**shark - whale**	**squirrel - hamster**
fly - spider	**rabbit - mouse**	**mouse - rat**

2. Make a large chart based on the pattern that is shown below for a pair of animals. Complete at least two more pairs of animals - listing similarities and differences.

Animal Pairs	Alike	Different	
1. horse - zebra	- hoofs on feet - mane - runs fast	- horse has no stripes	- zebra has stripes

Questions

1. Which animal traits might put it in a group with other animals? _____

_____

2. Which traits might put an animal in a different group? _____

_____

The Animal Kingdom

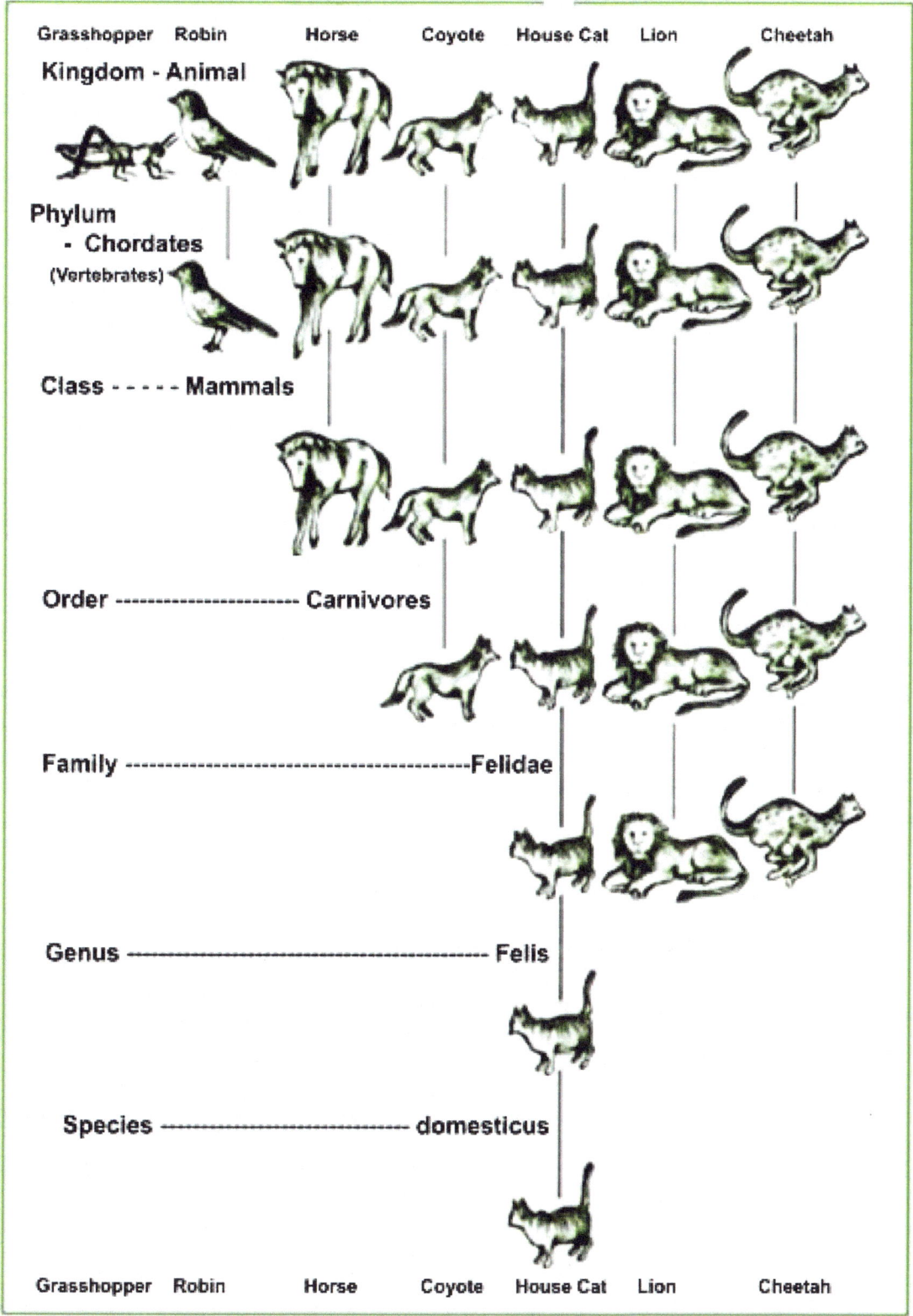

Study the chart.
- How is a cat related to a grasshopper?
- Name three ways that a cat is related to a horse.

[Interesting facts: Lions are in the genus Panthera - cats that roar.
Cheetahs are in the genus Acinonyx - cats that can not fully retract (pull in) their claws.]

Groups and Names of Organisms

Linnaeus' system gives names to organisms. It also is a way to **classify** organisms. When we classify things, we put them into groups. Biologists have long classified living things by their parts (**structures**). In recent decades, scientists have also been comparing the DNA of organisms.

A group of organisms with all of the same structures make up a **species**. Species (SPEE-sheez) are alike in so many ways that they are able to successfully reproduce. Species are grouped together in a **genus** (JEE-nus). *Felis* is a genus. It includes catlike animals.

A **scientific name** is made up of the genus and the species of an organism. *Felis domesticus* is the scientific name of a house cat. The first part of the name is the genus. This part of the name is always capitalized. The second part is the species, which is never capitalized. Both words of the scientific name must either be underlined or italicized.

Each living thing has its own scientific name, and this name is the same everywhere in the world.

Some catlike animals are not grouped in the genus *Felis*. They are placed in other genera. (**Genera** is the plural of genus). The cheetah is *Acinonyx jubatus*. Biologists place similar genera (JEN-ur-uh) in a larger group called a **family**.

Similar families are grouped into an **order**. The cat family is in the same order as the dog family. This is because cats and dogs are alike in many ways. For example, both have sharp teeth that help them eat meat (since both are carnivores).

A group of similar orders forms a **class**. Mammals is a class. The mammal class contains cats, dogs, horses, mice, and even people. All birds form another class.

Classes are put together in still larger groups called **phyla** (FY-luh). (Phyla is the plural of **phylum**). Mammals and birds are in the same phylum. Insects are in another phylum. Earthworms are in still another phylum.

Mammals, insects and worms are all animals. They are part of the animal **kingdom**. All organisms are grouped into one of six large kingdoms. The two kingdoms that you know best are the plant kingdom and the animal kingdom. Members of the plant kingdom make their own food. Members of the animal kingdom can't make food; they must eat food. Microbes and other simple living things belong to other kingdoms.

Finally, the largest grouping of organisms is called a **domain**. All kingdoms of living things belong to one of three huge domains.

REVIEW — U-2 L-4

I. In each blank, write the word that fits best. Choose from the words below.

| classify | structures | species | genus | domain | phylum |

Biologists group, or _____, all organisms. Members of the same

species have the same parts, or _____. In each genus, there can be

several _____. A _____ is the largest group of organisms.

II. Circle the word (between the brackets) that makes each statement correct.

    A. A genus contains more organisms than a [kingdom / species].

    B. In *Homo sapiens*, the genus name is [*Homo* / *sapiens*].

    C. An order is a [smaller / larger] group than a phylum.

III. Why should scientists use scientific names of organisms in their reports?

What Are the Simplest Living Things? U-2 L-5

Exploring Science / Historical Events

Big Challenges / Tiny Helpers. Fossil fuels, like coal, oil, and natural gas have made our lives easier. We have long used these fuels to produce the energy to run our cars and to build materials in our factories. But years of burning fossil fuels has left us with major troubles. In Unit 1 Lesson 11 you learned of **global warming**.

In addition, how we _get_ fossil fuels has caused serious problems. Removing mountains to get coal destroys ecosystems and pollutes waterways. Drilling for oil that is located under oceans carries high risks of leaks. In 2010, an oil drilling tower in the Gulf of Mexico exploded. Over 200 million gallons of oil was released into the gulf (and eleven workers were killed).

A giant offshore oil drilling platform

The ways that we _move_ fuel is also troublesome. In 1989, a ship called the **Exxon Valdez** was carrying crude oil from the Alaska pipeline. The ship ran aground, ripping a hole in its hull. Over 10 million gallons of oil poured out and soaked the beaches.

Beach polluted with spilled oil

The thick, sticky mess killed many animals and ruined the habitat for thousands of birds, fish and other wildlife. Workers tried to rescue some of the animals that were covered with the ooze. With detergent, some of the oil was painstakingly washed off of animals (as well as boulders) on the shore.

Washing oil from an oil spill victim

Scientists also came up with an idea to reduce the amount of long term damage. They put microbes to work. In any environment there are millions of microbes, even along Alaskan beaches. Some microbes help to break down dead plants and animals. The scientists added a special kind of fertilizer to the sand. This helped microbes to multiply faster. Some microbes ate a great deal of the oil. They continued to eat the oil for five months. Using these tiny helpers was a partial success.

Crude oil is made of carbon and hydrogen compounds called **hydrocarbons**. As the microbes ate, scientists hoped that most of their wastes would be harmless carbon dioxide and water - the same materials that you release when you breathe out. They were sad to learn that some of the oil-eating microbes gave off other harmful molecules.

Unfortunately, there have been many spills and leaks besides the infamous ones in 1989 and 2010.

Scientists will keep seeking ways that microbes can help us. However, we must find safer ways to _get_ and to _move_ fuels. And, we must get more energy without using fossil fuels.

➤ What are other sources of energy?

Classification of Simple Living Things

Simple living things are not all alike, but nearly all are made of only one cell. The tiniest types (too small to see with a regular microscope) have existed the longest. In these microbes, the DNA just floats in the cytoplasm; there is no nucleus that holds the DNA. These "no-nucleus," single-celled organisms come in two huge categories called **domains** - the domain <u>bacteria</u>, and the domain <u>archaea</u>.

<u>DOMAIN BACTERIA</u>. You have no doubt heard of bacteria. While most bacteria must eat to stay alive, some are able to make their own food (much like plants do). In fact, one group of these food-makers, called the **cyanobacteria**, changed the entire early atmosphere of the earth. They made the planet's first oxygen! From making oxygen to breaking down animal wastes, many bacteria certainly help humans. But a few types cause us serious problems.

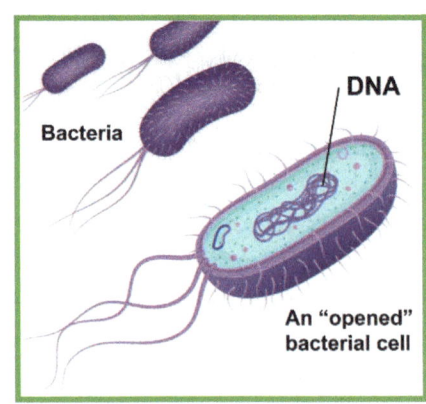

Bacteria with long thin DNA not "wrapped" in nucleus.

DOMAIN ARCHAEA. Members of this second domain of "no-nucleus," single-celled organisms are able to live in extreme conditions. These include environments similar to those that scientists think existed on Earth billions of years ago. The name **archaea** comes from the word archaic, which means ancient.

DOMAIN EUKARYA. In an earlier lesson you learned that the cells of many living things (like plants and animals) *DO* have a nucleus. All organisms whose cells <u>have</u> a nucleus are in the domain **eukarya** - which means "true nucleus."

<u>KINGDOM PROTIST</u>. Even in the domain eukarya there are some organisms that consist of only one cell. Scientists group most of the single-celled members of the domain eukarya in one <u>kingdom</u> - the **protists**.

Although protists are single-celled organisms, most are large enough to see under a normal microscope. They are exciting to see! The ones that move and eat are called **protozoans** (which means "early animals"). Some of these are **parasites** - living off of (but not actually eating) other living things. The blob-like ameba (shown in Unit 2 Lesson 2) and the slipper-shaped paramecium (shown at the right and in Unit 7, Lesson 1) are protozoans.

The protists that make their own food are called **algae**. Like plants, algae have cell walls. In fact, scientists believe that early forms of algae are the ancestors of plants. Among the most beautiful algae, with their two glass-like shells, are the **diatoms**. Kelp, made of very long thin sheets of algal cells, is a very important type of algae in the ocean; "kelp forests" provide food and shelter for many marine (ocean) animals. In the common pond algae called spirogyra (below), chloroplasts are arranged in a spiral.

DOMAIN Eukarya - Cells with <u>DNA in a nucleus</u>
KINGDOM: Protist

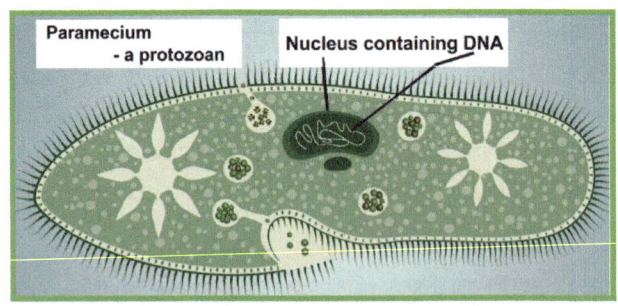

A typical <u>protozoan</u> (animal-like)

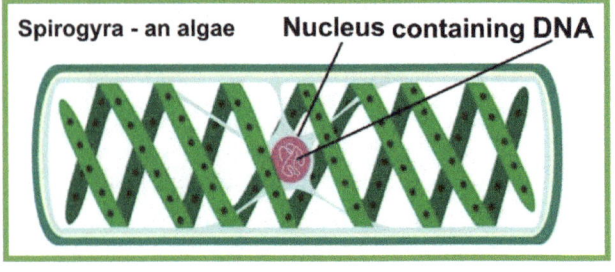

A typical <u>algae</u> (plant-like)

KINGDOM FUNGI. Fungi don't move, and they have cell walls; so fungi were long classified as plants. However, unlike plants, fungi's cell walls do not contain cellulose. Another key difference - plants make their own food; fungi do not. Most fungi live on dead matter. Some fungi live off of other living things.

Fungi vary greatly in size. **Yeasts** (shown in Unit 7, Lesson 1) are single-celled fungi. They sometimes form chains. Yeasts are used in making bread and alcohol. **Molds,** like bread mold, are fungi that have many cells. In the early 1990s scientists found a single, huge underground fungus covering an area larger than a thousand football fields!

A fungus that grows underground may send up parts called **mushrooms**. Many mushrooms are eaten by people all over the world. Others are quite poisonous.

WHAT ABOUT VIRUSES? When was the last time that you were sick with a cold or the flu? Blame a virus!

You likely remember people getting vaccines to protect themselves from the **viruses** that cause COVID-19, the measles, or the flu. Other infamous viruses are **HIV** (the cause of AIDS), and **HPV** (the cause of some cancers).

Viruses are extremely small - much tinier than bacteria. They are not made of cells. They have no cytoplasm. They even lack a nucleus. But viruses have either DNA or a similar molecule called RNA.

When they are alone, viruses look like crystals, and they show no life functions. Unfortunately, when viruses enter another organism, they can do one thing very well - reproduce. (See Unit 10, Lesson 1 for details).

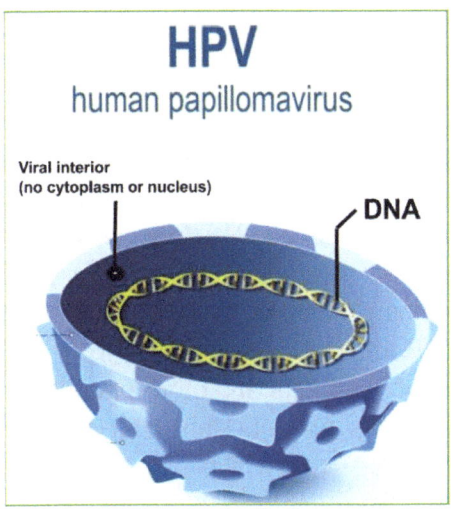

Bottom half of HPV virus showing DNA inside

So, are viruses alive? Most scientists say no. Therefore, scientists do not classify viruses in any domain.

------------------------------ **Review** ------------------------------ U-2 L-5

I. Write the number of the item in **list B** that *best fits* each item in **list A**.
Be careful; each choice must be used only once.

List A
A. ____ viruses
B. ____ algae
C. ____ bacteria
D. ____ fungi
E. ____ protists

List B
1. have no definite nucleus
2. have no cells
3. most live on dead matter
4. make their own food
5. include protozoans

II. For each organism,
Write **B/Ar** if it fits into either the **domain** bacteria OR the **domain** archaea.
Write **P** if it fits into the **kingdom** protist. Write **F** if it fits into the **kingdom** fungi.

A. ____ bacteria   C. ____ bread mold   E. ____ diatoms   G. ____ yeast
B. ____ mushroom   D. ____ algae   F. ____ ameba

51

What Are the Main Groups of Plants? U-2 L-6

Exploring Science / Historical Steps

<u>A Fruit-Salad Tree: Fact or Fake</u>? Picture in your mind a sort of fruit-salad tree. Most of its branches have peaches. But one branch has apricots. Another branch has plums. Fact or fake? Strange as it may seem, such a tree really can be grown.

The fruit grower starts with a young peach tree. Two of its branches are cut off. A twig from an apricot tree is joined, or **grafted**, onto one cut. A twig from a plum tree is grafted onto the other cut. Each twig grows into a branch that bears its own kind of fruit.

In the early part of the 20th century, **Luther Burbank** was the "plant wizard" of California. As a boy, Burbank lived on a farm where he learned to love plants. He also learned about grafting, since grafting was used on his family farm to grow apples. Breeding better kinds of plants became Burbank's life work. He developed hundreds of new fruits and other plants.

Grafting was an important part of Burbank's work. Suppose a new type of fruit had to be grown in large numbers. Often, it could only be reproduced by grafting onto another tree, called the **stock**. Sometimes, but not always, a graft would "take." When a graft takes, the stock accepts the graft as part of itself.

Luther Burbank created many new kinds of plants by grafting.

Plum and apricot grafts "take" on a peach stock. But a pear graft does not "take" on an apple stock. Why? Here's a clue. Plums, apricots, and peaches all belong to the same genus of plants. Pears and apples are not grouped into the same genus. Usually, grafts are more likely to "take" between closely related plants. When seeking whether a graft will "take," it helps to know how the plants are classified.

➤ Lemons and oranges both belong to the genus *Citrus*. Growing an orange graft on a lemon stock is probably [possible / impossible].

The Plant Kingdom

In the plant kingdom there are two broad groups of plants. Each group is called a **phylum**. The grouping depends on how a plant takes up water.

All plants need water. Some plants have <u>no</u> **veins**, which are tubes that carry water. The term **vascular** (VAS-kyuh-lur) means "having tubes," so plants with <u>no</u> veins belong to the phylum of **<u>nonvascular plants</u>**. These are the mosses and liverworts (LIV-ur-wurts). These plants have leaf-*like*, root-*like*, and stem-*like* parts. But because these parts have no veins, they are not *true* leaves, stems, or roots.

Phylum of <u>nonvascular</u> plants: <u>moss</u> on rotted log

You can see many kinds of mosses in the woods today. Mosses grow close to the ground where it is damp, and on the shady parts of trees. Liverworts also grow where it is dark and moist. Their flat, leaf-like parts grow on damp ground or on wet rocks - often beside streams.

Plants with veins belong to the phylum of **vascular plants**. Ferns, grasses, shrubs, and trees are vascular plants. All have true leaves, true stems, and true roots.

There are three main classes of vascular plants. **Ferns** make up one class. Another class contains the cone-bearing plants, or **conifers** (KUH-nuh-furs). A pine tree is a conifer. So is a giant redwood tree. The third class of vascular plants contains the **flowering plants**. A dandelion, a rose bush, and a red oak tree are all flowering plants.

Plants with veins are certainly able to grow tall!
Shown: "General Sherman" (a giant sequoia tree in western USA) is the world's largest tree by volume.

The three main classes of vascular plants:

1) the class of ferns / cinnamon ferns near log

2) the class of conifers / spruce tree twig and cone

3) the class of flowering plants / dogwood tree

REVIEW — U-2 L-6

I. In each blank, write the word that fits best. Choose from the words below.

nonvascular   vascular   veins   grafts

Some plants have tubes called _____ that carry water. Plants that have no veins are _____. Plants with veins are _____ .

II. Write the letter of the group to which each plant belongs. The pictures and captions of this lesson will help you find the answers.

A.   a. **Vascular plants**   b. **Nonvascular plants**

(1) ___ grass   (2) ___ moss   (3) ___ fern   (4) ___ liverwort

B.   a. **Cone-bearing plant**   b. **Flowering plant**   c. **Fern**

(1) ___ pine   (3) ___ dogwood tree   (5) ___ rose

(2) ___ cinnamon fern   (4) ___ sequoia tree

III. Why do mosses and liverworts never grow tall? Hint: What are they 'missing' that would help them grow?

53

What Are the Main Groups of Animals? U-2 L-7

Exploring Science / Historical Steps

The Beast That Wears a Girdle. In 1983, scientists discovered the first animal known that survives without oxygen! (There are many microbes that do so). The odd-looking little "beast" has spines around its head. It can pull its tube-like mouth in and out. As a baby, it swims with its "toes." As an adult, it has no toes. Thick plates of skin go around its middle. The animal is microscopic, and lives in the sea.

This "beast" had never been seen. The scientists who found it named it **Loricifera**. That means "girdle wearer." A girdle is a piece of stretchable, tube-like clothing that squeezes the middle of the body.

When these animals grow, they **molt**. This means that they shed their skin.

Scientists find and report about 6,000 new species a year. Why was this one special?

Because, it did not fit into *any* known phylum! (As you know, a phylum is a major group in a kingdom). This little girdle wearer fit into an entirely new phylum - all by itself. Only once before in the 20th century had a new phylum been added to the animal kingdom.

Surprisingly, in 2006, another new animal phylum was discovered and named! This animal is worm-like, about 2.5 centimeters long. It is now named *Xenoturbella*.

There are now over 30 phyla in the animal kingdom. The biodiversity of our planet is truly astounding, and we have much left to learn about it!

➤ Which statement is more likely to be true?

A. No new phyla of animals will be found in the future.

B. Some of the new species found in the future may be placed in new phyla.

Tiny members of the Animal Phylum
- **Loricifera** -
discovered in 1983

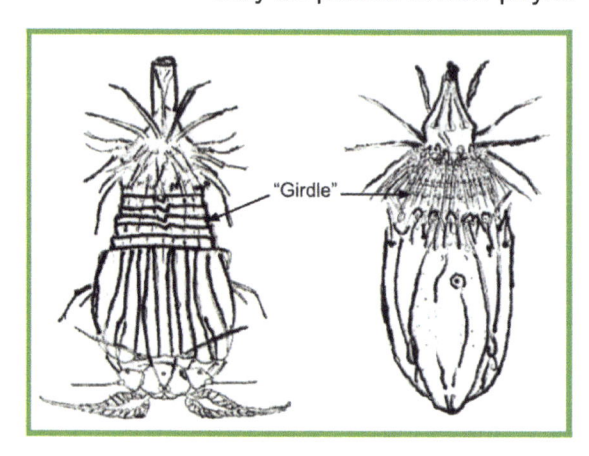

The Animal Kingdom

If you have a pet cat or dog, rub its back. You can feel its backbone. A backbone is made up of many small bones. Each bone is a **vertebrae** (VUR-tuh-bruh). Animals that have backbones are called **vertebrates** (VUR-tuh-bruhts). Do you think that you are a vertebrate? How can you tell? Do you have any other vertebrate pets?

Most phyla of animals have no backbones. These are **invertebrates**. The girdle wearer *Loricifera* and the worm-like *Xenoturbella* are invertebrates. The invertebrates that you know best are the insects and the worms. Some others may surprise you. You may have thought that they were not even animals!

Table 1 Phyla of Some Animals

Phylum	Description	Examples
Porifera (pore bearers)	Live on objects in water; filter food from water that flows in tiny pores and out of the large holes	bath sponge
Cnidaria (stinging tentacles)	Live attached or free in water; tentacles catch food; one opening for food and wastes	jellyfish; sea anemone; corals
Flatworms	Live free or on other organisms; one opening for food and wastes	planarian
Roundworms	Live free or on other organisms; food passes in mouth, and wastes pass out other end (anus)	nematode
Segmented Worms (segments)	Body in sections	earthworm
Mollusks (soft bodied)	Soft bodies; many with shells, organs, and organ systems	squid; octopus; snail; clam
Echinoderms (Spiny-skinned Animals)	Tube feet and spines; one opening for food and wastes	starfish; sea urchins
Arthropods (jointed-legs)	Hard outside covering (exoskeleton); jointed-legs	insects (crickets); crustaceans (crabs); spiders
Chordates (primarily the Vertebrates)	Animals with backbones	fish; amphibians; reptiles; birds; mammals

55

Table 2 Classes of Vertebrates

Class	Description	Examples
Fish	Scales on body; live in water; most lay eggs	shark; trout; guppy
Amphibians	Young live in water; adults live on land; smooth skin; lay jelly-like eggs in water	frog; salamander
Reptiles	Scales; live on land; (air-breathing); most lay leathery eggs (on land)	turtle; snake; lizard; crocodile
Birds	Feathers; most have wings and can fly; hard-shelled eggs; most live on land; (air-breathing)	eagle; robin; penguin
Mammals	Hair on body; produce milk for young	rabbit; gorilla; dolphin; whale

Table 3 Some Orders of Mammals

Order	Description	Examples
Egg Layers	Lay eggs	platypus
Marsupials	Carry young in pouch	kangaroo, opossum
Flying Mammals	Web of skin between toes	bat
Toothless Mammals	No teeth in front; armor plated body	sloth, armadillo
Rodents	Teeth in front shaped like chisels	squirrel, chipmunk

Table 3 Some Orders of Mammals

Table 3 (continued)   Some Orders of Mammals

Order	Description	Examples	
Sea Mammals	Live in ocean; flippers	whale; dolphin seal; walrus	
Trunk-nosed Mammals	Trunk noses and tusks	elephant	
Hoofed Mammals	Hoof as toenail	cow; sheep	
Carnivores	Meat-eaters	dog; tiger	
Primates	Walk on hind legs; front facing eyes; relatively large brains	monkey; lemur gorilla; orangutan human	

The tables on the previous pages list groups of animals. You will find phyla and classes. You will also find several orders of the class of mammals. Can you find yourself in one of those orders?

➤ To Do Yourself — How Can You Group Vertebrate Animals?

You will need: Pencil and paper

1. Make five columns on your paper. Head these with the Classes of animals - fish, amphibian, reptile, bird, and mammal.

2. Make a list of animals that you are familiar with. Start with the following list: frog, eagle, camel, penguin, dolphin, snake, pig, turkey, guppy, ape, deer.

3. Use the **Key to Vertebrate Classes**. For each animal in your list, answer the questions in the key. Write the name of the animal in the correct column.

Key to Vertebrate Classes

A. Does the animal have feathers?
   1. If YES, it is a BIRD.   2. If NO, go to B.

B. Does the animal have fins?
   1. If YES, it is a FISH.   2. If NO, go to C.

C. Does the animal have hair or fur?
   1. If YES, it is a MAMMAL.   2. If NO, go to D.

D. Does the animal have scales?
   1. If YES, it is a REPTILE.   2. If NO, it is an AMPHIBIAN.

Review — U2 L7

I. In each blank, write the word that fits best. Choose from the words below.

vertebra   vertebrate   Loricifera   invertebrate   mammal

Each bone in a backbone is a _____. An animal without a backbone is an_____. An animal with a backbone is a _____.

II. Write the letter of the group to which each animal belongs.

A. a. vertebrates   b. invertebrates

   (1) ___ bird   (3) ___ snake   (5) ___ sponge
   (2) ___ spider   (4) ___ flatworm

B. a. mammal   b. mollusk   c. crustacean   d. cnidaria   e. fish

   (1) ___ clam   (3) ___ whale   (5) ___ jellyfish
   (2) ___ crab   (4) ___ trout

C. a. amphibians   c. primate   e. joint-legged animal
   b. segmented worm   d. rodent

   (1) ___ earthworm   (3) ___ rat   (5) ___ frog
   (2) ___ gorilla   (4) ___ centipede

Unit 2 -- Review What You Know ---

A. Unscramble the groups of letters to make science words. Write the words in the blanks.

1. YTCOPLMAS / _____
   = part of a cell between the nucleus and the cell membrane
2. CENLSUU / _____ = center of control in a cell
3. RHOOPTASCLL / _____
   = green part found in plant cells but not in animal cells
4. GNOAR / _____ = group of tissues that does a special job
5. IRVSU / _____ = a microbe that has no cytoplasm
6. AVRSCLUA / _____ = plants that have water tubes

B. Write the ending that best completes each statement.

1. Compared to a microscope, a hand lens' magnifying power is …
   a. larger   b. smaller   c. the same   1. _____

2. Microbes that have chlorophyll include
   a. bacteria   b. algae   c. viruses   2. _____

3. A cell's DNA is found in its
   a. vacuole   b. cell membrane   c. nucleus   3. _____

4. A tissue is made up of a group of
   a. cells   b. organs   c. systems   4. _____

5. In *Panthera leo*, the name for a lion, *leo* is the
   a. phylum   b. genus   c. species   5. _____

6. Organisms classified in the fungi kingdom include
   a. bread molds   b. malaria germs   c. green algae   6. _____

7. Plants that are nonvascular include
   a. conifers   b. mosses   c. roses   7. _____

8. Among the vertebrate animals are
   a. crabs   b. earthworms   c. dogs   8. _____

9. The "skin" of a blood cell is its
   a. cell membrane   b. cell wall   c. cytoplasm   9. _____

10. A system is made up of a group of
    a. organisms   b. organs   c. phyla   10. _____

11. A part in both plant and animal cells that stores water
    is the   a. cell wall   b. vacuole   c. chloroplast   11. _____

12. An animal without a backbone is
   a. a vertebra   **b.** a vertebrate   **c.** an invertebrate   12. _____

13. The stiff woody matter in a cell wall is
   a. cellulose   **b.** DNA   **c.** chlorophyll   13. _____

14. In *Homo sapiens*, the name for a human, *Homo* is the
   a. class   **b.** family   **c.** genus   14. _____

15. The largest group into which living things are classified
   is a   **a.** domain   **b.** species   **c.** phylum   15. _____

C. Apply What You Know

1. Study the drawings below of different kinds of living things. Answer the questions on the next page.

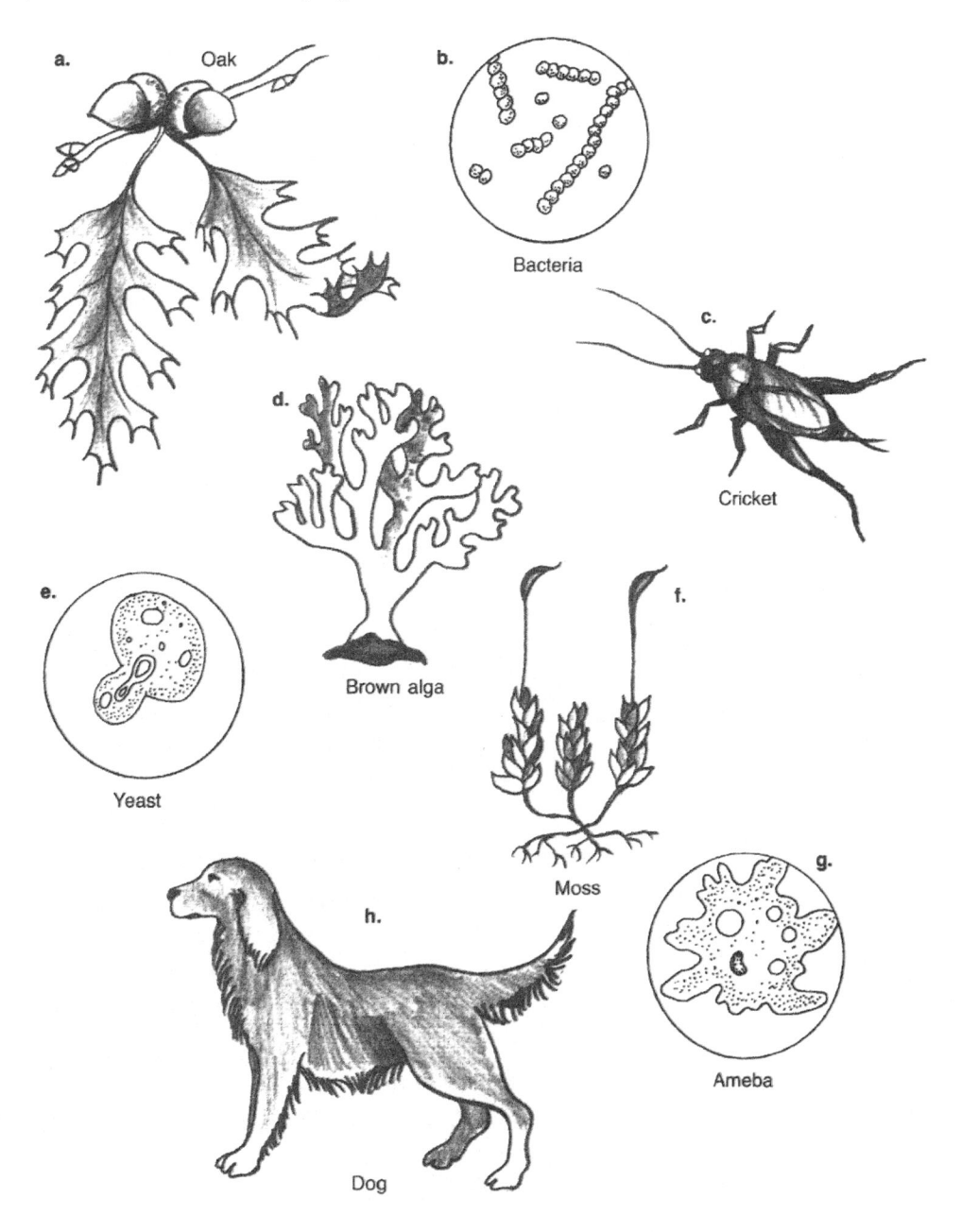

a. Oak
b. Bacteria
c. Cricket
d. Brown alga
e. Yeast
f. Moss
g. Ameba
h. Dog

61

Write the kingdom to which each organism in the drawing belongs. Use these kingdoms:

bacteria   **protist**   **fungi**   **plant**   **animal**

a. _____   e. _____

b. _____   f. _____

c. _____   g. _____

d. _____   h. _____

2. Again view the previous page. Fill in the blanks with the correct word.

    Which drawing shows a vertebrate? _____

    An invertebrate? _____

    A vascular plant? _____ A nonvascular plant? _____

3. Study the drawings of cells below. Match the labels for each numbered part.
Use these labels:

    A. cytoplasm   **B.** cell wall   **C.** cell membrane   **D.** nucleus   **E.** chloroplast

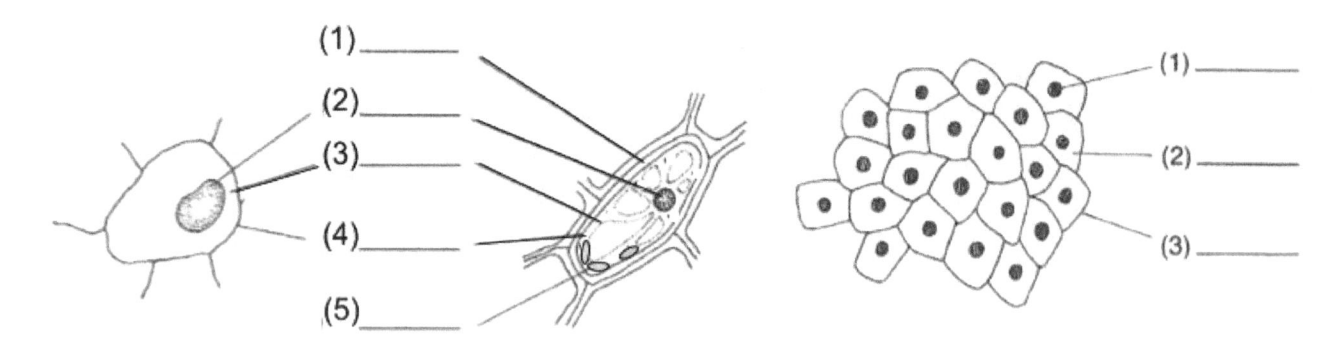

(1) _____
(2) _____
(3) _____
(4) _____
(5) _____

(1) _____
(2) _____
(3) _____

4. Fill in the blanks below with one of the following: • left   • center   • right

    Which drawing above shows: a plant cell? _____
                                 an animal cell? _____
                                 a tissue? _____

D. Find out more.

1. Obtain adult assistance (since this activity involves boiling water). Make a model of a cell. Place 1 packet of plain gelatin in a 2-cup heat-proof measuring cup. Fill to the 1-cup mark with boiling water. Stir to dissolve the gelatin. Fill to the 1 ½-cup mark with cold water. Add a teaspoon of cologne. Stir to mix. Pour the gelatin mixture into a small (sandwich-size) zip top bag. Place a marble in the bag. Close tightly. Let stand overnight. What does the plastic stand for? The gelatin? The marble?

    Place the bag into a jar. Add enough warm water to cover the bottom half of the bag. Let stand for a few hours. Take the cell model out of the dish. Smell the water. Explain what you observe.

2. Make a collection of examples of living things from some or all of the six kingdoms. Try to find the names of each one and make a labeled display.

Careers in Life Science

Plants in the Human World. By the end of Unit 2, you will know more about plants. Perhaps someday you will make a living using some of this knowledge.

Wherever there are people, there are plants. All of these plants must be arranged, grown, and cared for. So there are many jobs for people who provide the trees, flowers, lawns, and shrubs - where people live, work, and play.

Gardeners and Landscape Maintenance Contractors. To be a gardener, you should certainly like working outdoors. Do you have a "green thumb?" That's a help, too. You can learn gardening through on-the-job training. Gardeners plant seeds, do grafting, and care for the plants that they grow.

After a year or so of technical school, you can start a career as a landscape maintenance contractor. This job carries more responsibility than that of a gardener. You can become your own boss and run your own business. Landscape maintenance contractors are called in by landscape architects to carry out their plans.

Landscape Architects. A landscape architect plans and designs outdoor settings of all kinds. There are parks, gardens, golf courses, sports fields and arenas. There are the grounds of hotels, schools, resorts, hospitals, factories, office buildings, and housing developments. There are parkways, freeways, and airports. Can you think of other places where people build, or move the land, and thus need the services of landscape architects?

Some landscape architects specialize in the use of native plants. In order to protect the local biodiversity, it is vital that invasive plants not be used!

A landscape architect combines art, engineering, and science. Talent in these areas - and in math - is needed. So is college training for four or more years.

Landscape maintenance contractors and gardeners and enjoy outdoor, physical work that involves knowledge of plants and growth seasons.

After graduation, the person must gain work experience for a period of time. Then, in many states, she or he may become licensed as a full-fledged landscape architect.

Scenes like this result from the work of landscape architects - people who combine art and science.

From gardener to landscape architect, everyone in the field of landscaping helps to make our world more beautiful and easier to live in.

UNIT THREE

FOOD FOR LIFE

Copyright © 2008. For more information about The Healthy Eating Pyramid, please see The Nutrition Source, Department of Nutrition, Harvard T.H. Chan School of Public Health, www.thenutritionsource.org, and Eat, Drink, and Be Healthy, by Walter C. Willett, M.D., and Patrick J. Skerrett (2005), Free Press/Simon & Schuster Inc.

What Chemicals Make Up Living Things? U-3 L-1

Exploring Science / Historical Steps

It's Alive? Is this old movie scene familiar to you? Dr. Frankenstein is working in a lab full of flasks, test tubes, and wires. The "monster" lies unmoving on the table. Suddenly, the doctor throws a switch. Lightning flashes. Smoke swirls through the air. You see the monster rising. The doctor says, "It's alive!"

For years people have wondered if life did come from nonliving matter. In 1953, scientists **Stanley Miller** and **Harold Urey** completed a now famous experiment. They filled large flasks with water, ammonia, hydrogen, and methane. They believed that these chemicals were common on the early earth. Next, they heated and cooled the mixture. They also gave it jolts of electricity to imitate lightning.

After a while, some of the chemicals had combined. They formed a "soup" of substances that are found in all living things. These substances were amino (uh-MEE-noh) acids.

Amino acids are the building blocks that make up proteins. And proteins are found throughout your cells.

A few years later, other researchers found that some simple molecules can naturally form little spheres. Scientists viewed these tiny spheres with powerful microscopes. The spheres attached to one another and formed a network. Sometimes they even sent electric signals to one another when they were stimulated by light!

These microscopic spheres (called microspheres) are not living things. But the way that they attach to one another may be a clue. Could this be how the first cells of living things formed?

➤ Scientists think that the chemicals of life were made from...
A. living chemicals   B. nonliving chemicals
C. the earth's sea as it is today.

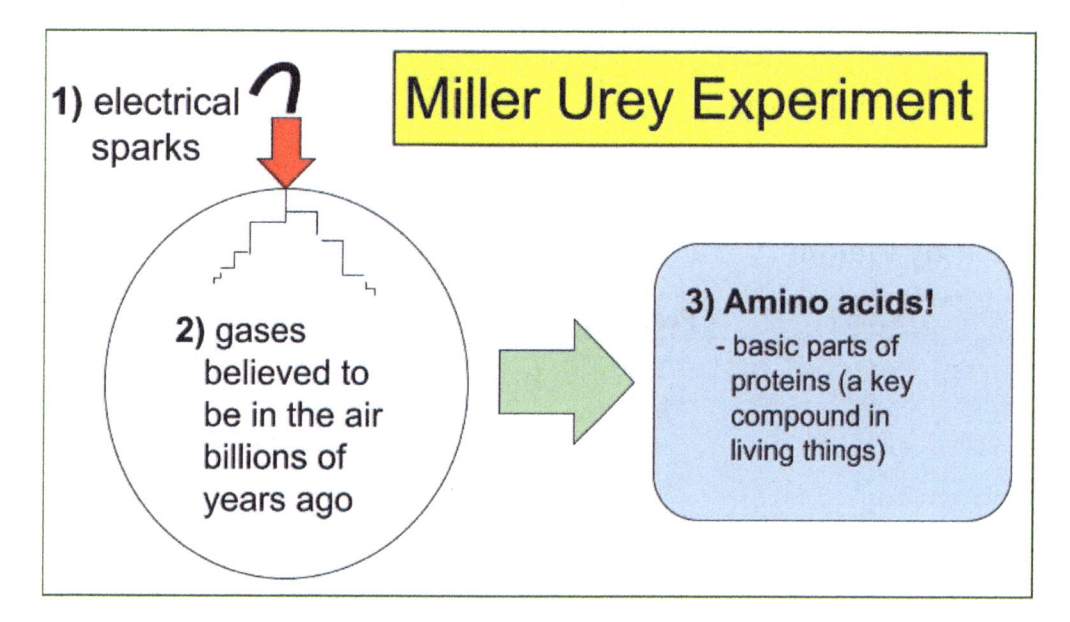

Miller Urey Experiment

1) electrical sparks
2) gases believed to be in the air billions of years ago
3) Amino acids!
   - basic parts of proteins (a key compound in living things)

The Chemicals of Life

What are these chemicals of life? What chemicals are in your body? A chemist would say that your body is made of **matter**. Matter includes all material things, living or nonliving. Matter can be a solid, a liquid, or a gas.

All matter is made up of **atoms**. The term **element** is used to describe the atoms that are identical to each other. There are about 100 different elements on Earth. You have likely seen a chart that organizes all of the elements - called the **periodic table**.

Oxygen is an element. So is hydrogen. Both are gases that you can't see, taste, or smell. These elements are some of the most common chemicals of life.

Carbon is another element. The black substance on a burnt marshmallow is carbon. So is the soot rising above the flame of a candle or match. Carbon, like oxygen and hydrogen, is an element found in very large numbers in all living things.

Some atoms are able to join together to form **molecules**. A molecule is the smallest amount of a substance that contains the properties of that substance. If two or more atoms of the _same_ element join together, we call the result a molecule of that element. For example, oxygen gas (that we breathe) is made up of two atoms of oxygen joined together, so we call this joined pair of oxygen atoms an oxygen molecule.

On the other hand, if the atoms of two (or more) _different_ kinds of elements join together, the result is not only called a molecule, it is also called a **compound**. Surprisingly, compounds are _not_ like the elements from which they are made. Water, for example, is a compound. When hydrogen gas and oxygen gas combine in a special way, water forms. Water is certainly unlike the two different gases that combined to form it!

Of course, water is one of the key compounds of life. Would you have guessed that over 3/4 of your weight is due to water? If you know how much you weigh, you can easily estimate how much water is in your body.

Scientists use a shorthand way of writing the names of elements and compounds. Each element has a **symbol**. A capital **C** is the symbol for carbon. Capital **H** is the symbol for hydrogen. Capital **O** is the symbol for oxygen.

Some elements have a capital letter followed by a small letter as their symbol. Here are some examples: **Fe** is the symbol for iron; **Na** is the symbol for sodium; **Cl** is the symbol for chlorine.

A **formula** is the way that you write the symbols for a compound. The formula for water, H_2O, tells you that a single molecule of water contains two hydrogen atoms and one oxygen atom. The small number 2 in the formula tells you this.

The formula for carbon dioxide is CO_2. Can you describe what this formula means? In other words, how many of each element join to make the compound called carbon dioxide?

The formula for salt (sodium chloride) is **NaCl**; each sodium atom joins with one chlorine atom.

Elements in Living Things / by Weight

Element	Symbol	Percentage
Oxygen	O	76.0%
Carbon	C	10.5%
Hydrogen	H	10.0%
Nitrogen	N	2.5%
Phosphorus	P	0.3%
Potassium	K	0.3%
Sulfur	S	0.2%
Chlorine	Cl	0.1%
Sodium	Na	0.04%
Calcium	Ca	0.02%
Magnesium	Mg	0.02%
Iron	Fe	0.01%

Compounds in Living Things / by Weight

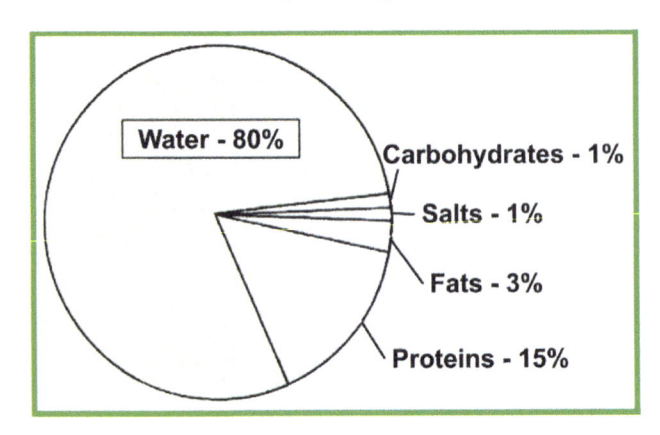

- Water - 80%
- Carbohydrates - 1%
- Salts - 1%
- Fats - 3%
- Proteins - 15%

> **To Do Yourself**   **What Chemicals Are in Sugar?**

You will need:

Old metal spoon, ¼ spoonful of sugar, pot holder, heat source

1. Under an adult's supervision, use a potholder to hold a spoonful of sugar over a flame. Be sure to use a potholder, as the metal will get very hot.
2. Observe and record what happens to the sugar.

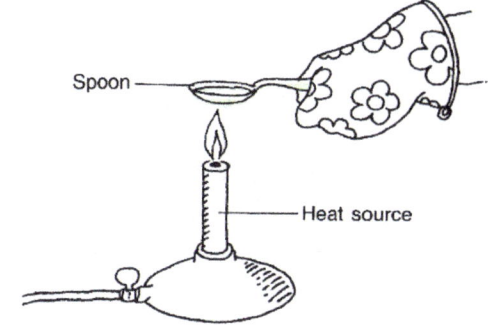

Questions

1. What first happened to the sugar?_____

2. What were your next observations?_____

3. What is the black material?_____

---------------------------------- **Review** ---------------------------------- U3 L1

I. In each blank, write the word that fits best. Choose from the words below.

elements	compounds	symbols	water
matter	formulas	molecule	scientist

There are about 100 different _____ on Earth. *Different*

elements often join together to form _____. The smallest piece of

any substance that still has the properties of that substance is called a

_____. Scientists use_____

to write the names of the elements. They use _____to write

the names of compounds. A key compound of life is _____.

II. Observe the graph and table in this lesson. Circle the answer to the following questions.

   A. The <u>element</u> that makes up most of the weight of the body is…
       (1) oxygen   **(2) carbon**   **(3) hydrogen**

   B. The <u>compound</u> that makes up most of the weight of the body is….
       (1) carbohydrate   **(2) fat**   **(3) water**

III. A scientist writes **H₂O**. Explain in a few sentences what this means, and include in your answer all of the following words: **element, compound, symbol, formula**.

How Do Green Cells Make Food? U-3 L-2

Exploring Science

Algae at Work. Have you had a milkshake at a fast-food restaurant lately? It might surprise you to know that it was thickened with a substance that comes from red algae. In fact, a quick search online for "foods with algae" will likely surprise you. Many foods contain some algae.

Algae of various types are also harvested from the sea, dried, and eaten every day by millions of people, especially in Japan.

Algae are plant-like organisms that grow in ponds, lakes, streams, and the ocean. They can grow very quickly, and are very important producers on our planet. In other words, algae start many of the Earth's food chains. In the wild, they are eaten by many organisms, from microscopic protists to quite large animals.

Producers Fight Global Warming. In this lesson, you will learn details about how plant-like protists (like algae) and actual plants (like trees) make food. A key point is that they *take in* carbon dioxide. So algae and plants help to lower the amount of carbon dioxide in our atmosphere!

Obviously, to fight global warming we must protect our oceans and waterways so that algae will grow well. And we must protect the land so that plants, especially trees, grow and reproduce.

➤ Since algae behave much like plants, in which specific aquatic habitat do you think more algae will be found?
   A. near the bottom. Explain.
   B. near the surface. Explain.

Kelp forests provide food (and shelter) for many marine organisms - all the while removing carbon dioxide from the ocean! This allows the ocean water to remove more carbon dioxide from the atmosphere.

Tropical rainforests are the most biodiverse land ecosystems. They remove enormous amounts of carbon dioxide from the atmosphere.

How Green Cells Make Food

How do algae and plants make food? To find out, let's look inside of the cells of the most common producers, those that are green.

Unlike animal cells, green cells contain small, usually oval, structures called **chloroplasts** (KLOWR-uh-plasts). These hold the special green compound, **chlorophyll** (KLOWR-uh-fil). Cells need chlorophyll in order to make food.

A green cell with chlorophyll uses three "ingredients" to produce food - water, carbon dioxide (from the air or the water), and light.

Chlorophyll traps light. It uses light's energy to change carbon dioxide and water into food. The light energy is changed into energy that is stored in a simple sugar.

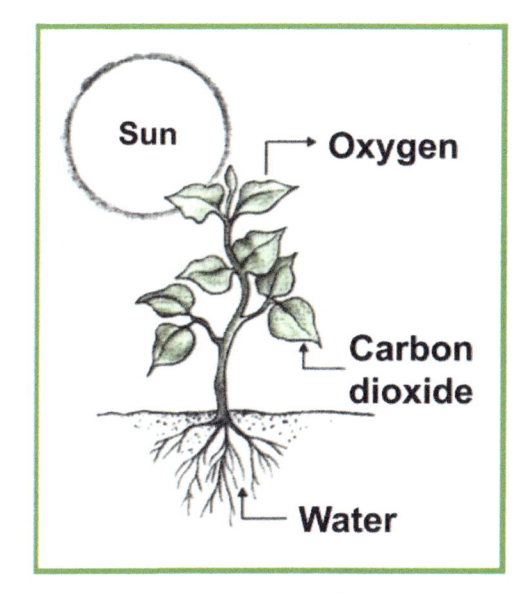

Green plants need three things to make food: <u>light</u>, <u>water</u>, and <u>carbon dioxide</u>.

▶ An Adult Demonstration   How Can the Green in a Leaf be Found?

You will need:

An adult to help you, a pair of googles, a spinach leaf, a saucepan, an electric hot plate, water, two test tubes, two corks for the test tubes, a test-tube rack, a well ventilated area, and alcohol

1. This lab **must** be done with an adult.
2. Tear a spinach leaf into small bits.
3. Put on your goggles. Boil the bits in a pan of water for a few minutes. Turn off the heat source; now work far from heat for the rest of this lab.
4. Place some boiled leaf bits in the empty test tube.
5. Fill this tube one-fourth full of alcohol.
6. Cork the tube and shake it every so often.
7. Allow the tube to cool for several minutes.
8. Pour the liquid into an empty test tube. Cork both tubes. Observe the liquid.
9. Observe the remains of the leaf bits.
10. Record your observations.

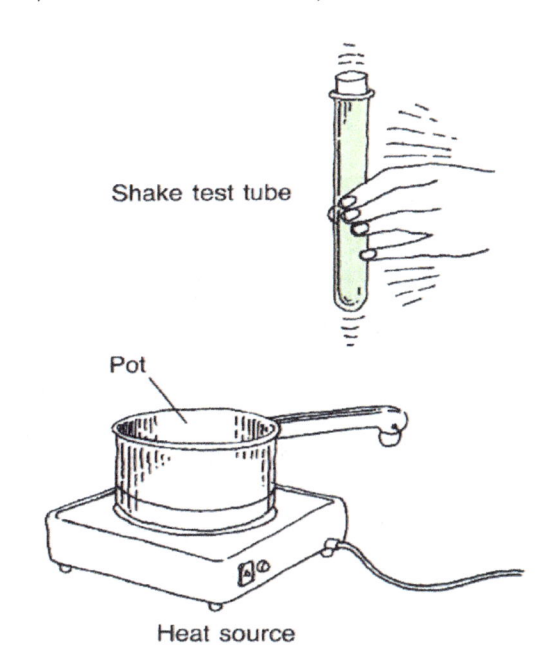

Questions

1. At the beginning, and then at the end, what color was the liquid ?_____/_____
2. At the beginning, and then at the end, what color were the leaf bits?_____/_____
3. What do you think caused the color changes?_____

Changing carbon dioxide and water into sugar in the presence of sunlight is called **photosynthesis** (foh-tuh-SIN-thih-sus). *Photo* means "light." In photosynthesis, the light is usually from the sun. *Synthesis* means "putting together." Carbon dioxide and water are put together using light energy - to make sugar! Oxygen is given off during the process.

We can show this process by using an **equation**. The top equation below uses words to tell the story. The same equation below it uses formulas. See if you can follow both equations. The arrows in the equations mean "makes" or "gives." Notice the numbers in front of the formulas for some compounds. These numbers tell how many of each compound is used.

Carbon dioxide + Water + Light energy ----------with-------------> chlorophyll   Sugar (stored energy) + Oxygen

$6 CO_2$ + $6 H_2O$ + Light energy ----------with-------------> chlorophyll $C_6H_{12}O_6$ + $6 O_2$

------------------------------------ Review ------------------------------------ U3 L2

I. In each blank, write the word that fits best. Choose from the words below.

| energy | chlorophyll | carbon dioxide | chloroplasts |
| oxygen | sugar | light | photosynthesis |

The green substance in a plant cell is _____ . A food that a plant makes is _____ . The plant makes food by the process of _____ . To make food, a plant needs carbon dioxide, water, and _____ . Chlorophyll in a green organism can be found in _____ . Animals eat green plants and other green organisms to get _____ . Photosynthesis not only makes food, it helps fight global warming by removing _____ from the atmosphere.

II. Circle the answer that best completes each sentence.

A. Chlorophyll uses light energy to change carbon dioxide and water into …
   (a) nitrogen   **(b)** food   **(c)** chloroplasts

B. Plants and other green organisms store energy in …
   (a) sunlight   **(b)** sugar   **(c)** carbon dioxide

III. Explain why some animals need to eat green plants.

How Do Living Things Get Energy? U-3 L-3

Exploring Science

Food for the Long Run. The first race of the cross-country season is about to begin. The starter's gun fires, and they're off! When you get ready for a race, you need to eat the correct food. But not everyone has the same idea of what food is correct.

Aiden has heard that sugar gives you quick energy. He eats candy on the day of the race. Maria thinks that a lot of meat makes muscles strong. She eats a steak dinner the night before the race. Mateo believes that vitamin pills give you energy. He takes some of these before the race.

Aiden, Maria, and Mateo all have ideas that people believed to be true for a long time. But Aliyah has a different idea. She read an article by some food scientists. They wrote that eating starch can help runners endure longer. So Aliyah has a meal of pasta and bread the night before the race. Both of these foods are rich in starch.

After running half of the race, Aiden, Maria, and Mateo suddenly become very tired. In runner's talk, they "hit the wall." They all slow down. But Aliyah keeps up her strong pace. She has enough energy to finish well.

➤ You are going on a bicycle ride that will last for several hours. The best food to eat before the ride is probably (bread / candy).

Getting Energy from Food

You need energy for living - for play, for exercise, and even for thinking. Where do you get all of this energy?

Living things get energy from **fuels**. (FYOO-uls). These fuels are food, such as sugar. To get energy, the sugar is "burned."

When fuels burn, they combine with oxygen. Wood is a fuel that is used to heat a campsite. When wood burns, it combines with oxygen. Some of the energy it gives off is light, and some of the energy is heat.

In your cells, sugar is the fuel. When sugar "burns," it combines with oxygen. There is no flame in this type of burning. You use some of the energy given off in your cells for living. And some of this energy is heat energy. Your body warmth comes from "burning" sugar.

Where does this sugar come from? It comes from the food that you eat. Your body breaks down the food into molecules that the cells can use. These molecules are then transported to your cells. Inside of the cells, the molecules are "burned."

During this "burning," carbon dioxide and water are formed. The energy that was stored in the sugar molecules is now stored in very small battery-like molecules called **ATP**. This kind of "burning" is called **cell respiration** (res-puh-RAY-shun).

One way to understand cell respiration is to put it in the form of an equation. Below are two ways to show the cell respiration equation.

Sugar + Oxygen -----inside-----> mitochondria   Carbon dioxide + Water + Energy

$C_6H_{12}O_6$ + $6\ O_2$ -----inside-----> mitochondria $6\ CO_2$ + $6\ H_2O$ + ATP

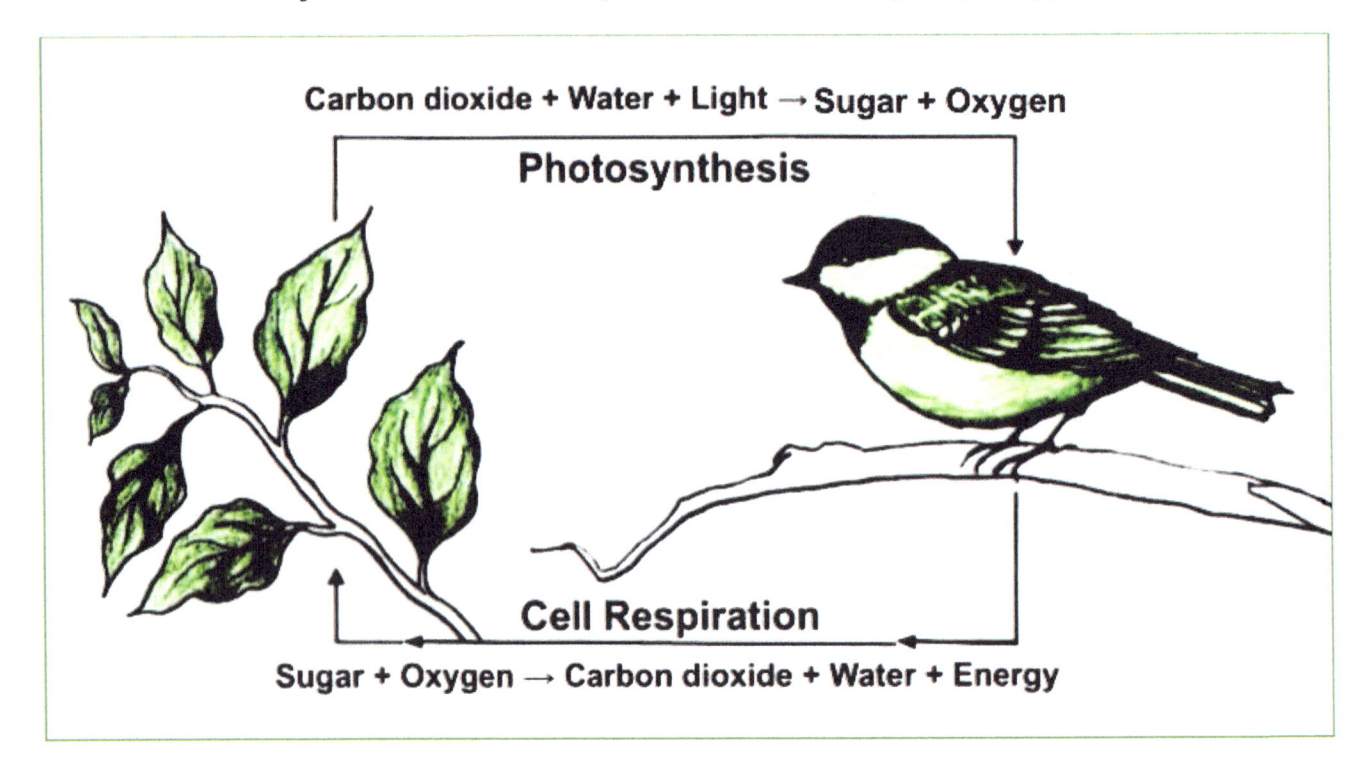

Photosynthesis and cell respiration are, in many ways, opposites.

Carbon dioxide + Water + Light → Sugar + Oxygen

Photosynthesis

Cell Respiration

Sugar + Oxygen → Carbon dioxide + Water + Energy

Opposites - but Not Really! The diagram above should remind you of the oxygen / carbon dioxide cycle from Unit 1 Lesson 5. Yes, photosynthesis and cell respiration are, in many ways, opposites.

There's a "catch" though. Plants don't simply *make* food. They must also *use* some of that food for themselves! To use food, plants carry out the same equation that animals do - cell respiration!

Lucky for animals (like us), plants nearly always carry out more photosynthesis than cell respiration. They make more food than they "burn." Some plants store this excess food in their roots, stems, or leaves. Most plants also store some of the excess food in their seeds and fruits. Do you see why animals (like us) should feel lucky?

Did you notice the specific site (inside of cells) where cell respiration happens? Look again at the arrow of the equation. Cell respiration happens inside of oval-shaped structures called **mitochondria**. (See Unit 2, Lesson 2). Obviously, mitochondria must be present in both plant and animal cells.

➤ **Trick question:**
Which organisms - plants or animals - depend on the equation below for energy?

Sugar + Oxygen -----inside-----> Carbon dioxide + Water + Energy
                 mitochondria

$C_6H_{12}O_6$ + 6 O_2 -----inside-----> 6 CO_2 + 6 H_2O + **ATP**
                         mitochondria

➤ To Do Yourself   Does Energy Make Heat?

You will need: A thermometer

1. Tuck a thermometer under your armpit. Keep it there for two minutes. Record the temperature to the right of Trial 1 under "Starting Temp."
2. Run 50 steps in place.
3. Obtain (and record) your temperature.
4. Do 10 jumping jacks.
5. Obtain (and record) your temperature.
6. Take a long break, and repeat all of the steps above. Record your results to the right of Trial #2.
7. Take another long break, and repeat all of the steps for the third time. Record your results to the right of Trial #3.

Exercise / Temperature Activity

Trial #	Starting Temp	Temp after steps	Temp after jumps
1	_____	_____	_____
2	_____	_____	_____
3	_____	_____	_____

Questions

1. What kind of temperature change did you observe? _____
2. What caused the temperature change? _____
3. What fuel was partially changed into heat? _____

REVIEW   U-3 L-3

I. In each blank, write the word that fits best. Choose from the words below.

chlorophyll   cell respiration   carbon dioxide   photosynthesis
oxygen   energy   mitochondria   food   ATP

The energy that we need to move, play, and simply stay alive comes from our

_____. Inside both plant and animal cells, food molecules are moved to

structures called _____, where they are "burned." The energy

obtained from this "burning" is trapped in battery-like molecules called _____. This

entire process is called _____. When food is

burned in cells, _____ and water are given off, along with

energy. In many ways, this process is the opposite of _____,

which happens in plant cells, but not in animal cells.

II. Explain how photosynthesis is different from cell respiration. Be sure to mention which applies to plants and which to animals. [Hint: Be careful on the second part!]

73

What Are the Parts of Plants?

U-3 L-4

Exploring Science / Historical Steps

The Plant and the Mouse. In the 1700s everyone knew that animals need air to live. But no one knew why. **Joseph Priestley**, of England, wanted to find out. He also thought that plants change the air in some way that is useful to animals.

To test his ideas, Priestley used large glass jars that had open bottoms and a knob on top. He placed a lively mouse in one empty jar that contained only air. The mouse soon became drowsy and seemed to go to sleep. The mouse had used up the part of the air that it needed; we now know that this part was oxygen.

Next, Priestley placed a mint plant under one empty jar, and kept this set-up in a sunny spot for a few days. He then quickly slid a mouse in the jar with the plant. The mouse stayed lively much longer than before. Why?

As the plant made food, it gave off oxygen. So the air around the plant had extra oxygen in it. This helped the mouse stay lively longer.

The leaf is the part of a plant that makes food. It is also the part of the plant that gives off oxygen. But a leaf by itself usually dies. It depends on the other parts of a plant to "keep going."

➤ Suppose that Priestley had kept the jar with the plant in the dark. What would have happened to a mouse put in this jar? Explain.

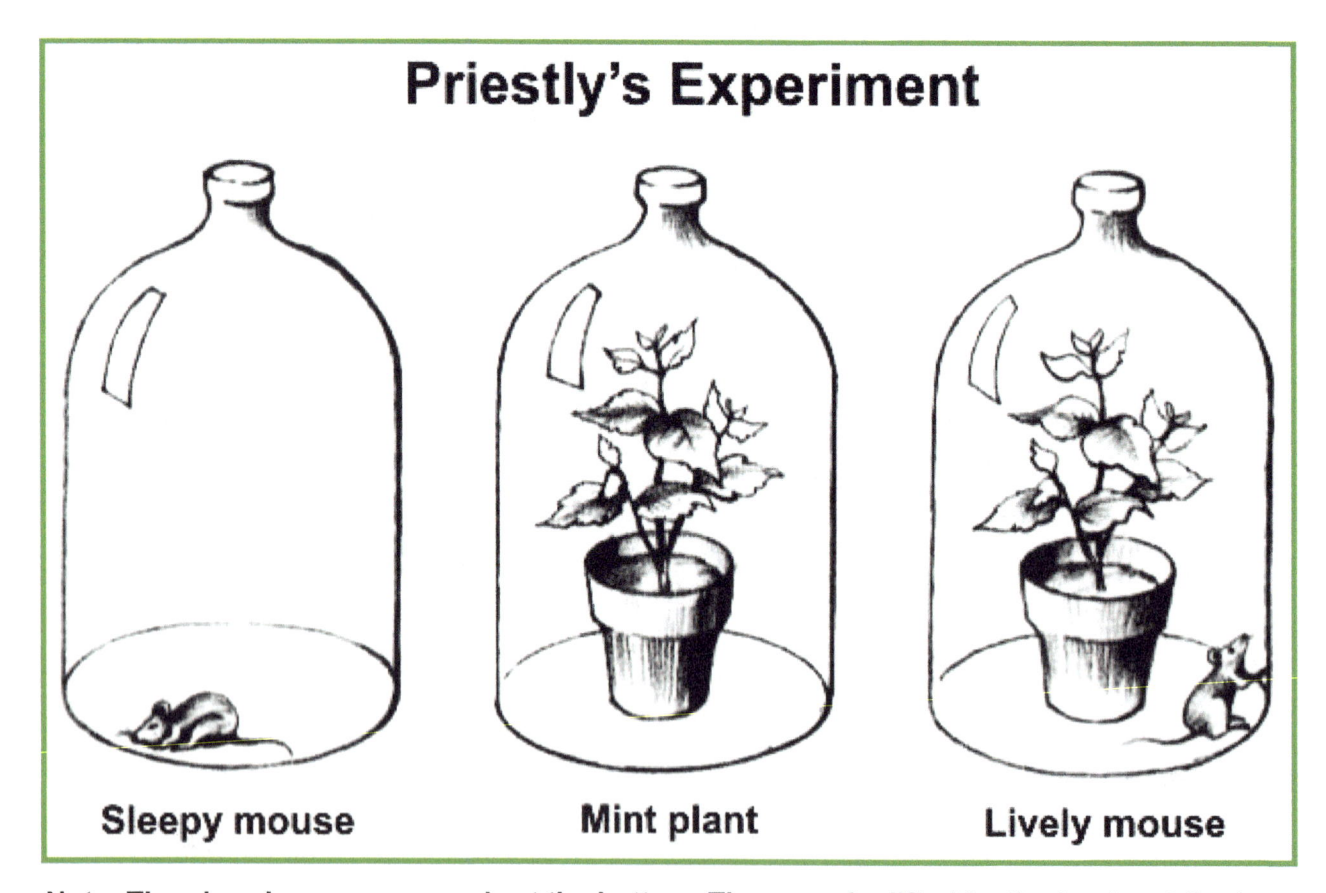

Priestly's Experiment

Sleepy mouse   Mint plant   Lively mouse

Note: The glass jars are open only at the bottom. They may be lifted by the knobs at the top.

Parts of Plants, the Food-Makers

The main parts of most plants are the leaves, stems and roots. Each of these parts has special jobs to do. But one job that all of these plant structures often do is store food.

Leaves. The main job of the leaf is to *make* food. Most of the chlorophyll in a plant is in the cells of the leaf. Here, in the leaves, green cells make food by photosynthesis. Some leaves also store a good amount of food. Many vegetables, such as lettuce, cabbage, and brussels sprouts, store food in their leaves.

Stems. A plant's stem holds up its leaves to the sun so that they may make food. It also connects the leaves with the roots. Some plants have a **nonwoody stem**. Most of a nonwoody stem is made up of a soft tissue called **pith**. The job of pith is to store food. We get table sugar from the pith of a plant called sugar cane.

Tissues in the stem move food and water to other parts of the plant. There are two kinds of fluid-moving tissue. **Xylem** (ZY-lem) moves water and minerals *up* through the stem.

Phloem (FLOH-em) moves food *down* through the stem.

Trees and shrubs have **woody** stems. The xylem and phloem grow in layers. Between the xylem and phloem layers of a woody stem is a layer called the **cambium**. (KAM-bee-um). The cambium is a growth tissue. As it grows, the cambium makes new xylem and phloem. Old xylem becomes wood.

You can see the layers of xylem in a woody stem if you look at the cross-section of a tree. The xylem produced each year looks like rings. The light part of a xylem ring is produced in the spring. These cells grow large since the conditions of growth are good in the spring. The dark part of a xylem ring is made in the summer. Since it is hotter and drier, these cells grow smaller. By counting the xylem rings (called **annual rings**) in the stem, you can tell the age of a tree.

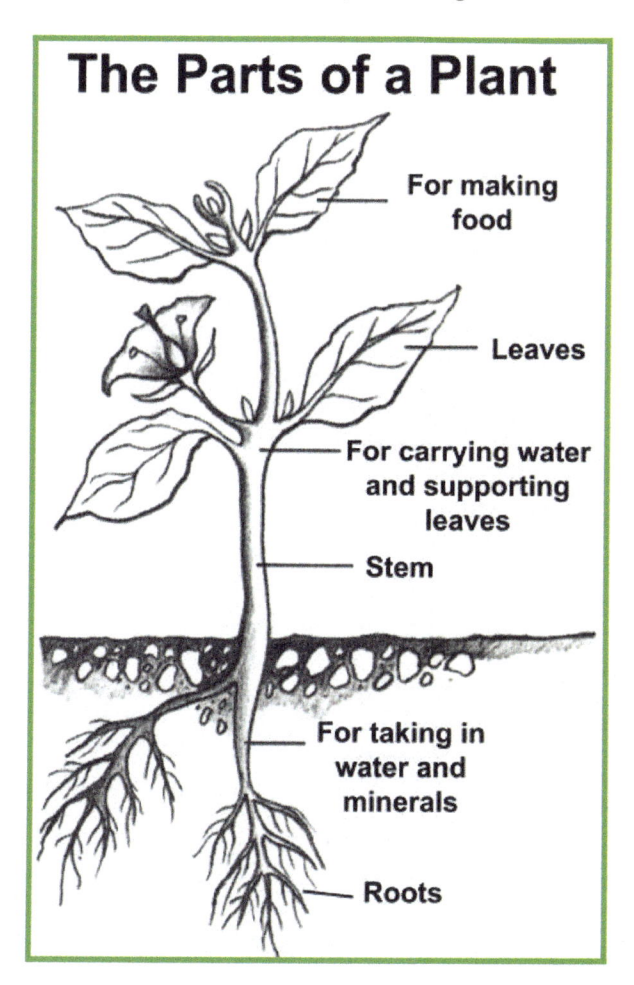

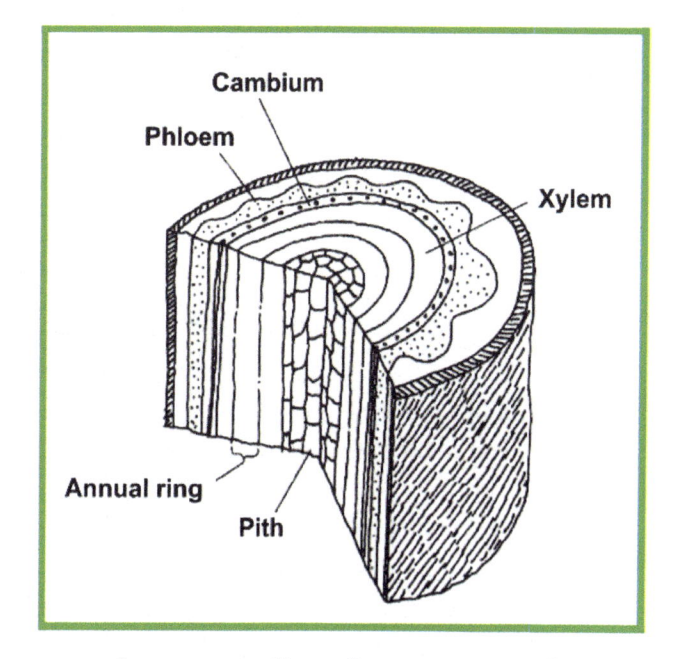

The Parts of a Plant

- For making food
- Leaves
- For carrying water and supporting leaves
- Stem
- For taking in water and minerals
- Roots

A cross-section of a young woody stem. Are you able to tell its age? [Hint: Note the label "annual ring."]

As the tree ages, the pith will be replaced by xylem tissue.

Roots. Roots anchor the plant to the ground. Both xylem and phloem are at the center of each root. Water and minerals move from the soil into the root. Along the outside of roots, very thin, hair-like portions of root cells stick out into the soil. These are called root hairs. It is the **root hairs** that actually take in the water and minerals.

Some plants have one large root, called a **taproot**; it stores a great deal of food. Carrots and beets are among these plants. Their roots can grow very long and take in water from deep in the ground.

Other plants have thread-like roots that branch over and over. These plants include grasses, strawberries, and pansies. Instead of growing deep, their roots spread out and take in water over a wide area.

The roots of still other plants grow in the air. These include orchids, which grow where the air is moist.

Finally, a few plants, like mistletoe, send roots *into* other plants and "steal" nutrients! In other words, some plants are **parasites** - organisms that live off of other living things.

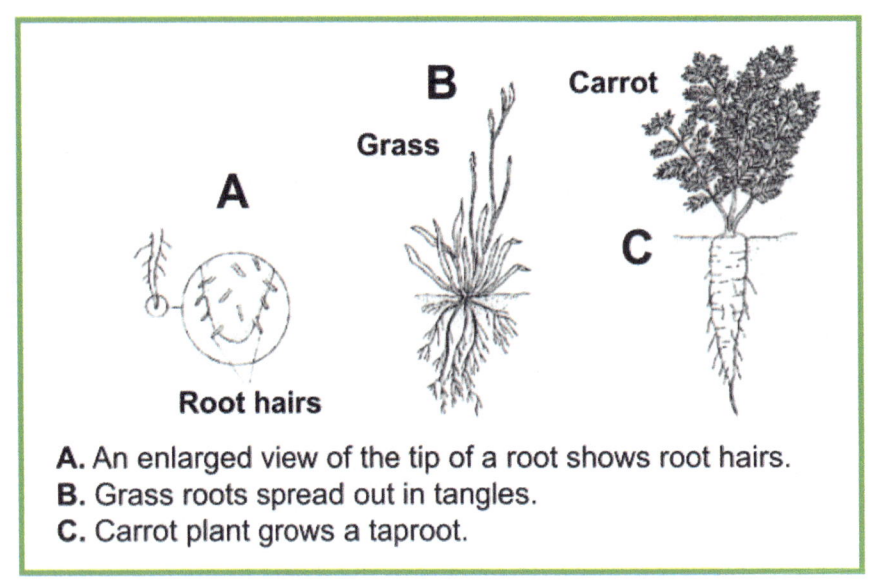

A. An enlarged view of the tip of a root shows root hairs.
B. Grass roots spread out in tangles.
C. Carrot plant grows a taproot.

➤ To Do Yourself — Do Foods from Plants Contain Sugar?

You will need:

Sugar-test tablets; test tubes or jars; water; food samples such as raisins, corn, banana, butter, apple, egg white

1. Crush a food sample and place it in a test tube or jar. Add some water and label the tube or jar.
2. Add a sugar-test tablet to the test tube.
3. Observe the color that forms. A green, yellow, or orange/red color shows that sugar is present.
4. Record your observations in a table.
5. Test each food sample for sugar.

Questions

1. Which foods contain sugar?_____

2. Where does the sugar in the food come from?_____

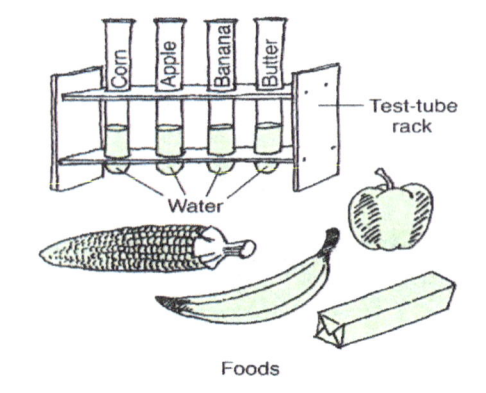

REVIEW — U-3 L-4

I. In each blank, write the word that fits best. Choose from the words below.

cambium   nonwoody   woody   pith   phloem
xylem   leaves   stems   root hairs

A plant tissue that moves food is _____. A plant tissue that moves water and minerals is _____. Soft plant stems are sometimes called _____ stems. In a nonwoody stem, food is stored in the _____. Trees have _____ stems. New cells in a woody stem are produced from the growth layer, called the _____ tissue.

Water and minerals enter a plant through _____.

II. Write the **number** of the plant part next to the job that it does.

A. ____ becomes wood   1. phloem
B. ____ anchors plant   2. xylem
C. ____ holds up leaves   3. leaf
D. ____ makes food   4. stem
E. ____ moves food down through the stem.   5. root

III. The veins of a plant have both xylem and phloem tissue. What are the jobs of the veins? Explain your answer.

_____

IV. Write the names of the tree parts on the blanks. Look carefully at the arrows. On the left, the arrow points upward. On the right, the arrow points downward.

What Foods Give Us Energy?

U-3 L-5

Exploring Science

Fuel Foods. Here's a challenge. First, carefully read this entire lesson. After reading the lesson, ask yourself what you might draw (in the space below) to help you remember this information.

Here are some ideas: Select one *common* food from each of the three numbered sections. Focus on these common foods in completing your drawing. This should help you to remember the *category* of each of these **fuel** foods (foods that are sources of energy).

Humor is welcome! Perhaps design a nutrient superhero, or a strange type of car or plane? Be sure to include labels, particularly the terms for the three fuel nutrient categories: **carbohydrates; fats; proteins**. Finally, add a title and a brief caption to your drawing.

Useful Materials in Foods

Nutrients (NOO-tree-unts) are the useful materials in foods. Most foods contain several types of nutrients. There are actually six types of nutrients. However, in this lesson we discuss the three types that our cells can "burn" to get energy; that is, these three nutrients can be used as our cells' **fuel**.

1) CARBOHYDRATES. Carbohydrates (often simply called "**carbs**") are our body's preferred source of energy. In other words, many carbohydrates are a *good* fuel for our cells to burn in order to obtain energy. Carbohydrates come in two categories, **simple** and **complex**.

Simple carbohydrates. We usually call these sugars. Sweet foods are, of course, rich in simple carbohydrates. Examples of foods that are high in simple carbohydrates are honey, ice cream, cake, candy, cookies, pie, and some fruits.

Complex carbohydrates. We usually call these **starches**. Some foods rich in starches are pasta, bread, potatoes, cereals, rice, and beans.

Another type of complex carbohydrate is **fiber**. Fiber is simply another name for cellulose, the key compound of the cell walls of plants. Human cells are not able to burn this particular complex carbohydrate. However, fiber helps to move food through our digestive system. Since fiber is cellulose, all plants that we eat contain some fiber. Particularly good sources of fiber are whole grains and fruits (especially prunes). You may have heard an older person share that fiber helps to "keep them regular." Can you guess what that means?

One last thing about fiber. Some species of microbes *are* able to break down cellulose. These special microbes live in the digestive system of herbivores (animals that eat only plants). The chewing by herbivores (such as cows and sheep) breaks the plants' cell walls into small bits. The microbes then chemically change the cellulose into other, smaller molecules. These smaller molecules are then used by the cells of the herbivore's body.

If you eat **meat** or dairy, be sure to thank fiber-eating microbes. They make it possible for plant-eating animals (like cows) to digest the cellulose in grass - and then to produce milk and grow their muscles (which we call meat).

Simple carbohydrates

Complex carbohydrates

High fiber foods (Humans can not digest the cellulose).

2) FATS. Fat is another nutrient that our cells are able to burn as an energy source. We sometimes eat fats directly. Plus, if we eat more carbohydrates or protein than our cells can burn, our cells actually turn these compounds into fats.

Why does our body do this? Fat molecules store *more* energy (in the same amount of space) than the other nutrients.

Also, stored fat in the skin acts like a winter coat; it slows the loss of body heat. For example, blubber (the fat of whales and seals) helps these animals stay warm in cold ocean waters.

Animals' bodies also use fat to cushion some organs. A layer of fat keeps your eyes from bumping into the bones of your eye sockets (the bony cups around your eyes).

Oils (liquid fat from plants)

Solid fats (such as the white in bacon)

Let's look at the three key types of fats:

Oils. Oils are simply plant fats that are liquid at room temperature. These oils may be squeezed from plant parts such as fruits, seeds, and nuts. Examples are olive oil, soybean oil, and peanut oil

Solid fats. Solid fats are fats that are not liquid at room temperature. These include the white fat around a slice of meat, and the fats that are naturally a part of eggs or milk.

Cholesterol. Cholesterol is a special type of fat. Our body makes its own cholesterol. It is needed in order to produce important molecules like hormones (See Unit 6, Lesson 6).

However, eating too much meat can overload us with cholesterol. This often leads to problems later in life. (See Unit 10, Lesson 3). Plant foods do not contain cholesterol.

---

3) PROTEINS. Proteins build and heal tissues, especially our muscle tissue. **Enzymes** are an additional important type of protein; these special proteins allow our cells' chemical reactions to occur, and to happen at the correct speed.

As with carbohydrates and fats, our cells are able to burn protein molecules for energy. However, our body does not "want" to burn this nutrient as a fuel. It prefers to use protein for the reasons stated above.

Both plant-based and animal-based foods contain protein. All proteins are made up of building blocks called **amino acids** (uh-MEE-noh ASS-ids). There are 20 kinds of amino acids, and our body can make all but 9 of these 20. We must get these nine amino acids from our food.

As stated above, meat is simply the muscle of an animal (such as a chicken, pig, or cow). Meat contains all of the amino acids that we need.

Very few plants have all of the needed amino acids. However, it is easy to get these amino acids from plants by simply eating a variety of plants - particularly seeds, beans and nuts. **Vegetarians** (people who avoid meat) make certain to eat many different plant foods. They also often eat foods made from mushrooms - a very good source of protein.

If you ask a vegetarian why they don't eat meat, you will likely hear one or more of the following responses: 1) they want to avoid excess cholesterol; remember, plants have no cholesterol; 2) they prefer that animals not be killed as food; 3) they want to reduce the amount of the Earth's land and water that is used to produce meat.

Humans across the globe depend on seeds (like grains, beans and nuts) to produce nutritious foods - including breads, tortillas, pastas, and soups.

80

> **To Do Yourself**   **What Foods Contain Protein?**

You will need

Biuret solution; medicine dropper; water; test tubes or jars; food samples such as peanuts, powdered milk, maple syrup, hamburger meat, egg white, cheese, gelatin

1. Choose a food sample and crush it. Place it in the test tube.
2. Add several drops of water. Add several drops of Biuret solution.
3. Record any color changes. A purple color shows that protein is present. Record your observations in a table.
4. Test each food sample for protein.

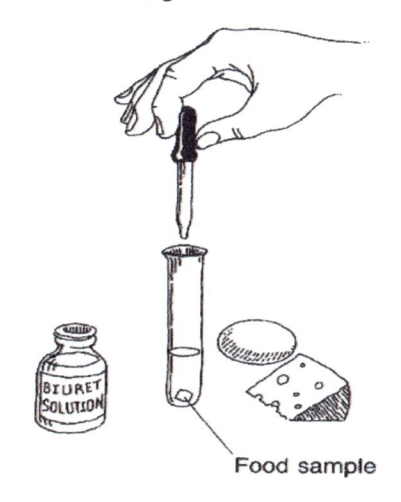

Questions

1. What are some foods that contain protein? _____

2. How could you find out if a food also contains starch or sugar? _____

_____

REVIEW — U-3 L-5

I. A). What is meant by a nutrient? _____
   B). List the three main groups of nutrients that organisms are able to use as <u>fuels</u>.

_____

II. Write the <u>number</u> from column **B** that matches the phrase from column **A**.
    Be careful. All answers are used only once.

A

A. ____ some microbes *can* digest it; animals *can't*
B. ____ controls the speed of chemical reactions
C. ____ can be used for energy if necessary
D. ____ cushions organs and keeps the body warm
E. ____ is not present in plants
F. ____ can be changed to fat

B

1. enzymes
2. fats
3. proteins
4. carbohydrates
5. fiber (cellulose)
6. cholesterol

III. A **fad diet** is a diet that becomes popular for a while, but often is determined to cause health problems. How might a fad diet of only rice and water make people sick?

_____

_____

81

Why Do We Need Vitamins and Minerals? U-3 L-6

Exploring Science / Historical Steps

The Story of Vitamin A. During World War I, many children in Denmark came down with a terrible eye disease. At first, their eyes became dry. Then their eyelids became swollen and red. Tragically, many children eventually went blind.

Dr. C.E. Bloch set out to find a cure for this disease. In his search, he read about the work of **Dr. E.V. McCollum**, an American scientist. McCollum fed rats what scientists then thought was a good diet. Yet the rats stopped growing and got sick. Like the Danish children, the rats came down with the eye disease. Then McCollum added milkfat to the diet of the rats. Soon their eye disease was cured.

What special substance was in the milkfat that cured the eye disease? After months of research, McCollum found out that the special material was Vitamin A.

Dr. Bloch knew that Danish children did not have enough milkfat in their diets. Because of the war, the children ate no butter and drank only skim (fat free) milk. This caused the children to get sick. Bloch added whole milk and butter to the diets of the sick children. The children got well!

Today, in America, vitamin A is added to all skim and low-fat milk. We also have other foods that are rich in vitamin A. It should be easy to get enough vitamin A from eating the proper foods. Yet, some people still get too little vitamin A. These people often notice that they have dry skin or hair.

Too little vitamin A also causes night blindness.

On the other hand, too much vitamin A can make you sick. Signs of too much vitamin A are a sick stomach, loss of hair, and a skin rash.

➤ In his later work, McCollum studied the diets of cattle. He observed that cattle fed only grain can become blind. Cattle fed on both grain and green grass stayed healthy. McCollum concluded that...

A. grain is rich in vitamin A.
B. green grass is rich in vitamin A.

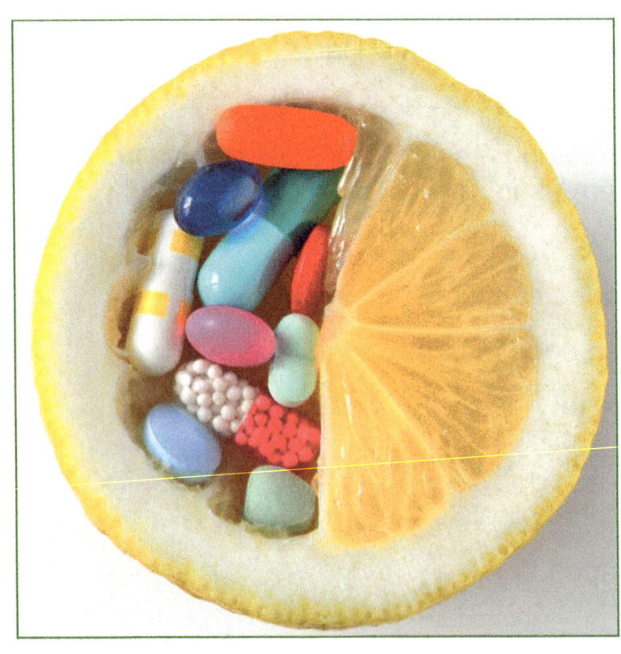

Should people take supplements? Many nutrition experts feel that only a few vitamins (such as vitamin D) need to be taken in pill form. They say that the best way to get almost all nutrients is through a varied diet - with *many* vegetables and grains.

Vitamins, Minerals, and Water

Vitamins, minerals and water are three types of nutrients that can *not* be burned for energy. In other words, these are not fuels. Our body does not make these nutrients; we get them from our foods. They help our body cells function properly.

To stay in good health, we need many kinds of vitamins and minerals. Foods have only small amounts of these nutrients in them. Yet, these small amounts have big effects on our health. A lack, or **deficiency** (di-FISH-un-see) of one of these nutrients can cause a disease. Night blindness is one example of a **deficiency disease**; it results from too little vitamin A.

Too much, or an **overdose** (OH-vur-dohs) of some vitamins and minerals can also cause illness. Overdoses usually come from taking large amounts of vitamins or minerals in the form of pills.

VITAMINS. Vitamins are compounds made of a variety of elements, though all contain carbon.

Some of the most common vitamins are the following: vitamin C helps us in many ways, one being to repair tissues; vitamin D helps our body use calcium in order to keep our bones and teeth healthy.

MINERALS. Minerals are simply specific elements that our body needs. A few of the most well known, and their key jobs, are the following: calcium (Ca) is needed for strong bones; iron (Fe) helps produce red blood cells; both sodium (Na) and potassium (K) are needed for healthy nerve tissue.

WATER. Do you recall what percent of your body (by weight) is made of water? Clean water is vital to staying healthy. For chemical reactions to occur inside of our cells, molecules must contact each other. These molecules float in water! Of course, our blood, which moves materials to all cells of our body, is mostly water.

Quite often, people wait too long to drink water, even when they feel thirsty. This is often the source of headaches. As a headache begins, try drinking some water. Staying **hydrated** is the term for keeping plenty of water

"I feel a headache coming on! What should I do?"

in your body. If you're **dehydrated**, you are too low on water.

Summary table

The following pages' table lists many foods that are rich in vitamins and minerals. Getting enough of certain vitamins and minerals prevents deficiency diseases. The names and **symptoms** (signs) of these diseases are also listed. The table also provides some important facts about certain vitamins and minerals.

Notice that many foods are rich in more than one nutrient. Whole grain breads are an example. They give us a group of vitamins called the B vitamins. Whole grain breads also give us vitamins E and K. Check the table to determine why these vitamins are needed in our body.

Can you find five vitamins and minerals that are found in milk? Eggs are also rich in several different vitamins. Which are they? How do these help our body? What are some signs of a deficiency of these vitamins?

Since different foods are rich in different vitamins and minerals, the key is to eat a variety of foods.

VITAMINS	SOME GOOD SOURCES	USES IN BODY
A	yellow, orange, or dark-green vegetables; liver; milk, butter, and margarine	helps keep skin and eyes healthy; helps ability to see at night; keeps bones and teeth healthy
C (ascorbic acid)	oranges, lemons, grapefruit, limes; tomatoes, potatoes; melons; strawberries; broccoli, cabbage	helps keep bones, teeth, and blood vessels healthy; aids clear skin
D	milk (fortified); liver; tuna, salmon; cod-liver oil; sunlight	keeps bones, teeth healthy; helps body use calcium
E	vegetable oils, margarine; wheat germ; whole-grain cereals and bread; leafy vegetables; beans	helps keep muscles and blood circulation healthy; protects body from damage by free oxygen
K	cabbage; cauliflower; green leafy vegetables; peas, beans; milk; eggs.	encourages normal blood clotting and possibly bone growth
B9 (folic acid or folate)	green leafy vegetables; wheat germ; whole-grain cereals and breads; liver; cabbage and broccoli; oranges, grapefruit	helps body use protein
B3 (niacin)	chicken; eggs; tuna; beef; whole-grain cereals and breads; nuts; dried beans and peas; liver	helps get energy from food and maintains skin, nerves, and digestive system
B2 (riboflavin)	milk; meat; dark-green leafy vegetables; eggs; whole-grain cereals and breads; dried beans and peas; liver	helps keep nerves, skin, and eyes healthy; helps body get energy from food
B1 (thiamin)	meat and fish; whole grain cereals and breads; beans and peas; eggs	helps keep nerves, skin, and eyes healthy; helps body get energy from carbohydrates
MINERALS		
Calcium	milk, yogurt, cheese; canned sardines and salmon eaten with the bones; dark-green leafy vegetables	keeps bones and teeth healthy; helps blood clot; encourages healthy muscle
Iron	meat or chicken; liver; raisins or other dried fruit; eggs; beans and peas; tomato juice	necessary for red blood cells that carry oxygen throughout the body
Iodine	seafood; iodized salt; seaweed	necessary for proper thyroid hormones (used in getting energy from food); aids growth
Potassium	bananas; oranges; dried fruits; meat; potatoes; peanut butter	keeps nerves and muscles healthy; helps regulate water in cells
Sodium	salt; milk; celery; beets; pickles	keeps nerves and muscles healthy; helps regulate water in cells

Deficiency Diseases and Signs	Helps protect against	Signs of Overdose
Night blindness (poor ability to see in dim light)	rough dry skin; dry hair; eye infections	skin rash; hair loss; dry skin and mouth; nausea; can be fatal
Scurvy (bleeding gums); loose teeth; dry, rough skin; weakness)	bruises; sore gums; perhaps colds	kidney and bladder stones
rickets (bowed legs, poor growth of bones and teeth)	soft bones; muscle disorders	too much calcium in blood test; nausea; kidney stones; deafness
signs of deficiency not observed in humans	certain eye diseases; leg pains; and possibly reduces odds of heart disease.	none known
poor blood clotting	bleeding and soft bones	turns skin yellow in babies
can cause a type of anemia (low red-blood-cell count)	certain birth defects and one form of heart attack	can mask a deficiency in another vitamin; otherwise no known signs
pellagra (skin disorders); diarrhea; confusion and other mental problems; swollen mouth	high levels of cholesterol that might cause heart disease	ulcer in intestine; high blood sugar
skin disorders (especially of the nose and mouth); eyes sensitive to light	poor growth; oily or scaly skin	none known
beriberi (weakness, leg cramps, confusion; loss of weight; partial loss of use of limbs)	tiredness	none known
soft bones that easily break, especially when older; poor teeth	loss of bone mass (osteoporosis)	sleepiness; tiredness; possibly kidney or bladder stones
anemia (paleness, weakness, tiredness); brittle fingernails	tiredness	damage to heart or liver; iron supplements can poison children accidentally
goiter (enlarged thyroid gland in neck); myxedema (tiredness; slow heart rate; weight gain)	heart disease; tiredness	none known
loss of water from cells; mood changes; poor heart beat; cramps and weakness; nausea and vomiting	loss of fluids or excess acidity or alkalinity in cells	muscle weakness, confusion, irregular heartbeat, difficulty in breathing
muscular weakness; headache; irregular heartbeat; confusion or even coma	low blood pressure; heat stroke	high blood pressure in some people

> **To Do Yourself**   **What Foods Contain The Mineral Iron?**

You will need:

Several small jars, water, ½ liter of strong tea, samples of juice such as apple, pineapple, orange, cranberry, grapefruit.

1. Fill each jar ¼ full of a juice sample. Clearly label the jars.
2. Add tea to each jar until it is half full.
3. Observe and record any changes. If a juice contains iron, this element will combine with the tea and form solid particles. These will settle to the bottom.

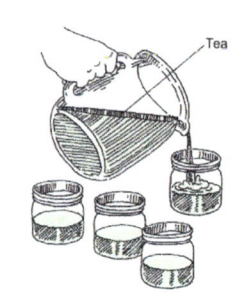

Questions

1. In just a few words, answer this: What is a test for iron in blood? _____

2. What are some foods that contain iron? _____

---------------------------------- **REVIEW** ---------------------------------- U-3 L-6

I. In each blank, write the word that fits best. Use each word below only once.

niacin   vitamins   deficiency disease   minerals   overdose

Foods contain small amounts of _____ and _____.

A lack of one of these nutrients can cause a(n) _____.

Taking too much of a vitamin is called a(n) _____. Whole grain breads are a good source of _____.

II. Use the tables to find the cure for each deficiency disease. Write the number from column **B** that matches column **A**.

A	B
A. ____ pellagra	1. iron
B. ____ goiter	2. vitamin B$_1$
C. ____ rickets	3. niacin
D. ____ scurvy	4. vitamin C
E. ____ night blindness	5. vitamin D
F. ____ anemia	6. iodine
G. ____ beriberi	7. vitamin A

III. Goiter, a swelling of the thyroid (making the neck appear large), is caused by a lack of iodine in the diet. It was once a common disease in places far from the sea.

A. Why do you think that this was so? _____

B. Why do you think that this is no longer true? _____

What Are Calories?   U-3 L-7

Exploring Science / Historical Steps

What's on the Bear's Menu? What does a hungry grizzly bear eat before settling down for a winter's sleep? Would you believe a feast of insect grubs? That's just one of the interesting things about grizzlies that **Dr. Chris Servheen** and his workers learned. Dr. Servheeen was hired in 1981 as the first head of the Grizzly Bear Recovery Program in Montana. He retired after 35 years.

In the early 1980s, grizzly bears were nearly gone from all but Alaska. Then Servheen's team tracked many of the remaining bears. They recorded how these giants travel from place to place. Learning some of the bears' habits allowed the team to reintroduce grizzlies to parts of the bears' former range.

Like all bears, grizzlies hibernate during the winter. All summer and fall, they fatten themselves up. Surprisingly, up on the mountainsides of the Rockies, insects are a big part of their diet. Insect larvae can be up to 40% protein - more nutritious than many other foods. Once the bears bulk up, they settle down for a long sleep. By spring, a grizzly has usually lost over 20% of his or her body weight - mostly by burning stored fat. No wonder she or he wakes up hungry as a bear!

In 1973 the Endangered Species Act was passed. Thanks to this law, and much hard work by people like Dr. Servheen, grizzlies are now listed as *threatened*, but not *endangered*. If we continue to protect the grizzly bears' habitat, they should be with us for a long time to come.

➤ The energy that a female bear needs during the *spring* and *summer* comes from food that
  A. she eats daily.   **B**. she has stored as fat.

A grizzly bear and her three cubs during good weather

Calories and Energy

How could bears live all winter without eating? During the fall, bears eat more food than they use. This is stored in their bodies as fat. In the winter, they burn that fat and obtain many **calories** (KAL-uh-rees) worth of energy.

Just what is a calorie? You see this word on food labels. A bread wrapper says, "Calories per serving -- 120." You may have heard that calories are "fattening." To lose weight, people say that you must cut down on calories. So what are they?

Food gives you energy to live. A calorie is simply a *measure* of food energy. It is not something that you can touch or feel, any more than a degree Celcius can be touched. (A degree is simply a *measure* of the temperature).

When your body "burns" one gram of sugar, starch, or protein you get four calories of energy. One gram of fat gives *nine* calories of energy. Do you see why fat is used as our bodies' energy storage molecule?

As you learned in Lesson 5, food that contains carbohydrates (sugars and starches), fat, or protein can be burned for energy. The amount of energy released when the food is burned is measured in calories. The number of calories that can be obtained from some common foods is shown in Table 1. In most cases, if a person's "input" of calories is more than their "output," they will gain weight. (This is not always the case, since a number of medical problems can interfere with how food is burned).

Of course, it is normal to gain weight while you are still growing. In addition, a full grown adult who exercises a great deal may add "muscle mass" - and gain weight while doing so.

If growing or adding muscle mass is not happening, a person who is gaining weight is very likely storing fat in their body.

In most cases, if a person's "output" of calories is more than their "input," they will lose weight. Like the mother grizzly bear, they will be "burning" stored food.

Table 1 Calories in Some Foods

Food	Amount	Calories
Apple	1 medium	75
Bread	average slice	70 - 120
Cornflakes	1 cup	80
Egg	1 (boiled)	75
Green peas	½ cup	55
Hamburger	1 medium size	250
Hamburger roll	1 average	110
Butter	1 teaspoon	33
Milk (skim)	8 ounces	80
Orange juice	½ cup	55
Swiss cheese	average slice	100

Table 2 Calories Used in Some Activities*

Activity	Calories per hour	Activity	Calories per hour
Sitting quietly	20	Bicycling (moderate speed)	110
Standing	20	Dancing	170
Eating	20	Playing ping-pong	200
Keyboarding rapidly	50	Violin playing	330
Dishwashing	50	Running	330
Walking	90	Bicycling (racing)	340
Carpentry, heavy	100	Swimming	360

*For a 46-kg (100-lb) person. If you weigh less, you use fewer calories per hour. If you weigh more, you use more.

REVIEW U-3 L-7

I. In each blank, write the word that fits best. Choose from the words below. Words may be used more than once.

fat	starch	protein	weight	winter
calories	sugar	energy	burning	

Food energy is measured in _____. One gram of _____, _____, or _____ provides four calories of _____. One gram of _____ provides nine calories of _____. If your "input" of calories equals your "output," your _____ stays the same. If your "input" is more than your "output," you will gain _____.

II. Use Table 1 to help you answer the following questions.

  A. Find the total number of calories in a breakfast that consists of the following: ½ cup orange juice, 1 cup cornflakes, ½ cup skim milk, 1 egg, 1 *small* slice of toast, and 2 teaspoons of butter. _____

  B. How many calories are in a lunch that consists of the following: hamburger on a roll, ½ tomato, 1 apple, 1 cup skim milk? _____

III. Circle the word (between the brackets) that makes each statement true.

  A. A person whose calorie <u>input</u> each day is 2,500 calories and whose calorie <u>output</u> is 2,000 will [gain / lose] weight.

  B. A person whose calorie <u>output</u> each day is 2,600 calories and whose calorie <u>input</u> is 2,000 will [gain / lose] weight.

IV. For a 46-kilogram (100 pound) person to "burn up" a 300 calorie snack, about how long would that person need to run - if they are running at 9 miles / hour? Show your work.

_____
_____
_____
_____

89

How Can You Balance Your Diet? U-3 L-8

Exploring Science

Tipping the Scales. Each year people in the United States spend billions of dollars on weight loss programs. Some people even try extreme diets (often called **fad diets**). Most of these are later found to be dangerous.

With so much media attention on body shape and image, some young people (girls more often than boys) become overly focused on their weight. This may lead to an **eating disorder**. This is a serious, but correctable condition. If a young person is losing weight instead of growing, they should seek adult help without delay.

At one point, the food industry had the public convinced that all fats were bad. Many low-fat and fat-free products were heavily advertised. Often these products were made with too much sugar. Since the body converts (changes) excess sugars into fats, the buyers did not lose weight.

Of course, eating too much of any nutrient is not healthy. It is true that, in America, many people struggle to keep from becoming overweight. **Obesity** means having way too much body fat. This condition makes the heart work harder to move blood to all of the body's cells.

Too much body fat may also upset our cells' ability to burn food. Even in young people, this can lead to a condition called **type 2 diabetes.** A person who has early signs of this serious disease may be able to halt it by eating better and exercising more.

Our nation has one of the highest rates of obesity and diabetes in the world. Why? First, it is very easy for Americans to obtain unhealthy foods and drinks. Meanwhile, in many neighborhoods, stores with healthy foods are too far from homes. Such neighborhoods are called **food deserts**. Second, the work and free time of many Americans involves very little movement. This means that we are often burning less fuel than we are eating.

While you are young, it is vital to get in the habit of eating well and staying active! If you truly must lose weight, keep eating a balanced diet, but eat smaller portions - and increase your physical activity.

A "prescription"(Rx) for a healthy body ...
- a <u>balanced diet</u> and <u>exercise</u>.

At least 8 hours of <u>sleep</u> is vital, too!

A Balanced Diet

To have a healthy body, you need to eat a balanced diet. A balanced diet supplies all of the nutrients you need. The study of how the body uses food is called **nutrition**. (new-TRISH-un). **Nutritionists** (scientists who study nutrition) have designed guidelines for eating well. Perhaps the best such guide is the **Healthy Eating Pyramid** that is shown below.

What foods should make up most of our diet? Check the bottom half of the pyramid. It's foods that are high in vitamins and minerals, like fruits, vegetables and **grains** (the seeds of "grass-like" plants such as wheat, corn, or rice). Grains are also great sources of carbs (fuel) and protein.

Fruits and vegetables also contain important nutrients called **antioxidants**. Antioxidants help to prevent some types of cancers.

The top of the Healthy Eating Pyramid shows that our diet should include only small amounts of red meat (like hamburger). Meat provides protein and iron (two important nutrients), but red meat is particularly high in cholesterol, which can cause blood problems. (See Unit 10, Lesson 3 for more on this topic).

HEALTHY EATING PYRAMID

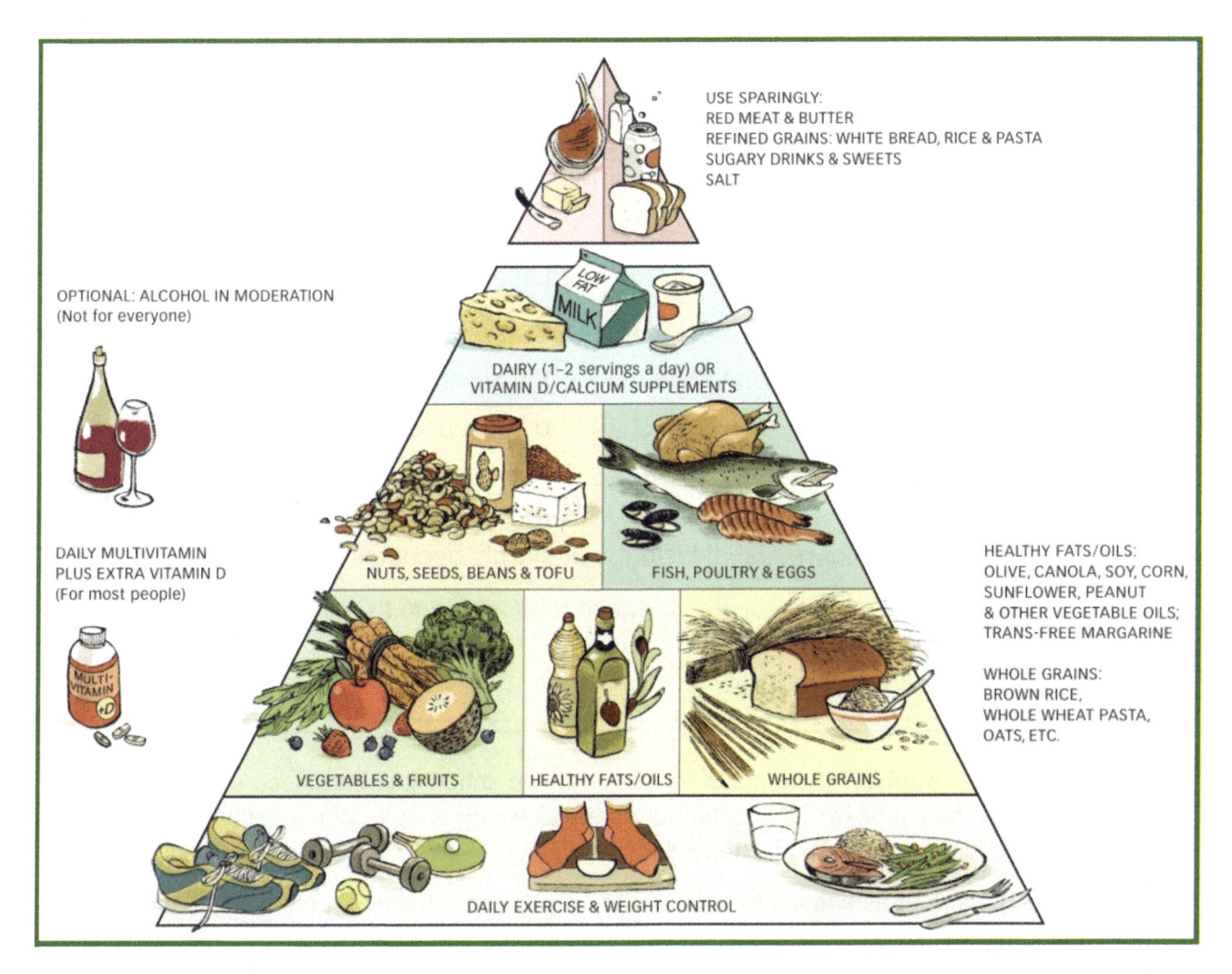

Copyright © 2008. For more information about The Healthy Eating Pyramid, please see The Nutrition Source, Department of Nutrition, Harvard T.H. Chan School of Public Health, www.thenutritionsource.org, and Eat, Drink, and Be Healthy, by Walter C. Willett, M.D., and Patrick J. Skerrett (2005), Free Press/Simon & Schuster Inc.

REVIEW — U-3 L-8

I. In each blank, write the word that fits best. Use each word below one time.

nutrition   balanced   Healthy Eating Pyramid
whole grains   red   nutrients

A _____ diet supplies all of the _____ that you need. The study of how the body uses food is called _____.

To help you make healthy food choices, nutritionists designed the

_____. As far as foods, the biggest area of the pyramid consists of fruits and vegetables, healthy oils, and _____.

The pyramid shows that, if we choose to eat meat, we should eat _____ meat only in small amounts.

II. Use the Healthy Eating Pyramid to help you answer the following questions. Place the <u>number</u> from column **B** in the spaces of column **A**. Each number is used one time.

A

- A. ____ milk, yogurt and cheese
- B. ____ poultry, fish
- C. ____ grains, fruits and vegetables
- D. ____ fats and sweets
- E. ____ healthy oils

B

1. should be used sparingly
2. from olives, peanuts, and sunflowers
3. good sources of vitamin D and calcium
4. the basis of a balanced diet
5. healthier animal-based foods (than red meat)

III. Circle one. A balanced diet is vital. If losing weight truly becomes necessary, it is best to

  A. leave out a food group   B. eat smaller amounts from all food groups

==

<u>**Are you sleeping enough**</u>? The brains of middle and high school students need at least 8 hours of sleep each night. In fact, a bit over 9 is better for nearly all young people. Why? At this age the brain is physically making big changes. Many connections between brain cells are being "re-organized."

In addition, in people of all ages, information that we hope to remember gets "sorted" during sleep. <u>Neurologists</u> (nervous system experts) describe this as moving information from <u>short term memory</u> to <u>long term memory</u>.

A key way to get a better night's sleep is to stay away from screens in the hour before bed. Screens give off specific wavelengths of light that make it harder to relax and fall asleep.

Unit 3 -- Review What You Know ---

A. Hidden in the puzzle below are the names of seven nutrients. Use the clues to help you find the names. Circle each name that you find in the puzzle. Then write each name on the line next to its clue.

Y	S	T	A	R	C	H	Z
W	A	S	B	M	C	P	Y
A	Q	U	P	I	R	R	X
T	D	G	E	N	F	O	G
E	F	A	T	E	H	T	I
R	J	R	K	R	L	E	M
U	V	I	T	A	M	I	N
I	N	O	Q	L	R	N	Z

Clues:

1. It has a sweet taste. _____
2. Butter is a rich source. _____
3. Calcium is one example. _____
4. Drink it if you're thirsty. _____
5. Niacin belongs to this group. _____
6. Complex carbohydrate found in bread. _____
7. Builds and repairs tissues. _____

B. <u>Write</u> the word (or words) that best completes each statement.

1. The formula CO_2 stands for
      a. an element   b. a compound   c. a mixture   1. _____

2. Most of the body is made up of
      a. sugar   b. starch   c. water   2. _____

3. For photosynthesis, a cell needs
      a. oxygen   b. light   c. heat   3. _____

4. During cell respiration, oxygen combines with
      a. sugar   b. water   c. nitrogen   4. _____

5. The tubes that carry water up through a plant's stem are
      a. xylem   b. phloem   c. pith   5. _____

6. Wood is tissue that once was
      a. pith   b. leaves   c. xylem   6. _____

7. A plant's growth tissue is
      a. chlorophyll   b. cambium   c. phloem   7. _____

8. A nutrient that provides calories is
   a. fiber   **b.** vitamin A   **c.** protein   8. _____

9. Included among the fats is
   a. starch   **b.** cholesterol   **c.** sugar   9. _____

10. The weight of a person whose calorie intake is lower than his or her output will
   a. increase   **b.** decrease   **c.** not change   10. _____

C. Apply What You Know

1. Study the drawing of the parts of a lunch that you might have: hero sandwich, salad, milk, and banana. Draw a line from each group on the pyramid to the food in the lunch that belongs to that group.

2. Below the one shown, draw another lunch (that has none of the same foods). Be sure that it includes food from most parts of the pyramid. Use a writing utensil of a different color to draw the connecting lines from the foods of this new lunch to the matching pyramid sections.

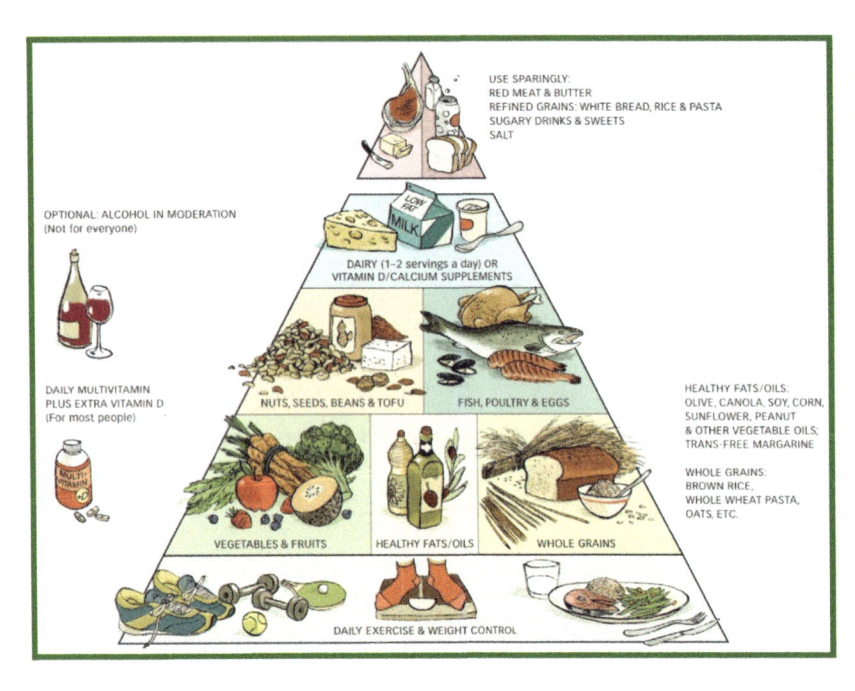

Copyright © 2008. For more information about The Healthy Eating Pyramid, please see The Nutrition Source, Department of Nutrition, Harvard T.H. Chan School of Public Health, www.thenutritionsource.org, and Eat, Drink, and Be Healthy, by Walter C. Willett, M.D., and Patrick J. Skerrett (2005), Free Press/Simon & Schuster Inc.

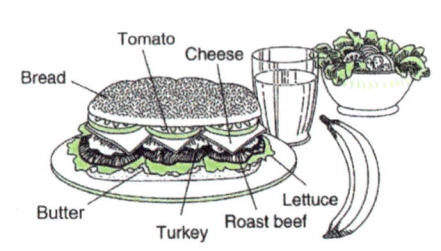

D. Find out more.

1. On Youtube, watch the seven minute video "*Can Science Explain The Origin of Life?*" produced by **Stated Clearly**. Discuss this video with a classmate or two.

2. A **complete protein** is one that supplies all nine of the amino acids that we need from food. An **incomplete protein** supplies only some of the nine amino acids. Animal proteins are complete. Most plant proteins are incomplete. Research how different incomplete protein foods can be combined to make complete proteins. In other words, what types of plants should be mixed in order to eat a meal that has all of the amino acids needed? Plan at least one vegetarian menu that provides complete proteins.

3. Collect labels from foods such as cereals, canned foods, bread, and so on. What do you learn from the Nutrition Facts listed on these labels?

SUMMING UP:
Review What You Have Learned So Far

A. Study the drawing. Use the choices below to write labels for the numbered parts below the drawing. No choice is used twice.

cell respiration   carbon dioxide   photosynthesis   oxygen
consumer   producer   decomposer

1. The apple tree's link in a food chain _____

2. The girl's link in a food chain. _____

3. Food-making process in the tree _____

4. "Burning" of food in the girl _____

5. Gas the tree gives off and the girl uses _____

6. Gas the girl gives off and the tree uses _____

B. Circle the word (or words) that makes each statement correct.

1. The cell parts in the tree's leaves that trap the sun's energy are [chloroplasts / vacuoles].

2. The clouds show part of the [oxygen-carbon dioxide / water] cycle.

3. One-celled microbes present in the soil include [bacteria / vertebrates].

4. The rocks, grass, air, and girl are all parts of the same [ecosystem / population].

5. The apple is a rich source of [vitamins / proteins].

6. The girl is classified in the genus [*Felis* / *Homo*].

7. The biome shown is of a [deciduous forest / arctic tundra].

8. The tree's [xylem / phloem] tissue forms woody rings in its trunk.

9. A nonvascular plant growing near the tree might be a [mushroom / moss].

10. Before testing a hypothesis about the tree, a scientist would make [observations / theories].

95

UNIT FOUR

DIGESTION AND TRANSPORT

What Does Your Digestive System Do? U-4 L-1

Exploring Science / Historical Steps

The Hug of Life. Kim Lee was 14 years old when she saved her brother Larry's life.

One day Kim and Larry were having lunch in a restaurant. While Larry was talking, he took a bite from his sandwich. Suddenly, he stopped talking and pointed to his throat. Kim asked what was wrong, but Larry could not speak. He also started to turn blue.

Kim could see that Larry was in danger. She knew at once what to do. In her first aid class, she had learned the "Heimlich (HYM-lik) Hug."

Kim stood behind Larry and put her arms around his waist. She made a fist with one hand and surrounded it with the other. A little above his belly button, she pressed her fists in firmly, the way that she had been taught. She heard a slight pop. Out came the piece of sandwich.

Larry took a deep breath. So did Kim, as she sighed with relief. Kim and Larry then went to see their family doctor to be sure that he was OK. Larry was just fine!

In 1972, **Dr. Henry Heimlich** invented the method that Kim used. Today, it is always a part of courses in first aid. It is often called the **Heimlich maneuver** or simply "abdominal thrusts."

The entrance to our air tube (trachea) and the entrance to our food tube (esophagus) are next to each other. As a result, choking on food is common. To prevent food from entering the air tube, a small door-like structure (the **epiglottis**) usually closes over the air tube as we swallow. Do you see why parents tell their children, "Don't talk with food in your mouth!"?

➤ What do you think is the main reason for taking a course in first aid before you try the Heimlich maneuver?

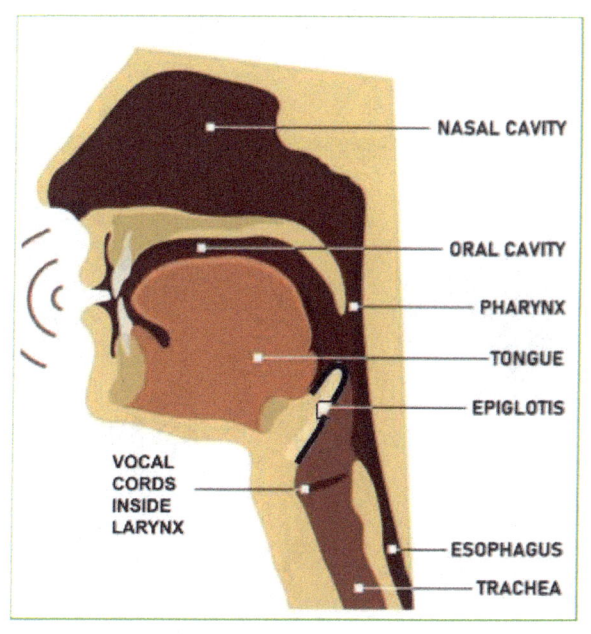

- Epiglottis in open position
- Risk of food entering trachea / blocking airway

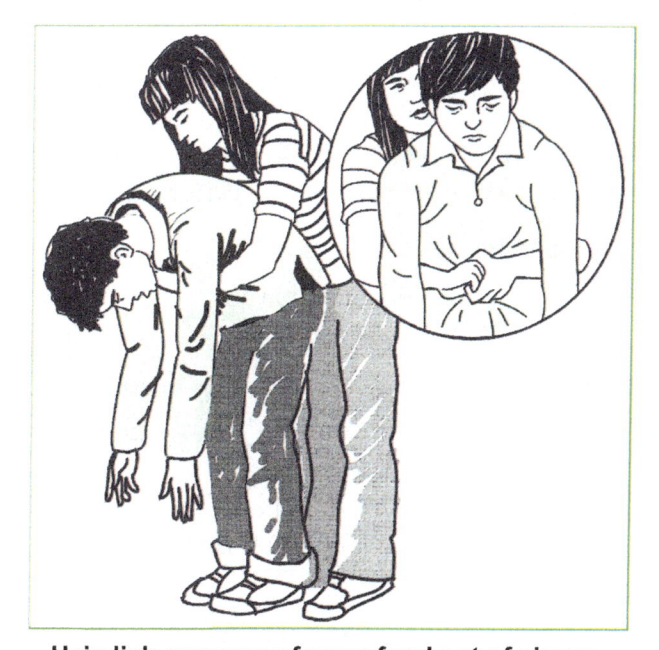

Heimlich maneuver forces food out of airway

Your Digestive System

Larry Lee's food bite "went down the wrong way." What happens to food when it goes down the *right* way? It gets digested.

Then, in a different form, the food moves to your blood. Finally it moves inside of your cells.

97

Chemically, foods are made up of compounds. Recall that the smallest amount of a compound is called a **molecule**. (MAHL-uh-kyool). Thus sugar, a nutrient, is made up of molecules. So are starch, fat, protein, and other nutrients.

Only small molecules can get into cells. During **digestion** (dih-JES-chun), your food is broken down into small molecules. Your cells can then use the food.

Your **digestive system** is a group of organs that break down food.

Find each part of the digestive system on the diagram as you continue to read. You take food into your **mouth**. When you swallow, the food goes into the **esophagus** (ih-SOF-uh-gus). The esophagus is a short tube that carries food into the top of the **stomach**.

The stomach is a bag-like organ shaped like a "J." Food leaves the stomach at its lower right side, and moves into the **small intestine** (in-TES-tin). The small intestine is long and coiled. It connects to a shorter and thicker tube, the **large intestine** (which ends with the short section called the **rectum**). Muscles in the walls of the esophagus, stomach and both intestines move the food along.

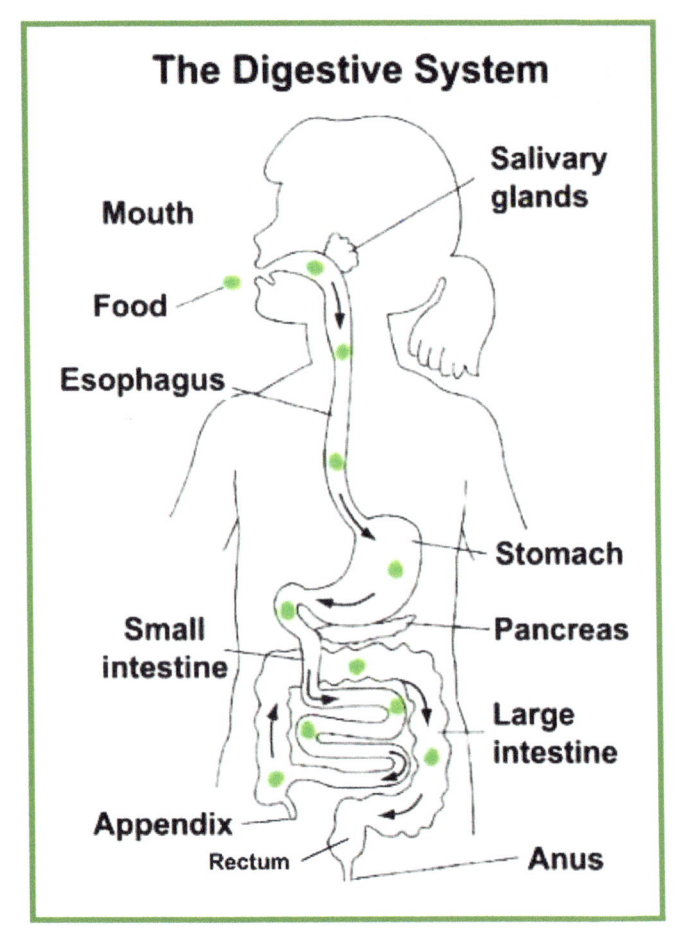

The Digestive System

Mouth, Salivary glands, Food, Esophagus, Stomach, Small intestine, Pancreas, Large intestine, Appendix, Rectum, Anus

Everything that you eat passes through your digestive system.

➤ To Do Yourself   Where Does Digestion Begin?

You will need:

Unsalted soda cracker, stopwatch or another timer.

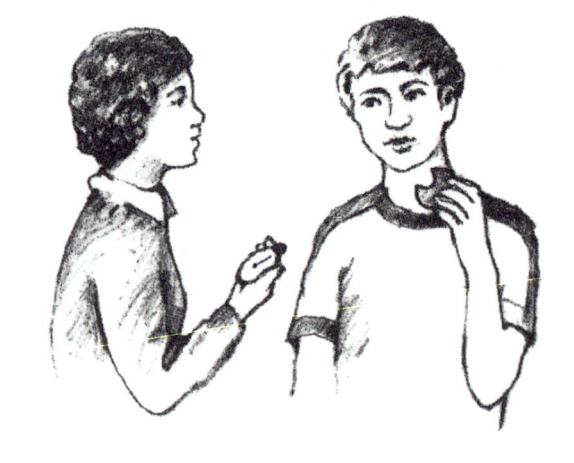

1. Take a bite of a cracker. Notice its taste.
2. Have a partner begin to time you as you start chewing.
3. Try not to swallow. Try to chew until the taste changes. When it does, have your partner stop timing. Record how long it took for the change to occur.

Questions

1. How did the cracker taste at first? How did it taste later?_____ / _____

2. What do you think caused this change?_____

3. Based on this activity, where in the body would you say that digestion begins? _____
_____

The food that your body cannot use collects in the large intestine as wastes. The wastes leave the body through an opening called the **anus**.

Attached to the large intestine is the **appendix** (uh-PEN-diks), The appendix is a worm-shaped sac; food does not pass through it. It has no known use in humans.

There are some organs, called glands, that work with the digestive system. These are the **salivary** (SAL-uh-ver-ee) **glands**, the **pancreas** (PAN-kree-us), and the **liver**. Salivary glands are near the mouth. The pancreas is below the stomach. The liver is to the right of the stomach.

Food does not pass through the glands. The glands add digestive juices to the food that is being broken down. Chemicals in these juices help to digest foods. Many of these chemicals are **enzymes** (EN-zyms). Enzymes are special proteins that help to break down large food molecules.

Each part of the digestive system has a specific job. You will learn more about these jobs in lessons to come. You will see what happens to your food in each part.

REVIEW — U-4 L-1

I. Use the choices below to write, in order, the names of the parts of the digestive system that a piece of food <u>passes through</u>. Start with the mouth. [You must *write out* the words.]

| small intestine | pancreas | appendix | large intestine | anus |
| stomach | esophagus | liver | salivary glands | |

A. ___mouth___

B. _____

C. _____

D. _____

E. _____

F. _____

II. Write the names of the organs described.

    A. Glands near your mouth _____

    B. Gland below your stomach _____

    C. Useless finger-like sac attached to your large intestine _____

    D. Gland to the right of your stomach _____

III. State briefly how food changes when it is digested. Use the word **molecule**.

_____

IV. Glands give off digestive juices. What do these juices contain that help digestion?

_____

V. Study the drawing at the top of the previous page. Where do digestive juices of the following glands <u>do their work</u>?

a) the salivary glands _____

b) the liver _____

c) the pancreas _____

99

How Does Digestion Begin? U-4 L-2

Exploring Science / Historical Steps

The Man With a Hole in His Stomach. In <u>1822</u>, **Alexis St. Martin**, a hunter in Michigan, accidentally shot himself. A U.S. Army physician, **Dr. William Beaumont**, was called in to save the man.

Under Beaumont's care, the hunter got well. But he was left with an opening into his stomach from the outside. In 1822, little was known about what happens to food in the stomach. Beaumont saw St. Martin's wound as an opportunity to study digestion, and Alexis agreed.

When Beaumont placed bread in Alexis' stomach, juice flowed from the stomach's walls!

Beaumont collected some of this juice. He put it in a test tube with some meat. After an hour, the meat fibers were swollen and separated. Within a few hours, the fibers turned into a liquid.

What Beaumont learned about the stomach is still useful. Dr. William Beaumont and Alexis St. Martin were pioneers in the study of digestion.

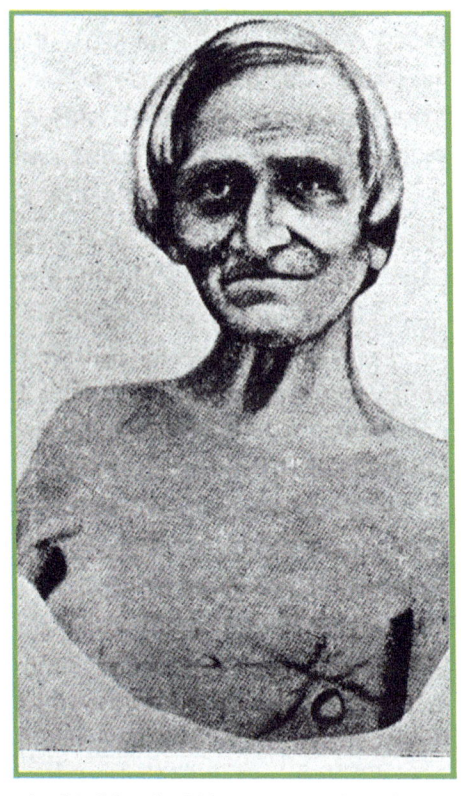

Alexis St. Martin / Note wound and opening

How Digestion Begins

Suppose that you eat a sandwich made of bread, meat, lettuce, and tomato. What happens to these foods? First, your teeth bite, tear, and grind the food. This chewing is part of mechanical (muh-KAN-ih-kul) digestion. **Mechanical digestion** breaks large pieces of food into small pieces.

At the same time, the salivary glands send **saliva** (suh-LY-vuh) into your mouth. Saliva wets food and makes it easier to swallow. Saliva also starts the chemical (KEM-ih-kul) digestion of starch. In **chemical digestion**, large molecules are broken down by enzymes into smaller molecules. There is an enzyme in saliva that acts on starch in the bread. It starts to break down the *starch* molecules into smaller *sugar* molecules.

After the food is chewed and mixed with saliva, your tongue pushes it into your throat. You swallow, and the food enters your esophagus. Wavelike movements of muscles in the wall of the esophagus push the food downward. These wavelike motions are called **peristalsis** (pair-i-STAL-sis).

From the esophagus, food enters the stomach. Digestion in the stomach is both chemical and mechanical. Lining the stomach are tiny **glands**. These glands give off **gastric** (GAS-trick) **juice**. This juice contains an enzyme and an acid. The acid helps this enzyme do its work. The enzyme works on the meat in your sandwich. Specifically, the enzyme starts to break down *protein* molecules. As you recall, meat contains proteins.

Peristalsis in the stomach wall mixes food with gastric juice. This motion helps to break down the food into smaller pieces. After one to three hours, your sandwich has become a thick liquid. It is ready to go from the stomach to the small intestine.

> **To Do Yourself**   How Does Saliva Help Digest Food?

You will need:

Starch solution, saliva, water, test tubes, iodine solution, dropper, teaspoon, sugar-test tablets

1. Collect some saliva in a test tube.
2. Place one drop of starch solution in each of two test tubes that are ¼ full of water.
3. Add a drop of iodine solution to each tube. The starch should turn dark blue.
4. Add the saliva to <u>one</u> tube. Add an equal amount of <u>water</u> to the other tube. Shake the tubes.
5. Add a sugar-test tablet to each tube. Record your observations.

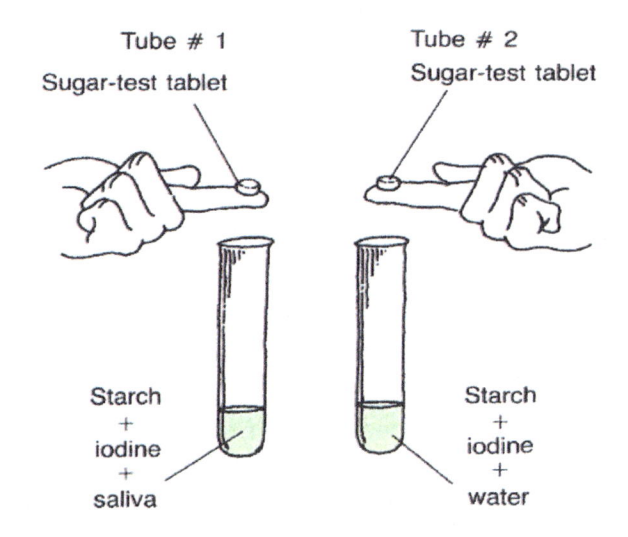

Questions

1. In which tube did the dark blue liquid turn clear? _____

2. Why did this happen? _____

3. What did the test for sugar show? _____

REVIEW — U-4 L-2

I. In each blank, write the word that fits best. Choose from the words below.

gastric   **saliva**   **chemical**   **mechanical**   **esophagus**

Chewing food is _____ digestion. Changing large molecules to

small molecules is _____ digestion. In the mouth, an enzyme in

_____ changes starch to sugar. In the stomach

_____ juice acts on proteins.

II. Circle the word (between the brackets) that makes each statement correct.

A. Glands in the stomach produce enzymes and [acid / saliva].

B. Digestion carried out by enzymes is [mechanical / chemical].

C. When food leaves the stomach its digestion is [only begun / finished].

III. Birds' stomachs have sections. One very muscular section, called the **gizzard**, has small stones in it that the bird has swallowed on purpose. Food passes through the **gizzard** before it goes into the intestine. What do you think is the job of the gizzard?

101

How Does Digestion End? U-4 L-3

Exploring Science / Historical Steps

How to Live Without Eating. In the 1960s a young surgeon, **Dr. Stanley Dudrick**, noticed a serious problem. Too many patients were dying after surgeries that seemed to go well. Why?

The digestive system works hard to do its job. It uses up a great deal of energy. Some patients were too weak to digest enough food after their operations. They were actually starving to death!

What if, for a short time, patients like these didn't need to digest food? Perhaps all of the needed nutrients could be put directly into the blood? Dr. Dudrick invented a machine that could steadily pump liquid food into a large vein near the heart.

This method is now widely used. It is especially helpful to patients who have serious digestive system problems. But the method is not risk free. It requires great care to keep germs from entering at the site where the food enters the large vein.

➤ A hospital patient with normal intestines is fed sugar, water and minerals through a vein. How does this differ from Dr. Dudick's method?

How Digestion Ends

Your small intestine is like a long coiled tube. Stretched out, it would be 6 or 7 meters long. Most digestion takes place in the small intestine.

Digestive juices from three places work on foods in the small intestine. These juices come from the liver, the pancreas, and the small intestine itself.

The liver makes a juice called **bile**. Bile is stored in the **gallbladder** (GALL-blad-ur). This is a small bag under the liver. When you eat fats, bile flows from the gallbladder into the small intestine. There, the bile breaks the fat into tiny droplets.

Imagine that you eat a roast beef sandwich. All meat contains some fat. After bile breaks the fats into small droplets, enzymes are better able to act on them.

The pancreas (a gland below the stomach) also makes enzymes. One enzyme from the pancreas digests fats. It breaks fats into molecules of **fatty acids** and **glycerol** (GLIS-uh-rohl).

The small intestine itself has tiny glands in its walls. These glands make a variety of enzymes. Some of these enzymes digest carbohydrates. Although starch begins to break down in the mouth, in the small intestine starch is broken down completely into simple sugars.

What about the protein in the beef? Protein molecules begin to be broken down in the stomach with the help of gastric juice.

Enzymes in the small intestine finish digesting proteins. Large protein molecules are broken down into small amino acid molecules.

Your sandwich has finally been completely changed into molecules that your body can use.

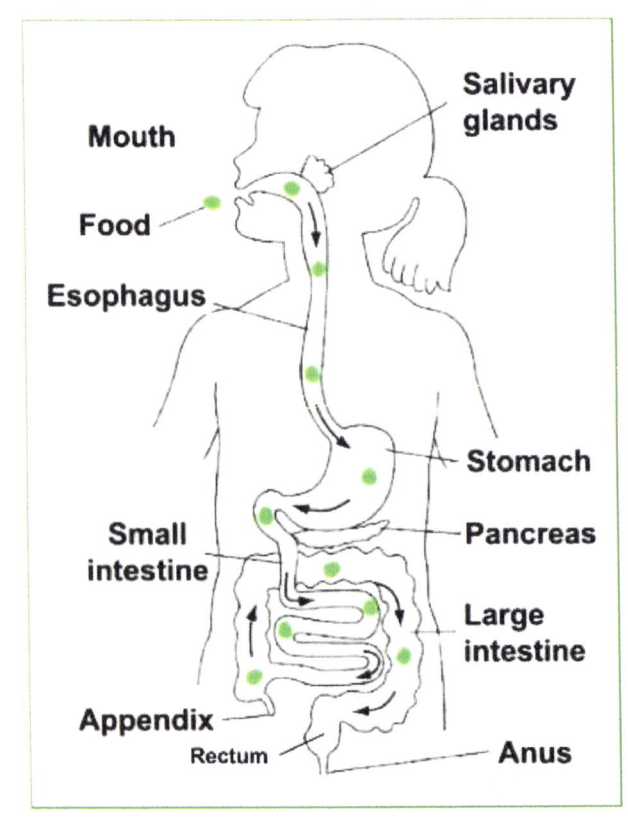

Most digestion occurs in the small intestine.

> **To Do Yourself**   **How Does Digestion Change Fats?**

You will need:

2 test tubes with stoppers, cooking oil, baking soda, dropper, teaspoon, water

1. Fill the two test tubes half full of water. Add a few drops of cooking oil to each tube.
2. Add one-fourth teaspoon of baking soda to one tube of oil and water.
3. Stopper each tube and shake it well. Record your observations.

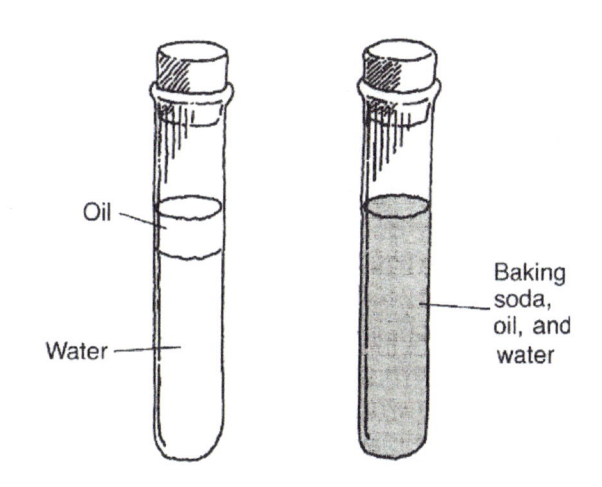

Questions

1. Which test tube has fats that have *not* been reduced to smaller droplets? _____

2. Which tube has fats that *have* been reduced to smaller pieces? _____

3. What digestive juice did the baking soda act like? _____

REVIEW   U-4 L-3

I. Complete the table below, which sums up digestion. Use words or phrases from the list to fill in the blanks. Some words or phrases will be used more than once.

| small intestine | amino acids | glycerol | stomach |
| mouth | fatty acids | simple sugars | large intestine |

Nutrients	Where Acted Upon	Final Forms
Carbohydrates	a. _____	c. _____
	b. _____	
Proteins	d. _____	f. _____
	e. _____	
Fats	g. _____	h. _____
		i. _____

II. Suppose that you eat a piece of butter. Describe briefly what happens to its fatty part (which is most of it) until it is completely digested.

_____

III. A person who has had their gallbladder removed is still able to digest fats. How? Explain.

103

What Happens After Food Is Digested? U-4 L-4

Exploring Science / Historical Steps

The King's Appendix. Edward was to be crowned king of England on June 25, 1902, and London was decorated for the big show. People had come from all over the world.

But 12 days before the ceremony was to start, Edward got sick. His stomach was upset. He had a fever and a sharp pain in his right side. On June 24, word came from the palace: the crowning could not take place.

Edward had **appendicitis**. This happens when undigested food and bacteria become trapped in the appendix, the small, finger-like organ attached to the large intestine. The appendix gets tender and starts to swell. If it bursts, the infection can spread to other parts of the body. The person may even die.

The doctors operated on Edward and removed his appendix. He recovered rapidly.

Note: Edward was finally crowned King of England on August 9, 1902.

An appendectomy (removal of the appendix) delayed his coronation, but saved King Edward's life.

➤ If someone shows signs of appendicitis, they should not take a laxative. Can you think why this is the case?

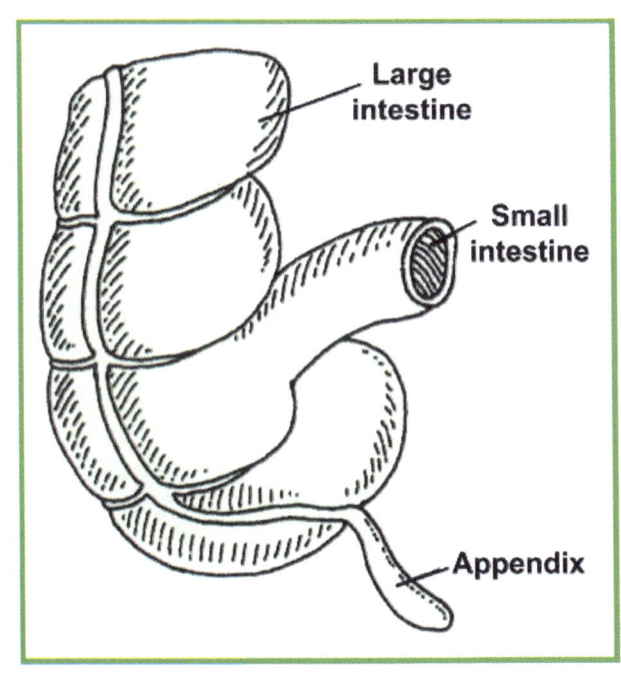

The appendix is located near the start of the large intestine, below the entrance of the small intestine. In most people, this is the lower right side of the abdomen. Hence, a doctor will often press here during an exam, and ask if you feel any tenderness.

After Digestion Is Complete

[To avoid confusion while you read this section, keep in mind that the same word is sometimes used to mean two different things.

The word "wall" is one example. When speaking about the outside layer of anything, writers often use the word "wall." Do not confuse this with the term "cell wall" which is the outer layer of plants' cells. Remember, plants have cell walls, but animal cells do not.]

You ate a sandwich and digested it. The digested food is in your small intestine. It has been changed into molecules of amino acids, simple sugars, glycerol, and fatty acids.

All of these molecules are now small enough to move through the intestine walls and then into your blood.

Observe the illustration on the next page. The walls of the small intestine are lined with tiny bumps called **villi** (VIL-eye). There are many very tiny blood vessels inside of villi. Digested food passes through the walls of the villi and then into the blood. This movement of food molecules into the blood of villi is called **absorption**. The blood next carries the food to all of the cells of your body. Each cell then takes in, or **absorbs**, the food that it needs.

Many *helpful* bacteria live in the large intestine. These bacteria cause further chemical breakdown of the wastes. This process also produces some vitamins for you!

In the large intestine, peristalsis moves the waste material along. The resulting "solid" waste is called **feces** (FEE-seez). Feces fill the last part of the large intestine (the **rectum**), then pass out of your body through the **anus**.

A BIT MORE ABOUT DIGESTION

If you eat food that contains germs, to rapidly get rid of it your body takes two unpleasant actions. First, you vomit. Second, the large intestine allows the material to pass by without removing water. In this case, the wastes are quite liquid; you have a condition called **diarrhea**.

A donut-shaped muscle surrounds the anus. Donut-shaped muscles are called **sphincters**. You have control over this particular sphincter.

Two other sphincter muscles are not under your control; one at the top and one at the bottom of the stomach. These keep food inside of the stomach until it's time for it to be released.

The stomach's top sphincter may weaken with age. Some partially digested (and acidic) food then moves upward through this sphincter and into the esophagus. This causes a burning sensation that is poorly named "*heart*burn." To deal with mild heartburn, people take **antacid** tablets ("against" the acid). If heartburn happens too often, a doctor should be seen.

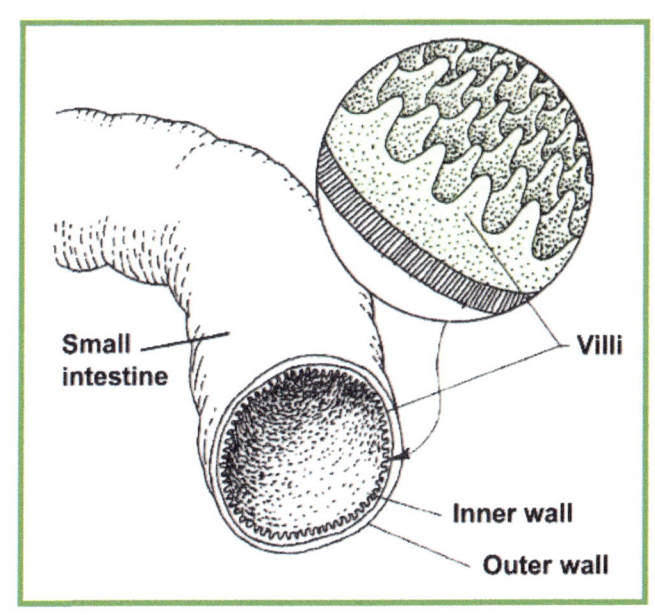

The enlarged view in the circle shows how thousands of villi line the inner wall of the small intestine.

It will be helpful to look back at the illustration in the last lesson as you complete the reading of this lesson.

Your sandwich includes lettuce and tomato, both of which contain fiber. Recall from Lesson 5 that your body cannot digest fiber, but it is an important part of your diet. Fiber keeps your intestines healthy. It makes it easier to move and then eliminate (get rid of) wastes.

Food parts that you cannot digest become wastes. These wastes pass into the large intestine. Here, much of the water in the undigested food is taken back into the blood. This makes the wastes more solid.

------- **REVIEW** ------- U-4 L-4

I. In each blank, write the word or words that match the statement. Use the words below.

   villi   absorb   feces   fiber   blood vessels   appendix

   A. _____ to take in
   B. _____ solid waste that gathers in the rectum
   C. _____ cannot be digested
   D. _____ found in villi
   E. _____ bumps in the small intestine

II. Circle the word or words (between brackets) that make each statement true.

   A. Water is absorbed in the [large / small] intestine.
   B. Bacteria in the large intestine are [always harmful / usually helpful].

III. It is not recommended (of course), but if you stood on your head after a meal, food could still move "down" your digestive system. Why?

What is the Work of the Blood? U-4 L-5

Exploring Science / Historical Steps

<u>Dr. Drew; "Father of the Blood Bank."</u> In <u>1940</u>, World War II was being fought in Europe. One August day that year, **Dr. Charles Drew**, an African American, got a cablegram from a doctor in Britain. The cable read: CAN YOU GET 5,000 UNITS OF DRIED PLASMA FOR TRANSFUSION?

In a **transfusion** (trans-FYOO-zhun) one person's blood is given to another. **Plasma** (PLAZ-muh) is the blood's liquid part. Drew had developed a way to dry blood plasma. When dried, plasma is easy to keep and to ship. Just before use, the dried plasma is mixed with water.

Drew answered Britain's call for help and many lives were saved. He became the head of the *Plasma for Britain* project. He also spoke up strongly against the appalling American practice of separating blood donations by race.

In peacetime, as in war, plasma saves lives. All hospitals use plasma.

➤ Donors at a blood bank are given fruit juice and sweets after they give blood. What two parts of the lost blood are partly replaced by these foods? Explain.

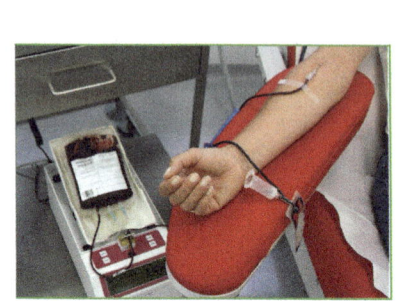

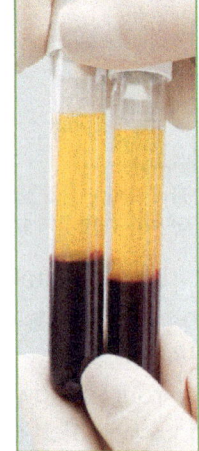

Left: A person donating blood
Right: Tubes of blood - after solid and liquid layers have been separated

The Work of the Blood

Has a doctor ever tested your blood? For many tests, a test tube is filled with blood. The tube is put in a machine (called a **centrifuge**) that spins very fast. When the spinning stops, the blood has been separated into two layers. At the top is the liquid part, called **plasma**. At the bottom is the "solid" part.

Plasma is 90 percent water. Dissolved in the plasma are digested foods, wastes from cells, and special chemicals. One chemical helps the blood to clot (stop flowing) when you get a cut. Other chemicals help to fight germs.

The solid part of the blood has three kinds of cells. **Red blood cells** are small and shaped like rounded plates. Often they are simply called **RBCs**. They have no nuclei. A protein called **hemoglobin** (HEE-muh-gloh-bin) in the red blood cells gives them their color. It is the hemoglobin that actually carries the oxygen to all of your body cells.

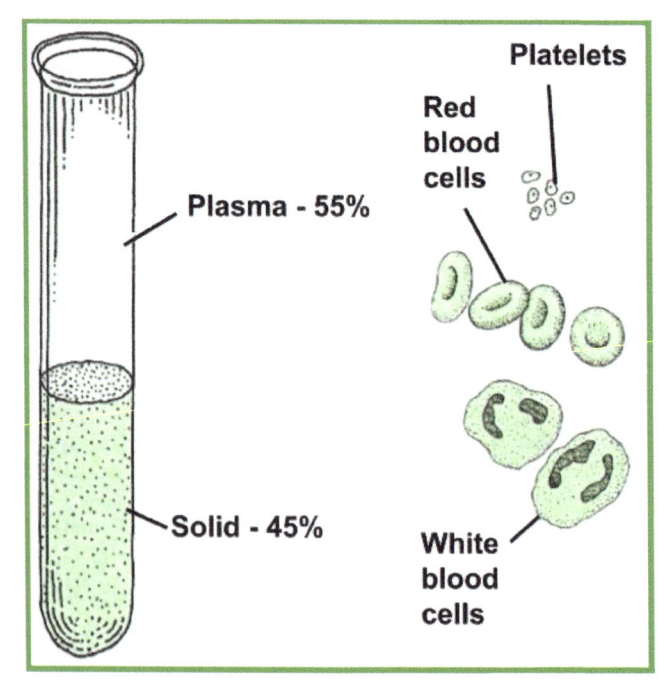

Plasma - 55%
Solid - 45%
Platelets
Red blood cells
White blood cells

Left: Tube of blood after being spun in centrifuge
Right: The solid part is made of red blood cells, white blood cells, and platelets.

White blood cells (often called **WBCs**) come in several different forms. These cells *do* have nuclei. There are many fewer white blood cells than red blood cells. The job of the white blood cells is to fight germs. You will learn more about these in Unit 10, Lesson 2.

Tiny blood cells called **platelets** (PLAYT-lits) are also in the solid part of the blood. Platelets help form blood clots. A clot keeps blood from flowing from a wound. When you are cut, the platelets give out a chemical. This chemical works with part of the plasma to make a thread-like net. This net catches some red blood cells, and together they form a clot. When the clot dries it is called a **scab**.

➤ To Do Yourself   What Are Some Parts of Blood?

You will need:

Prepared slide of human blood, microscope

1. Observe the prepared blood slide under low power. Then observe it under medium and high power.
2. Identify the red blood cells. Locate some of the stained white blood cells. Draw and label each kind of cell.
3. Select a very small section of the slide near a white blood cell. Count the number of red blood cells and white blood cells in this small section.

Questions

1. Which kind of cells are larger?_____

2. Which kind of cells are present in the largest numbers?_____

REVIEW — U-4 L-5

I. In each blank, write the word that fits best. Choose from the words below.

| platelets | plasma | hemoglobin | clot | white | red |

The liquid part of the blood is _____. Blood cells that fight germs are _____ blood cells. Rounded, plate-shaped cells without nuclei are _____ blood cells. The _____ in red blood cells carries oxygen. Tiny cells that help to form clots are _____.

II. Circle the word (between the brackets) that makes each statement correct.

  A. A net of threads in a clot forms partly from [plasma chemicals / white blood cells].

  B. A scab forms from a chemical from platelets and dried [white / red] blood cells.

III. Blood clots sometimes form inside of the body rather than where the skin is cut. Why could such a clot be harmful?

What Does Your Transport System Do? U-4 L-6

Exploring Science / Historical Steps

Birth of a Bloodstream. One night in the early 1900s, **Dr. Florence Sabin** worked late in her lab. She was the first female full professor at the famous John Hopkins Medical School in Baltimore, Maryland. She described that night as "the most exciting experience of my life."

What did she do? As you might guess, it was something that was new to science. She had been watching a small developing chicken embryo as it grew. Under her microscope, she saw tiny blood tubes forming. Blood cells then appeared, and the heart began to beat! She called what she saw "the birth of a bloodstream."

In many ways, a chick's heart, blood tubes, and blood are like a human's. Seeing how these parts develop inside of a chick embryo led to new knowledge about our own blood system. For example, Dr. Sabin showed where human blood cells are made. Most blood cells come from the center, or **marrow** (MEH-roh), of the long bones in our arms and legs.

During her career, Sabin also studied how certain blood cells help to fight a lung disease, **tuberculosis** (too-bur-kyuh-LOH-sis), that has killed many people. Some white blood cells can move like amebas. They can move out of the blood vessels and into spaces between the cells of the lungs. There they attack and kill the tuberculosis germs.

➤ What kinds of cells are probably found in a chick embryo's blood? Why do you think so?

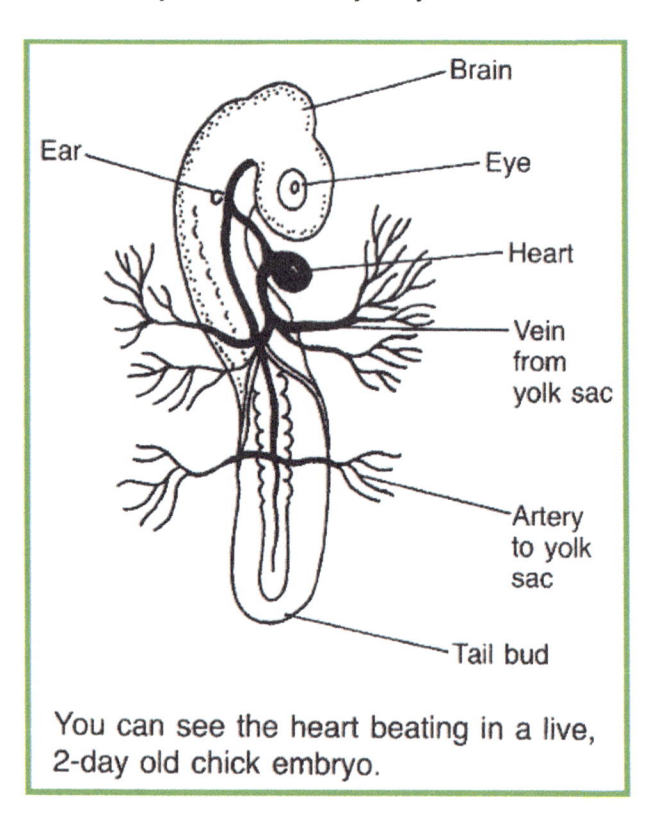

You can see the heart beating in a live, 2-day old chick embryo.

Your Transport System

The bloodstream in Dr. Florence Sabin's chick embryo was a transport system, much like a road. So, too, is your bloodstream. It transports food, oxygen, and other materials. The main parts of this system are the blood vessels, heart, and blood. The blood goes in a complete path or "circle," in the body. We say that it **circulates**. Because of this, the transport system has another name: the **circulatory system**.

Blood vessels are the tubes through which the blood moves. You have three kinds of blood vessels. **Arteries** (AR-tuh-rees) carry blood away from the heart. **Veins** (VAYNS) carry blood back to the heart. **Capillaries** (KAP-uh-leh-reez) are microscopic blood vessels that connect arteries with veins.

The largest blood vessels are arteries. And the largest artery in the human body is the **aorta** (ay-OR-tuh). The long aorta is located at the top of the heart and deep inside of the body's trunk.

In the veins, **valves** let blood flow in only one direction - toward the heart. The valves function like one-way doors.

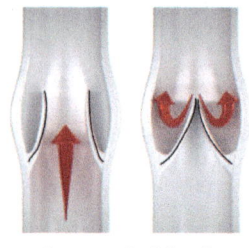

In a leg vein, a pulse sends blood upward through a valve. The valve then closes, preventing backflow. Blood in veins is also moved as our muscles contract - squeezing nearby veins. (A good reason to move!)

The Circulatory System

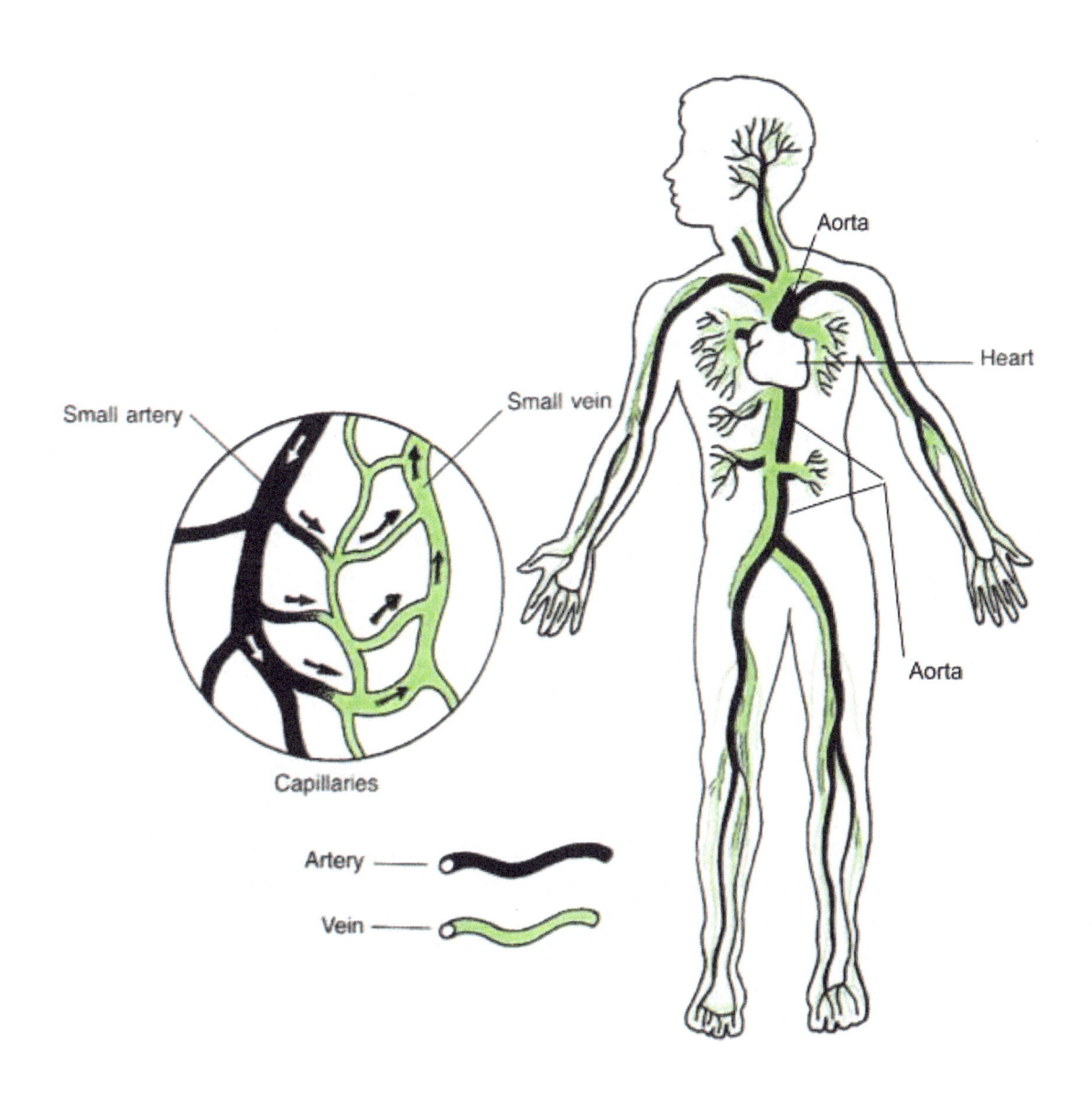

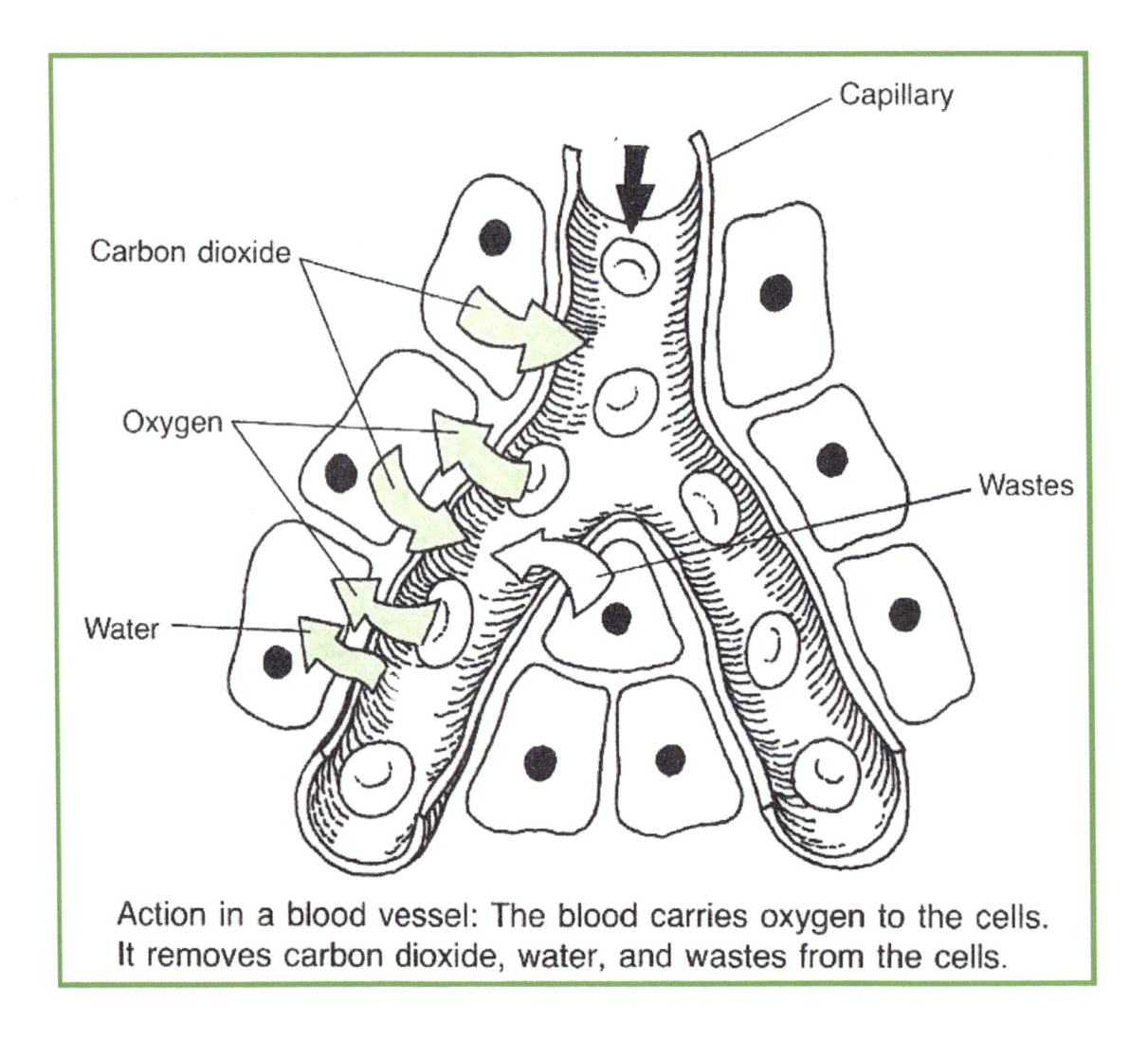

Action in a blood vessel: The blood carries oxygen to the cells. It removes carbon dioxide, water, and wastes from the cells.

You have millions of capillaries. These tiny blood tubes can be seen only with a microscope. They take the blood very close to every body cell.

The **heart** is a strong muscle; its job is to pump blood through the blood vessels. Your heart is about the size of a large fist. It lies in the center of your chest. The heart is actually two pumps in one. A wall of muscle divides the right pump from the left pump. Blood cannot flow from one side of the heart to the other.

Each side of the heart is divided into two chambers, or rooms. Each upper chamber is an **atrium** (AY-tree-um). Veins carry blood to each atrium. The blood flows through each atrium to a lower chamber, which is called a **ventricle** (VEN-trih-kul). The job of each ventricle is to send blood out of the heart.

Between each atrium and ventricle is a **valve**. These heart valves keep blood flowing in one direction - from an atrium to a ventricle. In other words, the valves prevent blood from flowing backward from a ventricle to an atrium.

Each time your heart beats, it **contracts** (squeezes together). This pushes blood out of the heart and into the arteries. The force of this contraction also makes the arteries "beat." They expand and contract as each "gush" of blood passes through them. This beating of the arteries is your **pulse**. Wherever an artery is close to the skin, you can feel your pulse.

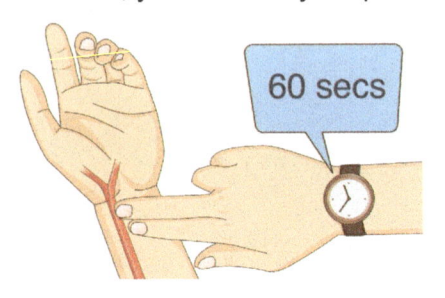

Press using two fingers (not your thumb) on the thumb side of the wrist (with the palm upward). The pulse rate is usually given per minute. To save time, check for 20 seconds and multiply by 3.

From upper body — **From lungs** — **From lower body**

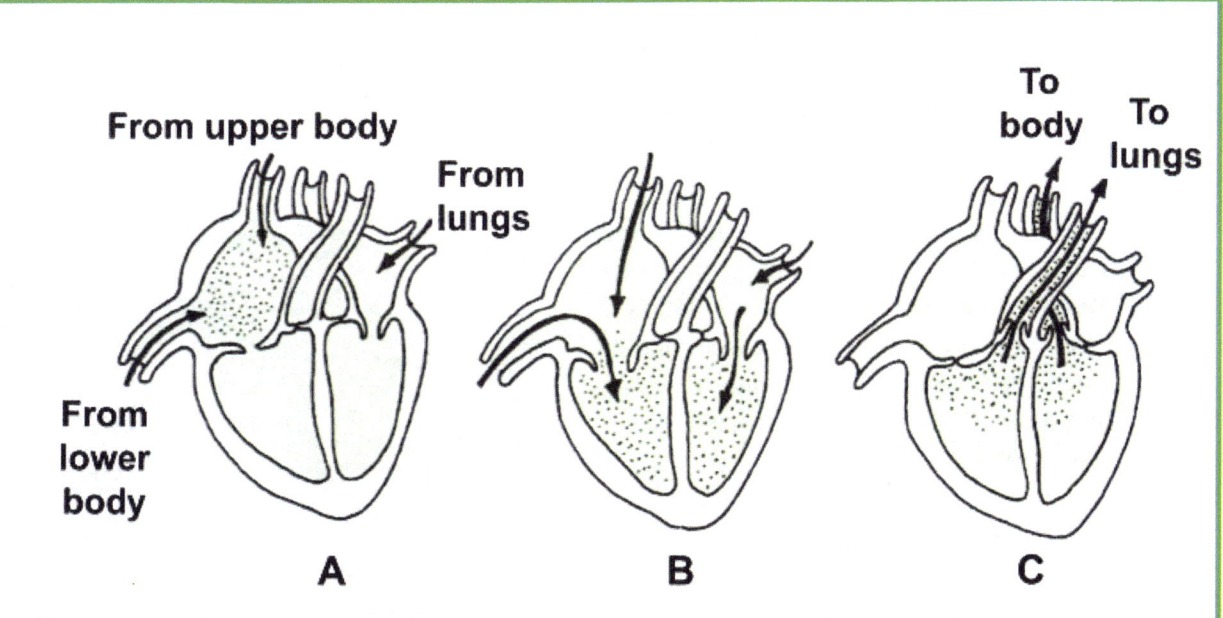

A   **B**   **C** (**To body**, **To lungs**)

In **A**, blood enters the atria (plural of atrium) from the body and lungs. In **B**, the atria force the blood into the ventricles. In **C**, the ventricles send the blood to the body and the lungs.

--- REVIEW --- U-4 L-6

I. Use the list of words below. Write the names of the parts of the transport (circulatory) system that match each clue below.:

    arteries   cell   capillaries   ventricle   veins   atrium   valve   heart

    A. _____ upper chamber in the heart
    B. _____ lower chamber in the heart
    C. _____ carry blood away from the heart
    D. _____ carry blood toward the heart
    E. _____ one-way "door" in the heart, or in a vein
    F. _____ smallest blood vessels in the body

II. Circle the word or words (between brackets) that make each statement true.

    A. A blood vessel that has a pulse is [an artery / a vein].

    B. As blood carries food and oxygen to cells, it [contracts / circulates].

    C. Blood flows from an atrium into a [vein / ventricle].

    D. Capillaries connect an artery with a [vein / atrium].

III. Unlike your veins, your arteries do not need valves. Can you explain why?

_____

_____

IV. The walls of arteries and veins are not quite the same. Which do you think are stronger? Explain your answer.

What is the Path of Your Blood? U-4 L-7

Exploring Science / Historical Steps

Harvey the "Circulator." In the 1660s in England, quacks (people who pretended to know about medicine) sold remedies during medicine shows. These shows went around, or circulated, through the country. The quacks were called "circulators." When **William Harvey**, a real doctor, first announced that the blood circulates, people laughed at him. They called him a "circulator," as if he were a quack, too.

Before Harvey began his experiments on the blood's path, people thought that the arteries were air tubes. They knew that the heart beats, but they had no idea why. They also knew that their veins had blood. But they thought that the veins were two-way "streets" - that the blood moved in both directions.

Harvey showed that the heart beats because it is a pump. He showed that both arteries and veins carry blood. And he said that the blood tubes are *one-way* "streets." It was clear to Harvey that blood moves in only one direction, and that it goes in "circles."

Harvey was sure that somehow blood from the arteries moves into the veins. But he could find no tubes linking arteries with veins. Still, he predicted that someday the links would be found.

People did not want to believe in anything that they could not see. This is a common bias. They did not believe Harvey's hypothesis. In 1661, only four years after Harvey died, **Marcello Malpighi**, in Italy, found the tubes that Harvey could not see. Malpighi had a tool that Harvey did not have - a microscope. As you have already learned, capillaries are microscopic (able to be seen only with a microscope). After that, everyone stopped laughing at Harvey.

➤ Circle the word (or words) that makes the statement true:
Harvey [could / could not] have predicted that the red blood cells carry oxygen to the body tissues. Explain your answer.

William Harvey discovered that blood circulates.

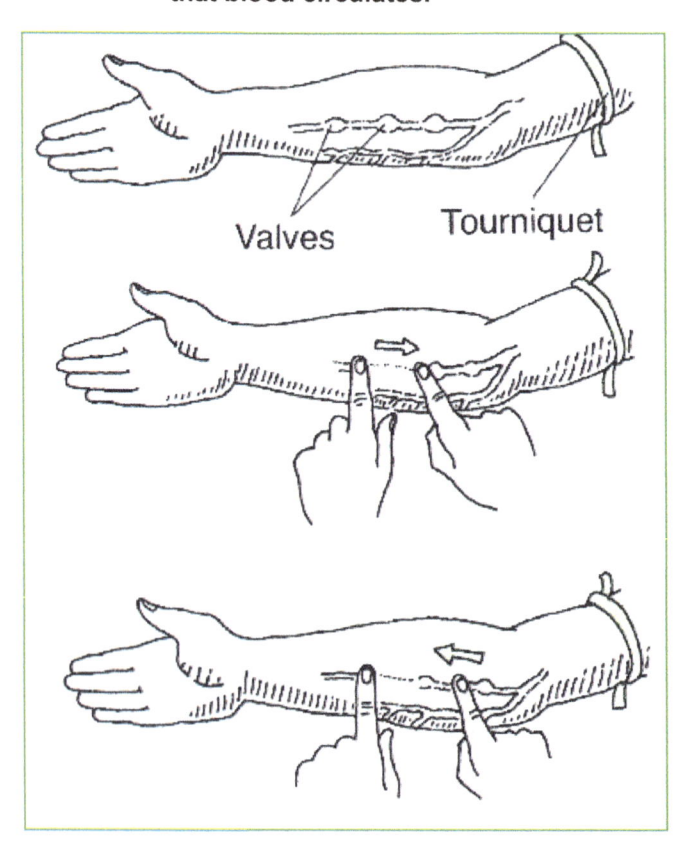

Harvey showed that the vein's valves work so that blood flows in one direction only - toward the heart.

The Path of the Blood

Because of Harvey's pioneer work, you can trace the path of your blood through your body. To do so, use the diagram shown below. The numbers of the following descriptions match the numbers in the diagram.

(1) Blood in the **right atrium** has come from your body cells. This blood is high in carbon dioxide and low in oxygen. It is *dark* red.

(2) The right atrium contracts. The blood flows through a one-way **valve** into your right ventricle.

(3) The **right ventricle** contracts. This pushes the blood into the arteries that go to your lungs.

(4) The blood goes into the capillaries of the **lungs**. There, carbon dioxide moves out of the blood and oxygen moves into the blood. Now your blood is low in carbon dioxide and high in oxygen. Your blood is now *bright* red.

(5) Through veins, the blood moves from the lungs into the **left atrium**.

(6) The left atrium contracts. The blood goes through a one-way **valve** into your left ventricle.

(7) The **left ventricle** contracts. This pumps the blood into your body's largest artery, the aorta.

(8) The **aorta** divides into other arteries. Some arteries take blood to your upper body. Other arteries take blood to your lower body.

(9) The arteries divide again and again. Blood moves from the smallest arteries into **capillaries** all over your body. There, oxygen passes from your blood into your body cells. Carbon dioxide passes from your body cells into the blood.

Other materials also pass between the blood and the body's cells. When blood goes through the capillaries of your small intestine, it picks up digested foods. The food molecules later pass from the blood into all cells of your body. Besides carbon dioxide, other cell wastes also pass from your cells into your blood.

From the capillaries, blood moves into the smallest veins. Small veins join together to form large **veins**. The largest veins take your blood back into your right atrium.

Your blood has completed a round trip - from the right atrium, back to the right atrium.

HOW YOUR HEART WORKS

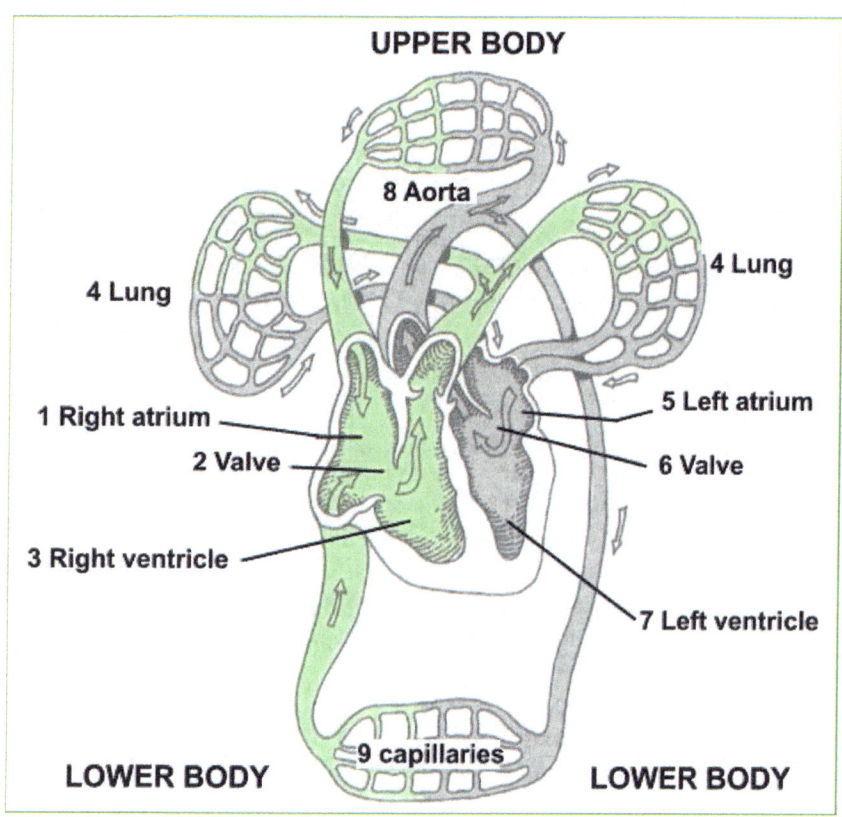

UPPER BODY

8 Aorta
4 Lung
4 Lung
1 Right atrium
5 Left atrium
2 Valve
6 Valve
3 Right ventricle
7 Left ventricle
9 capillaries
LOWER BODY
LOWER BODY

➤ To Do Yourself   What Is Inside of a Heart?

You will need:

Pig heart, sharp scissors or knife

1. Observe the pig heart. Find and count the blood vessels.
2. Cut the heart open. Your teacher may help you do this, or show it to you.
3. Count the chambers and find the valves.
4. Find the section that has the most muscle - the thickest walls.

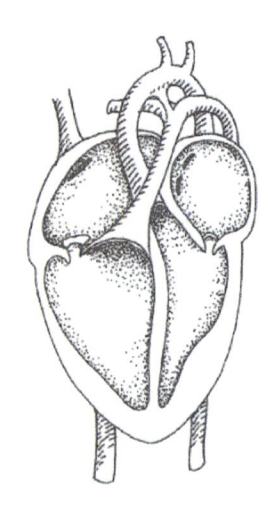

Questions

1. How many chambers does the heart have? What is the function of each?_____

_____

_____

2. What are the valves? What is their function?_____

_____

Helping weak hearts. In some people, the path of blood in the circulatory system may become blocked. In other people the heart itself may not pump blood properly. These people may be suffering from **heart disease**.

Some heart problems can be solved with heart surgery. Weak valves can now be replaced with valves made artificially, or with valves from pig hearts. Some people whose hearts are not beating properly have a special device inserted next to their hearts. (See the illustration on the next page). This device gives the heart an electric signal to start beating. It sets the pace for the heart to beat, so it is commonly called a **pacemaker**. Thousands of people each year have heart valve and pacemaker surgeries.

Sometimes a person's heart may have severe damage caused by disease or an inborn condition. If the condition is life-threatening, a person may have a **heart transplant**. During a heart transplant, a heart from a donor is placed inside the chest of a person with a damaged heart. Many thousands of Americans have failing hearts and are waiting for a transplant.

Sadly, there are many more people with damaged hearts than there are donor hearts. Each day in the U.S., 17 people die waiting for a heart transplant. Scientists have tried for decades to find ways to help desperate patients.

An artificial heart, called the Jarvik-7, was used in 1982. This made news globally. But blood clots formed in the device and the patient lived for only 112 days. The focus shifted to smaller devices called **VAD**s (ventricular assist devices). With VADs, the patient's heart is left in place. VADs help (assist) the left ventricle to move blood until a donor heart is available.

In the summer of 2021, a French company built a device that, for a short time, can replace both ventricles in the heart. This device gives people more time to wait for a heart transplant.

In early 2022, the DNA of a pig was changed slightly so that its heart would (hopefully) not be attacked by cells in a human body. The pig's heart was transplanted into a man who was about to die of a weak heart. If this method works well, there may soon be many people living with pig hearts!

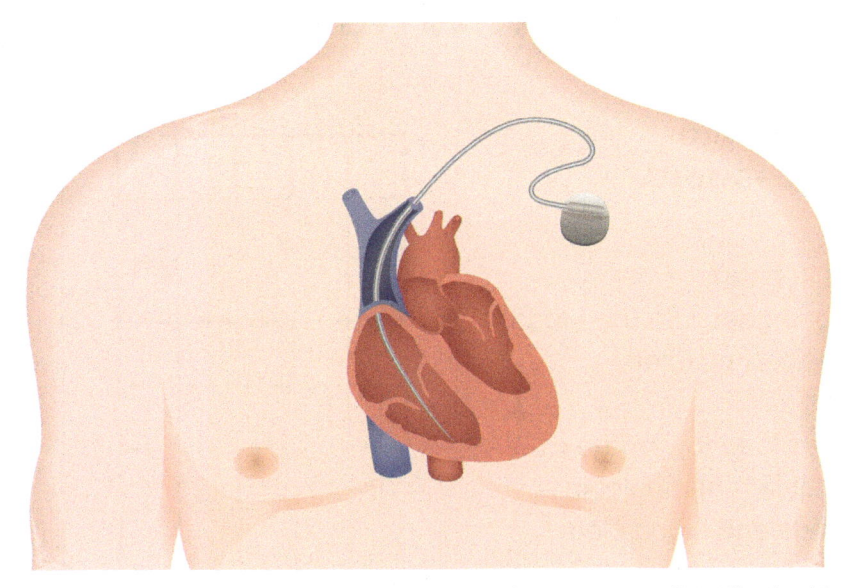

The pacemaker is placed under the chest muscle. A wire connects it to the inside of the right side of the heart. Electrical impulses from the pacemaker stimulate the heart to beat regularly.

REVIEW — U-4 L-7

I. Below is a list of the parts of the circulatory system. Begin with the right atrium and put the parts in the order in which blood passes through them. Record only <u>letters</u>.

    a. arteries to body
    b. capillaries in lungs
    c. veins from body
    d. arteries to lungs
    e. right ventricle
    f. left ventricle
    g. veins from lungs
    h. capillaries in body
    i. left atrium

1. __right atrium__
2. _____
3. _____
4. _____
5. _____
6. _____
7. _____
8. _____
9. _____
10. _____

II. Circle the word or words (between brackets) that make each statement true.

    A. The aorta is the body's largest [vein / artery].

    B. Carbon dioxide passes into the blood in the capillaries of the [body / lungs].

    C. Blood in the right ventricle is [low / high] in oxygen.

    D. The color of the blood in the left atrium is [dark / bright] red.

III. The blood that goes through arteries <u>to</u> the body cells is rich in oxygen. If this blood is mixed with blood that is low in oxygen, the skin can look blue. One cause of the color of **"blue babies"** is a hole between the right atrium and the left atrium.

    A. Explain why such babies are "blue."   **B.** How can an operation save these babies?

Unit 4 -- Review What You Know ---

A. Use the clues below to complete the crossword.

Across
1. Sending chamber in heart
5. Where digested food goes into the blood
6. Formed with the help of platelets
8. Liquid part of blood
10. Vessels that carry blood toward heart
11. Blood cells that contain hemoglobin
12. Dried net of threads and blood clots

Down
1. Keeps blood flowing in one direction
2. To go around in a circle
3. Digested fats become glycerol and _____ acids
4. To break down into small molecules
7. They produce juices that contain enzymes
9. Where feces leave the body

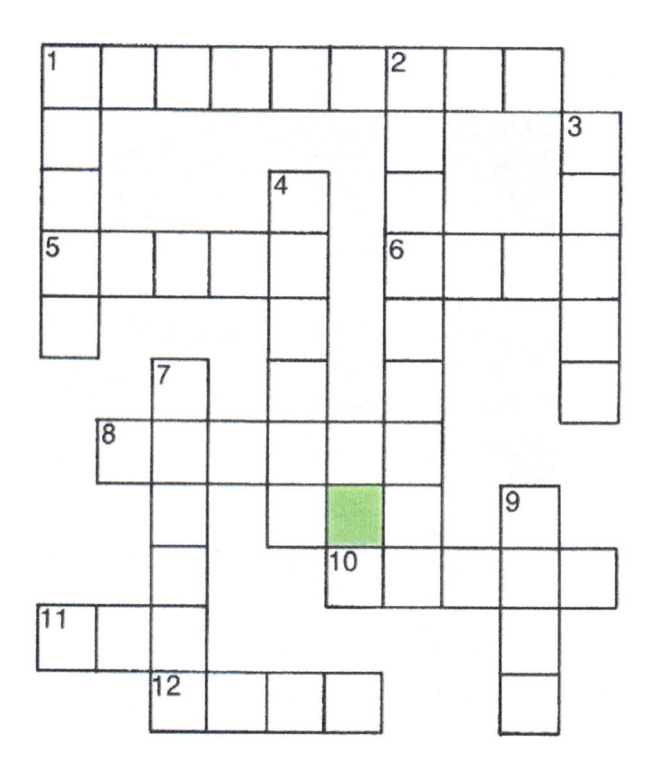

B. Write the word (or words) that best completes each statement.

1. A juice from the pancreas does its work in the
   a. esophagus   b. stomach   c. small intestine   1. _____

2. Gastric juice helps to digest
   a. fiber   b. meat   c. oil   2. _____

3. Bile is made in the
   a. liver   b. villi   c. appendix   3. _____

4. Saliva starts to break starch into
   a. amino acids   b. fatty acids   c. sugar   4. _____

5. Food and oxygen pass into cells through the walls of
   a. arteries   b. veins   c. capillaries   5. _____

6. The heart's receiving chambers are
   a. arteries   b. atria   c. aortas   6. _____

7. Carrying oxygen is the job of the
   a. red blood cells   b. white blood cells   c. platelets   7. _____

8. Amino acids are formed by the digestion of
   a. fats   b. proteins   c. starches   8. _____

9. Chewing food is a part of digestion called
   a. gastric   b. chemical   c. mechanical   9. _____

10. Blood returning to the heart from the lungs is
    a. high in CO_2 b. low in O_2 c. high in O_2 10. _____

11. The longest part of the digestive system is the
    a. large intestine   b. small intestine   c. esophagus   11. _____

12. The most numerous cells in the blood are the
    a. white blood cells   b. red blood cells   c. platelets   12. _____

13. The gallbladder stores a liquid that helps to digest
    a. proteins   b. starches   c. fats   13. _____

14. CO_2 passes out of the blood through the walls of
    a. veins   b. capillaries   c. the aorta   14. _____

15. Glycerol is formed from the digestion of
    a. fats   b. proteins   c. fiber   15. _____

C. Apply What You Know

1. Study the drawing below that includes numbered body parts. Use the <u>letters</u> to the left of the words below to match a label to the right of each number.

   A - liver   C - large intestine   E - heart   G - salivary glands
   B - esophagus   D - aorta   F - small intestine   H - stomach

2. In the drawing, name the **two** organ <u>systems</u> that are at least partially illustrated. The first letter of each system has been provided for you.
   [For now, ignore the blanks to the right of the word "numbers" (below).]

   systems → __D_____   __C_____

   numbers → ___ , ___ , ___ , ___ , ___ , ___   ___ , ___

   1 _____
   2 _____
   3 _____
   4 _____
   5 _____
   6 _____
   7 _____
   8 _____

3. Look back at question #2 (above). In the short blank spaces below each <u>system</u> that you named, make a <u>list</u> of the <u>numbers</u> of the parts that belong to each of the two systems. Each short blank should receive one number.

D. Find out more.

1. A. Obtain a fresh, whole fish. Using the internet or a source book from the library, find the following parts of its digestive system: pharynx (throat), stomach, intestine, liver, gallbladder. Also find the following parts of the circulatory system: heart (cut it open to find its chambers), blood vessels (the larger ones). What are some differences between the fish's organs and yours?

   B. Fill in the blanks with the names of the organs.

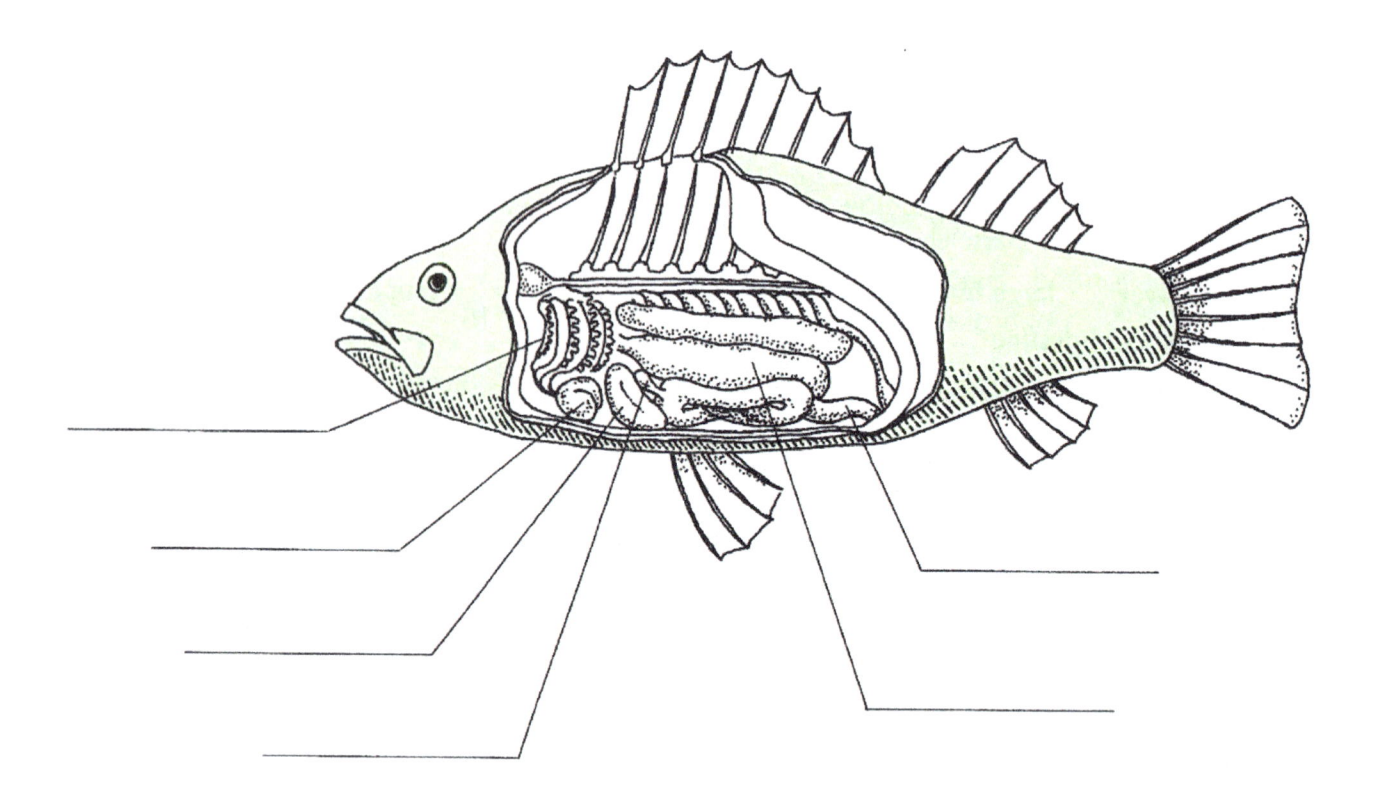

2. Plants as well as animals produce enzymes that digest foods. An enzyme that digests protein is found in papayas. A meat-tenderizer that you can find in a supermarket contains this enzyme. To show how this meat-tenderizer acts on the protein in egg white, try the activity described next.

   In one test tube, place bits of cooked egg white, water, and meat-tenderizer. In another test tube, place only the egg white and water. Let both tubes stand for several hours. Observe. Explain.

   Another protein-digesting enzyme is found in fresh pineapple juice. If the pineapple is cooked, as it is when it is canned, the enzyme is destroyed. Gelatin is a protein. People often add fruit to gelatin when making salads or desserts. Try the activity described next. Prepare one dish of gelatin to which fresh pineapple juice is added. Add juice from canned or cooked pineapple to a second dish of gelatin. What do you observe? Explain.

3. Research the basics of human **blood types**. Why are blood types important? Visit a blood bank or a hospital laboratory. Ask to see how blood types are determined. Ask a parent or your doctor about your blood type. Make a bulletin-board display to explain what you have discovered.

Careers in Life Science

Staying Active. Are you a sports fan? Join the crowd! So are millions of Americans. Do you play sports or exercise daily? Join an even bigger crowd! Visit any park, zoo, public pool, golf course, or tennis court and you'll find it busy with people exercising! The numbers are getting larger all the time. So are the numbers of people with careers related to physical activity.

Athletic trainer helps a man recuperate from an injury.

Athletic Trainers. Do you love both sports and science? Do you like helping people? Athletic trainers need these qualities. Their work helps both amateur and professional athletes perform better. Trainers help athletes prevent and recover from injury.

Some athletic trainers begin learning while still in high school. They work with experienced trainers. Both the coach and the team doctor depend on the trainers - as do the athletes themselves - for many special tasks. To become a certified athletic trainer, four years of college are required.

Physical Therapist. The same interests and talents that lead some people into careers as trainers lead others into careers as physical therapists. Athletes who have been injured, or who need surgery, are first treated by doctors. Then the physical therapist takes over. She or he plans and carries out treatments ordered by the doctor. The object: to relieve pain, restore the use of the injured parts, prevent further injury, and get the player back into action.

Some physical therapists specialize in working with older people. There is a strong demand for professionals who value helping older people stay physically active. But there is certainly a need for physical therapists who work well with other age groups, including young children.

In America, a degree in physical therapy is now a doctorate. After earning a college degree in a health-related field (such as biology), you will complete 3 years of specialized study. Being accepted to a physical therapy program is very competitive - and the pay once you graduate is quite good.

A physical therapist is a highly educated and trained professional. In addition to directly helping clients, most supervise physical therapy assistants.

You can become a physical therapist assistant with just two years of technical school. An assistant therapist works under the direction of a licensed physical therapist.

UNIT FIVE

BREATHING AND MOVEMENT

What Does Your Respiratory System Do? U-5 L-1

Exploring Science / Historical Steps

Breath of Thin Air. In 1981, scientist **Chris Pizzo** set a record for the world's highest frisbee throw. He did this while standing on top of the world's highest mountain, Mount Everest.

To reach this in style, Chris Pizzo took along ….

…..this! Actually, without the dog.

Of course, Pizzo didn't climb the mountain just to throw a frisbee. He was part of the American Medical Research Expedition. Over 300 people carried scientific equipment to a camp two-thirds of the way up Mt. Everest. The air there is very thin; there are fewer molecules in the air. For example, this air contains only half as much oxygen as at sea level. Human lungs must work harder to get the same amount of oxygen. Tests were done to learn how much work people can do while breathing so little oxygen.

Pizzo climbed the rest of the way to Everest's top. The air was so thin that Pizzo needed extra oxygen. He wore an oxygen mask.

He also took along a gas sampler. As he climbed into thinner air, Pizzo went without the mask for a few minutes at a time. Before he put the mask on again, he breathed into the sampler. He took the air samples back to the camp, where they were studied by other scientists.

The scientists found out just how thin the air at the top of Mt. Everest actually is. They also found out how little oxygen is needed to keep a person alive and well. Doctors used these findings to help people with breathing problems.

➤ Circle the word (between the brackets) that makes the sentence correct.

The native peoples who live high in the Andes Mountains of Peru have very large lungs and chests. To join them on a hike you would probably breathe [faster / slower] than they do.

Your Respiratory System

Recall that all of your cells are carrying out cell respiration. This process uses oxygen to burn food - so that your cells may obtain energy. Obviously, you need oxygen to stay alive. You get this oxygen from the air, which is about one-fifth oxygen.

The entire process of breathing is called **respiration**. Your breathing organs take air into, and let air out of, your body. These organs make up your **respiratory system** (shown on the next page). When you breathe in, or **inhale**, you take air into your nose. There the air becomes warm and moist. From your nose, air goes down your throat and into your **trachea** (also called the **windpipe**).

Rings of a stiff tissue called **cartilage** keep the trachea open. You can feel these cartilage rings if you rub up and down at the front of your neck.

In both the nose and the trachea, air is cleaned. **Cilia** (SIL-ee-uh), which are like little hairs, catch tiny bits of dust. **Mucus** (MYOO-kus), a sticky liquid, also traps dust. Sometimes the dust (and some mucus) is removed as we sneeze; hence covering a sneeze is vital!

The trachea branches into two air tubes called **bronchi** (BRONG-keye); the singular is **bronchus** (BRONG-kus). One bronchus goes to each lung.

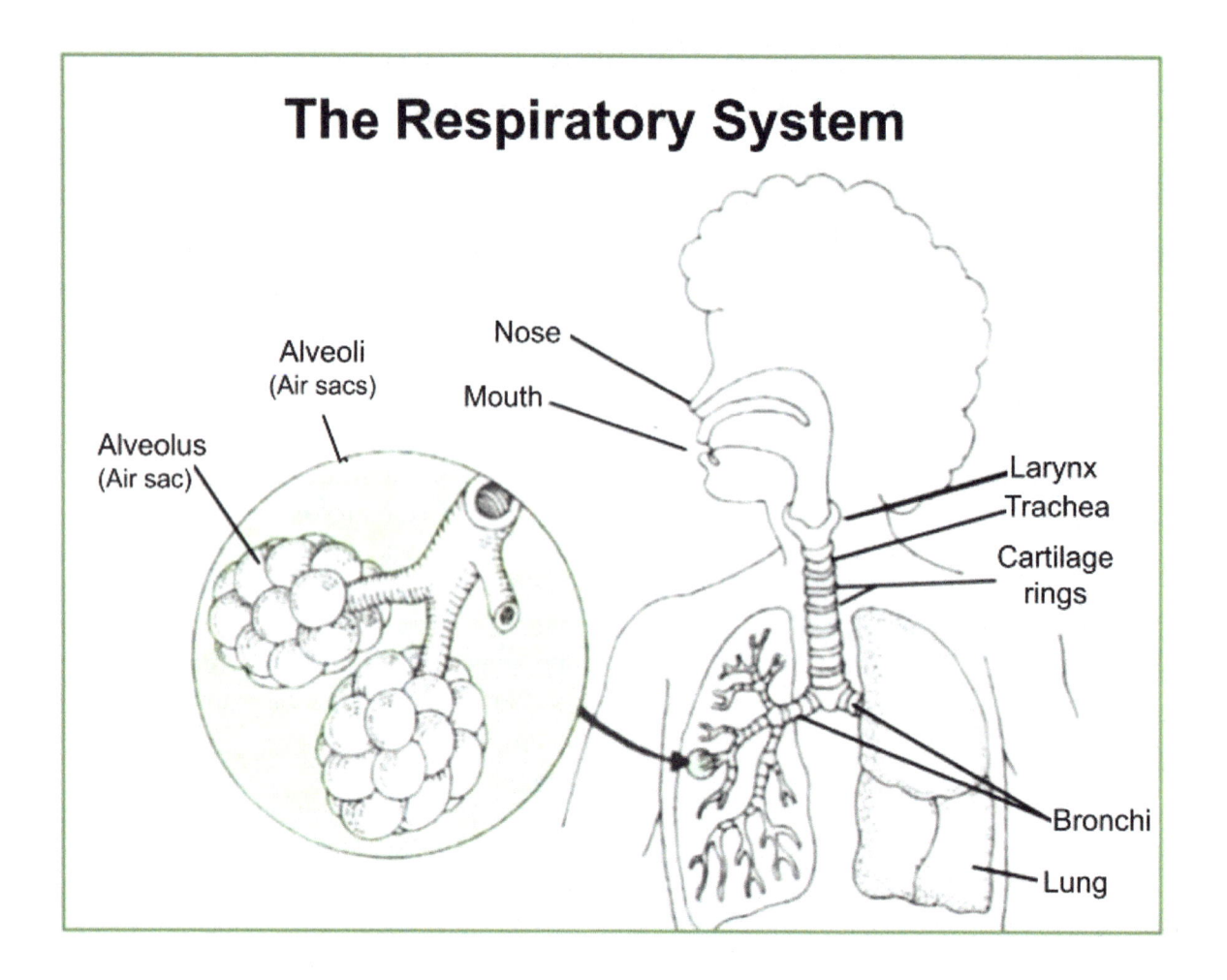

The Respiratory System

- Alveoli (Air sacs)
- Alveolus (Air sac)
- Nose
- Mouth
- Larynx
- Trachea
- Cartilage rings
- Bronchi
- Lung

➤ To Do Yourself — What is Your Lung Capacity?

You will need: rubber tubing, large plastic tub, large plastic jar, plastic saucer, measuring cup, water

1. Set up the bottle of water and the rubber tube as shown, but with no air in the jar.
2. Inhale and fill your lungs with air. Now exhale slowly into the rubber tube.
3. Keep the jar's opening underwater, Now slide the saucer under the mouth of the jar. Carefully, while keeping them together, take the jar and saucer from the tank, and then flip them over.
4. Remove the saucer. Now, fill the jar with water, using the measuring cup. Record how much water you use to fill the jar. This is the amount of air that you exhaled.
5. Repeat the entire procedure three more times and take an average.

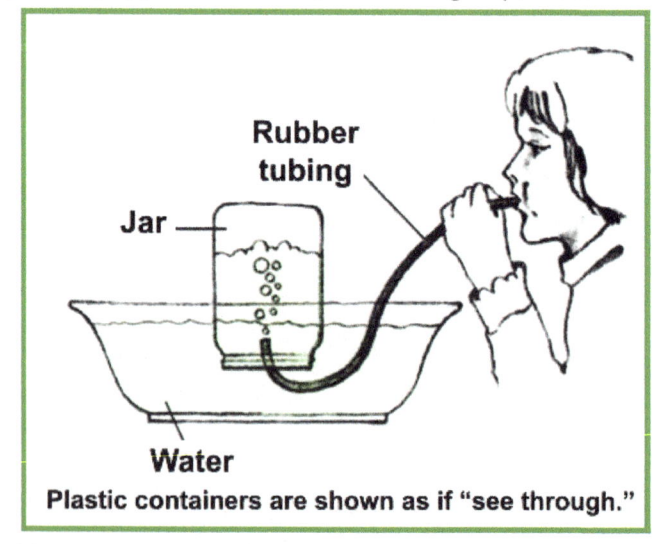

Rubber tubing, **Jar**, **Water**
Plastic containers are shown as if "see through."

Note: If you choose to use glass containers instead of plastic, obtain adult support.

Questions

1. What is the air capacity of your lungs? _____

2. How does this compare with others in your class (or your family)? _____

In the lung, each bronchus branches like a tree, into smaller and smaller tubes. The tubes end in millions of tiny air sacs called **alveoli** (al-VEE-oh-lie); singular **alveolus**.

Each alveolus is surrounded by capillaries (microscopic blood vessels). Oxygen moves from the air within each alveolus into the blood. At the same time, carbon dioxide moves from the blood into the alveolus. Some water also moves from the blood into the air sac.

When you breathe out, or **exhale**, the air in the alveoli passes back through the same tubes that it came in. But this exhaled air now has less oxygen and more carbon dioxide than the air that you inhaled. Are you clear why? There is also more water in the air that you exhale. Are you clear where this water comes from?

A Respiration Riddle: What organ won't work properly unless you are exhaling? A hint: You don't actually need this organ to breathe.

The organ, made of firm cartilage, is the **voice box** (also called the **larynx**). The larynx is located at the entrance to the trachea. It contains two elastic bands - your **vocal cords**. These bands vibrate as air, on its way out, rushes between them. Your voice is the result! [Sadly, smokers sometimes get cancer of the larynx. After surgery they are often unable to talk].

REVIEW — U-5 L-1

I. In each blank, write the word that fits best. Choose from the words below.

mucus   exhale   respiratory   cilia   inhale   larynx   oxygen

Your _____ system takes air into and lets air out of your body.

When you _____, you breathe in. When you

_____, you breathe out. Hairlike parts that clean the air are the

_____. A sticky liquid that traps dust is _____.

Because you have a _____, you are able to speak.

II. *Write* the names, in order, of the body parts through which oxygen passes during respiration. Start with the nose. Use these parts:

bronchi   throat   capillaries   trachea   alveoli

A. ____nose____   C. _____   E. _____

B. _____   D. _____   F. _____

III. Circle the word (between the brackets) that makes each statement correct.

A. Air that you exhale contains [more / less] carbon dioxide than the air you inhale.

B. A capillary that carries blood into the lungs contains [more / less] oxygen than the same capillary that carries blood out of the lungs.

IV. Smoking destroys cilia. How does this lead to lung diseases like cancer and emphysema?

_____

_____

What Happens When You Breathe?   U-5 L-2

Exploring Science / Historical Steps

The Boy Who Did Not Drown. On a cold January day in 1984, **Jimmy Tontlewicz** went sledding with his dad along the shore of Lake Michigan. Four-year-old Jimmy threw his sled onto the frozen lake. His dad walked on the ice to bring back the sled. Jimmy followed. He jumped onto the ice … and both of them fell through!

Jimmy's dad was rescued by passersby. It was at least 20 minutes before divers finally pulled Jimmy out, and there were no signs of life.

A few years before this happened, Jimmy would have been another boy who drowned. For many years scientists thought that no one could survive for more than 4 minutes under water. Fortunately, by 1984 it was known that some people survive even after an hour in cold water!

So the rescue team did not give up. They immediately used **CPR** (a method that anyone can learn) to help Jimmy's heart restart and to get him breathing again. In the hospital, doctors put to use special methods that they had recently learned. Soon, Jimmy was on his way to being well again.

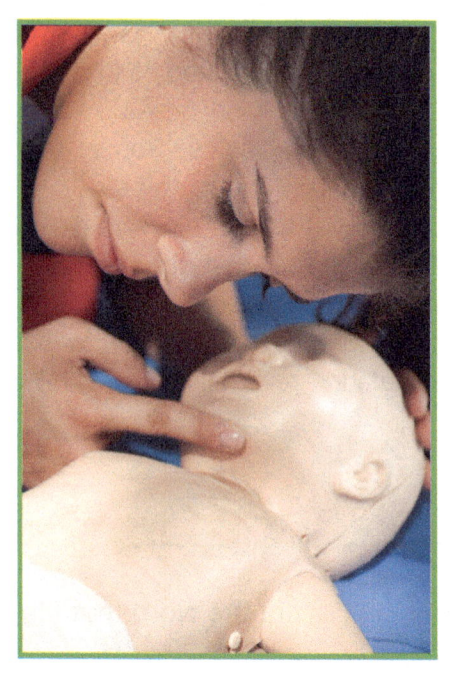

A woman learning CPR on a baby manikin

➤ Explain how Jimmy's case shows that the body can store oxygen for use at a later time.

➤ Want more? Search the "mammalian diving reflex" to learn why some "drowned" people are able to survive.

➤ To Do Yourself   How Can You Measure The Carbon Dioxide You Exhale?

You will need:

Bromothymol blue solution, straw, glass jar, timer (stopwatch)

1. Fill the jar half full with bromothymol blue solution.
2. Take a deep breath and exhale through the straw into the solution.
3. When enough carbon dioxide is bubbled through the bromothymol blue solution, it turns yellow. Have a partner time how long it takes for you to do this.

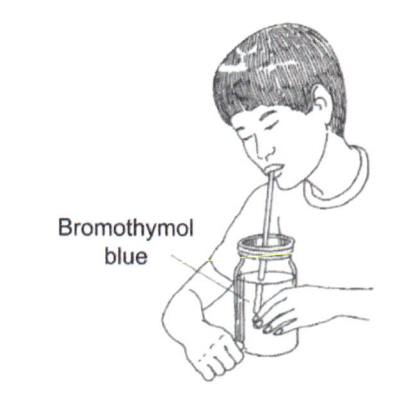

Bromothymol blue

Questions

1. How long did it take for the bromothymol blue to change to yellow? _____

2. How did this time compare to other students (or family members)? _____

Respiration and Breathing

When Jimmy went under the water, he could no longer take oxygen into his lungs. But there was already enough oxygen in his body to keep him alive for a while. His breathing and even his heartbeat stopped. But a process that keeps the cells alive did not stop. You have learned that this process is called **cell respiration**.

As a reminder, here are the major steps of cell respiration - the process used by nearly all living things: The cells take in both oxygen and food. When the food (a simple sugar) combines with oxygen (burns), energy is released. Carbon dioxide and water are given off.

Moment by moment, so that your cells receive and remove molecules, you breathe. This moves air into and out of your lungs. When the single word **respiration** is used, it means **breathing**. Muscles in your chest help you breathe. One of your breathing muscles is the **diaphragm** (DY-uh-fram). The diaphragm is like an elastic sheet. It lies below the lungs. At rest, the diaphragm is curved upward, like an upside-down bowl. When you **inhale** (breathe in), your diaphragm moves down and becomes less curved. At the same time, rib muscles move your ribs up and out. As a result, the space inside of your chest gets larger. Air rushes (from the outside) into your lungs to fill the larger space.

When you **exhale** (breathe out), your diaphragm moves up, becoming more curved. At the same time, your ribs move down and in. The space inside of your chest gets smaller. The air is pushed from your lungs to the outside.

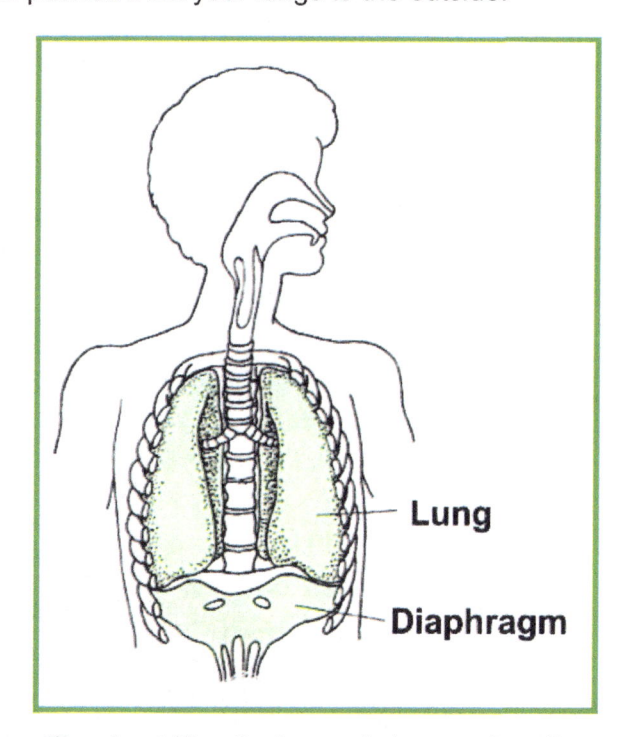

The sheet-like diaphragm helps you breathe.

REVIEW — U-5 L-2

I. In each blank, write the word that fits best. Choose from the words below.

    breathing   respiration   CPR   diaphragm   ribs   bronchi

In cell _____, sugar combines with oxygen. The <u>common</u> term for the process of moving air into and out of your lungs is _____.

Muscles move your _____ up and out when you breathe in. When your _____ moves up, air is pushed out of your lungs.

II. Circle the word or phrase (between the brackets) that makes each statement correct.

    A. The diaphragm becomes [more / less] curved when you inhale.

    B. When you exhale, your ribs move [up and out / down and in].

    C. A process that takes place in all body cells is [breathing / cell respiration].

III. Some bacteria can live only in areas <u>without</u> oxygen. Do you think that these bacteria carry out cell respiration? If not, how do you think that they survive? [Note: No need to worry. Just make a logical guess at the answer.]

How Does Your Body Remove Wastes? U-5 L-3

Exploring Science / Historical Steps

Heroes Still Needed! In recent decades, amazing progress has been made in medicine. Some young people even think that there is nothing left to discover in this field. Not true!

In 2018, Ken Tschanz, a 64 year-old farmer in Kentucky, was nearly a hero. Nearly? His younger brother John had survived a serious blood infection. But this badly damaged John's kidneys. Ken planned to donate one of his kidneys to John. Sadly, this never happened. Ken suffered a fatal stroke, and there was too little time for doctors to transplant a kidney.

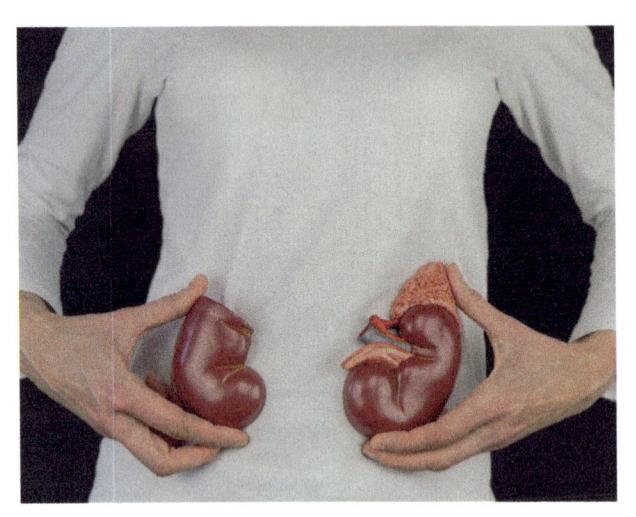

Actual size models of kidneys. The one on the right includes blood vessels and the triangular-shaped adrenal gland (at the top).

John soon began to use an **artificial kidney** machine. He also added his name to the waiting list for a donated kidney. As you can imagine, denoted kidneys are in short supply.

Kidneys continually "clean the blood." No machine has been designed that can match the human kidney. The two best designs have been in use since the 1960s!

Thousands of people with failing kidneys have been helped by the first design. It runs the patient's blood through a filtering process called **dialysis**. This takes four hours, and it must be done three times per week. These machines are very expensive and they require a skilled staff to operate. Thousands of people who need these machines will not live long enough to use one.

In the late 1960s, a creative, less costly method of cleaning the blood came into wide use. With this method, called TPD,* the patient's blood never leaves the body.

The **abdomen** (also called the "belly") is the area below the chest and above the waist. Lining the abdomen is a large thin sheet of tissue. It helps keep the digestive organs in place. The many blood vessels in this sheet can be "tricked" into removing unwanted molecules (sometimes called bodily wastes) from the blood.

To do this, about 2 liters of a specially designed fluid is pumped into a patient's abdomen. Some "waste" molecules leave the blood vessels and enter this fluid. Next, the fluid is drained out of the body. Three or four times each night, while the patient sleeps at home, a pumping machine completes the entire process. During the day, the patient's life is close to normal.

Soon after Ken died, John used blood dialysis. Later he used TPD. These methods can keep people like John alive for several years while they wait for a kidney transplant. However, great care must be taken to avoid dangerous infections. Happily, John did eventually get a kidney transplant - and it worked very well!

Obviously, heroes are still needed; heroes, like Ken, who are willing to donate a kidney, and heroes of science who will build better artificial kidneys! Maybe you or a classmate will be such a hero!

*TPD = tidal peritoneal dialysis

➤ For kidney patients, two significant advances occurred in the Fall of 2021.
1) A company began selling a dialysis machine that patients could use at home.
2) Scientists changed a bit of DNA in a pig. Later, the pig's kidney was transplanted into a human. For weeks the human's body reacted to the pig's kidney as if it was a human kidney!

Hopefully, these are steps toward a long term solution for those with failing kidneys.

How do you think people on kidney dialysis feel about these advances?

Getting Rid of Wastes

Your cells give off many wastes. These include carbon dioxide, water, salts, and urea. **Urea** (you-REE-uh) is formed when amino acids are broken down. All of these wastes must be removed from your body or you will get very sick. Getting rid of the cell's wastes is called **excretion** (ik-SKREE-shun).

The cells pass their wastes into the blood. The blood carries these wastes to three places - the lungs, the kidneys, and the skin. Excretion takes place through these organs.

• <u>LUNGS</u>. Your lungs remove carbon dioxide and some water from your body. When you breathe out, your lungs get rid of (**excrete (ik-SKREET)**) these wastes. If a person drinks alcohol, some of the alcohol will also be given off by the lungs. Police sometimes test an automobile driver's breath to see if they have been drinking alcohol.

• <u>KIDNEYS</u>. Water, urea, and other wastes from body cells are removed from the blood by your two kidneys. The kidneys are in the back part of the abdomen, just above your waist. Each is about the size of the palm of your hand.

In each kidney there are about one million microscopic filters. As the blood passes through these filters, the cells' wastes are removed. These wastes, plus water, form a liquid called **urine**. (YOOR-in).

The urine flows from the kidneys' microscopic filters into tiny tubes. These tubes join other tubes. Finally, urine leaves each kidney through a large tube called the **ureter** (YUR-uh-tur). The ureter from each kidney carries the urine to the **bladder** (sometimes called the **urinary bladder**). The bladder is an elastic sac that stores the urine until it is excreted. The urine leaves your body through a single tube called the **urethra** (yoo-REE-thruh).

At the base of the urethra (near the bladder) you have another sphincter muscle that is under your control. Once again, this is a sphincter that often becomes weaker as people age.

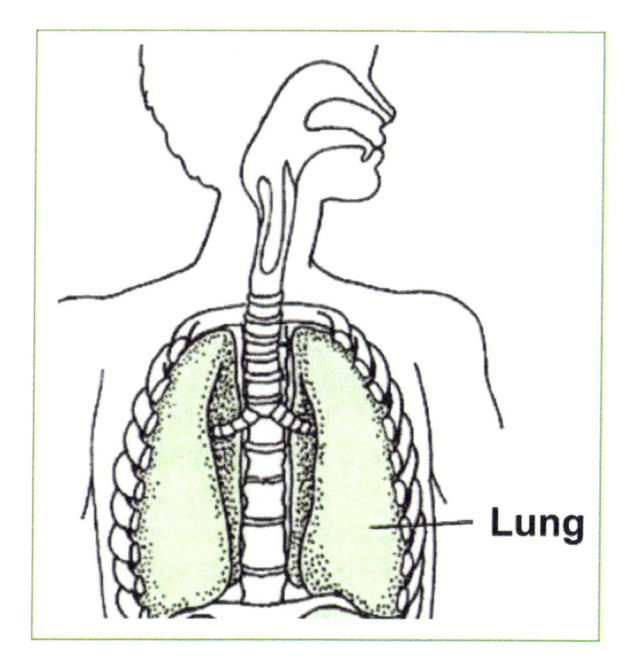

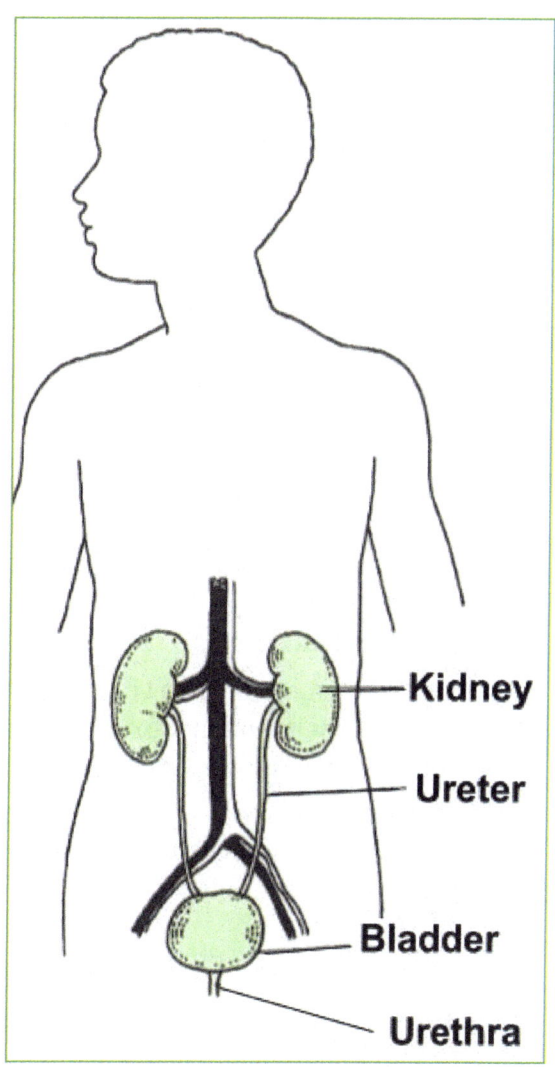

One kidney is located on each side of your spine, in your lower back.

127

- SKIN. Your skin covers and protects your body. It also functions as an organ of excretion. The skin contains approximately two million tiny **sweat glands**. Capillaries carry blood next to each sweat gland. (See the illustration). Water, salts, and some urea move from the capillaries into the sweat glands. Together, these wastes make up the liquid called **sweat**.

At the top of each sweat gland is an opening in the skin. This opening is called a **pore**. Pores release the sweat.

Your skin also helps to control your body's temperature. When your body is too warm, more blood flows to the sweat glands, and more water moves out of them and on to the pores. Sweat on the skin evaporates (turns into water vapor). This uses heat, so it helps to cool your body.

By the way, other glands in the skin make oils to keep the skin from drying out. When the pores of these oil glands get clogged, **acne** ("pimples") can occur. Hormones (discussed in Unit 6, Lesson 6) have much to do with the activity of oil glands. Some acne is normal for young people.

For excess acne, it is best to see a **dermatologist**, a doctor who specializes in skin care.

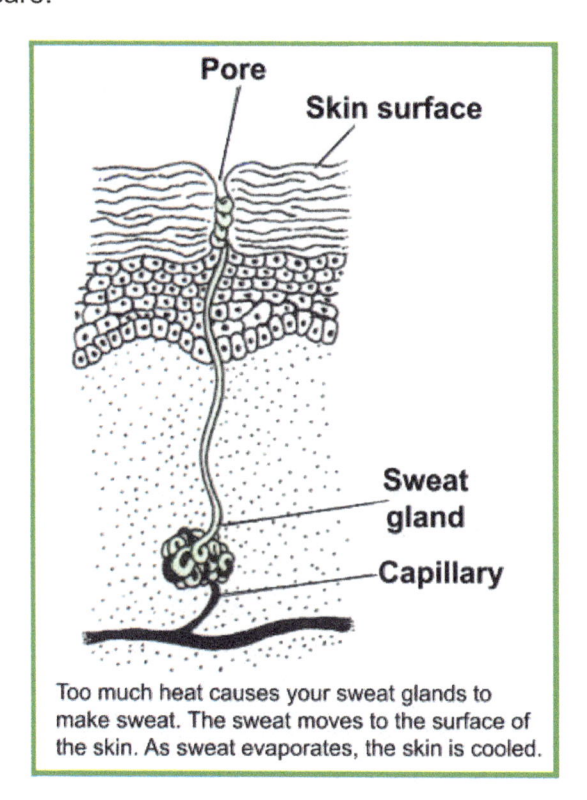

Too much heat causes your sweat glands to make sweat. The sweat moves to the surface of the skin. As sweat evaporates, the skin is cooled.

➤ To Do Yourself   How Does Sweating Cool the Body?

You will need:

Water, 2 thermometers, a piece of cardboard, cotton

1. Wet the cotton and use it to cover an area of your body.
2. Fan both the wet area and a dry area with the cardboard. Record (below) which area feels cooler.
3. Place one thermometer in a ball of wet cotton. Place the other thermometer next to it. Note the temperatures and record them below.
4. Fan both thermometers for several minutes. Again, note the temperatures and record them.

Questions

1. Which area of the skin was cooler? _____

2. Which thermometer showed a lower temperature? _____

3. Besides getting rid of wastes, why is sweating important to your body? _____
_____

REVIEW U-5 L-3

I. In each blank, write the word that fits best. Choose from the words below.

urea   bladder   pores   urethra   filter   sweat   ureter   excretion   urine

The process of getting rid of cells' wastes is _____. The kidneys _____ cells' wastes from the blood. The waste made by the kidneys is called _____. A tube called the _____ carries urine from each kidney to the _____. Urine passes from the bladder and then exits the body through the _____. The _____ glands excrete wastes through openings in the skin called _____.

II. Use the letter in front of each waste to record all of the wastes that are excreted through the listed organ. Some wastes will be used more than once.

    a. urea   **b.** salts   **c.** water   **d.** carbon dioxide

1. lungs   _____

2. kidneys _____

3. skin _____

III. When you go for a checkup, your urine is tested. If the doctor finds sugar in your urine, this can be a sign of disease. Can you suggest why?

_____

IV. Observe the illustration and the numbered organs. On the lines below, write the *names* of the numbered organs of excretion.

1. _____

2. _____

3. _____

4. _____

129

What Is Your Body's Framework? U-5 L4

Exploring Science / Historical Steps

New Faces for Old Bones. 50,000 years ago, a second group of humans existed with us on Earth. These were the **Neanderthals** (nee-AN-der-tahls). Their skulls and various other bones have been found in Europe and the Middle East.

Scientists consider Neanderthals to be in the same genus (*Homo*) as modern humans. Our scientific name is *Homo sapiens*. Theirs is *Homo neanderthalensis*. Some of the ancestors of these two groups of humans actually mated and produced children. As a result, all modern humans have a small percentage of Neanderthal DNA in our cells!

What did Neanderthals look like? People who are both scientists and artists are able to draw (or make models of) faces based on old bones. To make a model of a Neanderthal, a scientist-artist starts with a plaster copy of the bones. The bones and the copies have ridges and grooves on their surface. The ridges and grooves show where muscles were once attached. Deep grooves on a heavy bone mean a big muscle. Clues like this tell how the muscles might have looked.

Soon the scientist-artist has an idea of muscle sizes, positions, and shapes. The next step is to make clay versions of the muscles and add them to the Neanderthal bones. Clay is also used to shape the ears and the end of the nose. Next, a thin layer of clay "skin" is added over the whole model.

Finally, the model is painted. [See the result at the end of this lesson.] Scientists agree that Neanderthals looked much like us. However, they had a much flatter forehead, a larger nose (with a wider bridge), puffy cheeks, and a smaller chin. Neanderthals also had a slender area of bone underneath their eyebrows that stood out. This is called a "brow-ridge."

➤ What part of the model of a face would likely involve the most "guesswork"?

(A) tip of the nose; **(B)** chin; **(C)** forehead. Explain your answer.

➤ To Do Yourself   What Does A Bone Look Like?

You will need:

Chicken bones, a large beef bone sawed in half, a hand lens

1. Scrape the outside of the bones and describe what you find.
2. Look for cartilage at the end of the bones.
3. Use a hand lens to examine the inside of the beef bone. Draw and label the parts that you can identify.
4. Wash your hands with soap and water.

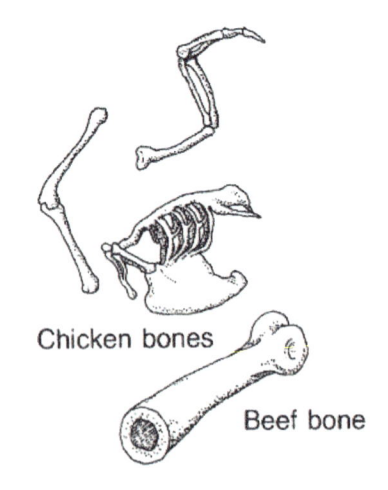

Chicken bones

Beef bone

Questions

1. How is the outside of the bone different from the inside? _____

2. How is cartilage different from bone? _____

Your Body's Framework

BONES. Like the Neanderthals of long ago, your body gets its shape from your bones. There are 206 bones in your body. Together, they make up your **skeleton** (SKEL-ih-tun).

The skeleton does much more than give shape to your body. It also provides the body's basic framework, much as wooden beams provide the framework for a house. Without the skeleton, you would not be able to stand upright.

Another job of your skeleton is protecting vital organs. Your skull protects your brain.

The Human Skeleton

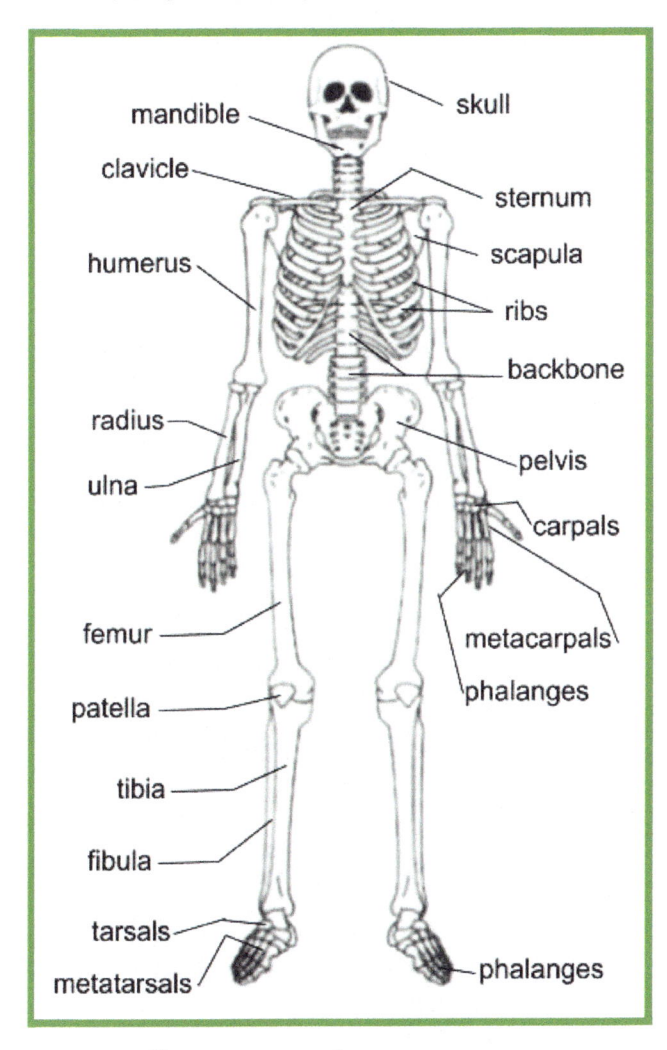

How are your lower arms similar to your lower legs?

Hint: Look at an entire arm and an entire leg. Now count the number of bones at each location. Do you see a pattern? The skeletons of other vertebrates follow a similar pattern! All have one large bone nearest the trunk, followed by two bones. Compare the bones of our hands and feet with those of other vertebrates to discover more similarities.

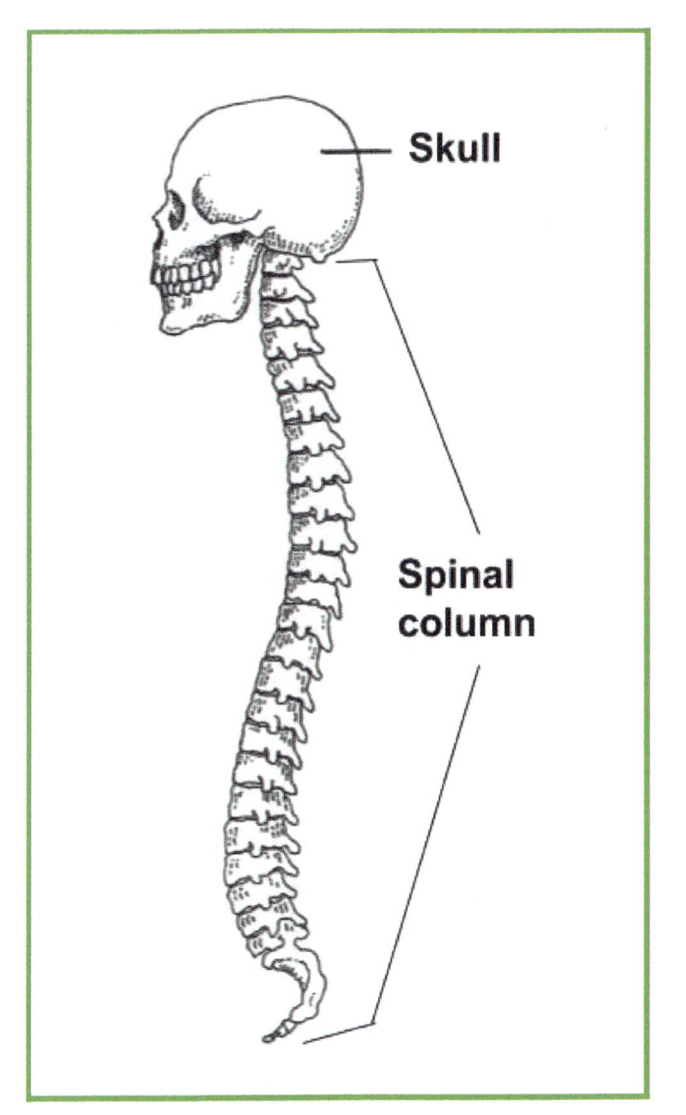

Two of the vertebrae in the spinal column are fused together. Can you find them?

Your breastbone (sternum) and ribs protect your heart and lungs. Your backbone (which is also called the **spinal column** (or the vertebral column), is made of 33 **vertebrae** and encloses and protects your **spinal cord**. The spinal cord carries messages between your brain and the rest of your body, and must be protected.

Some bones have the job of making new blood cells. This work goes on in the soft center (the **marrow**) of the long bones in your arms and legs.

Your skeleton also enables you to move your body. Muscles that are attached to bones work with the bones when you move. You need both muscles and bones to walk, stand, sit, write, eat, speak, play music and sports, and so on.

Kinds of Joints

Fixed joint

Hinge joint

Ball-and-socket joint

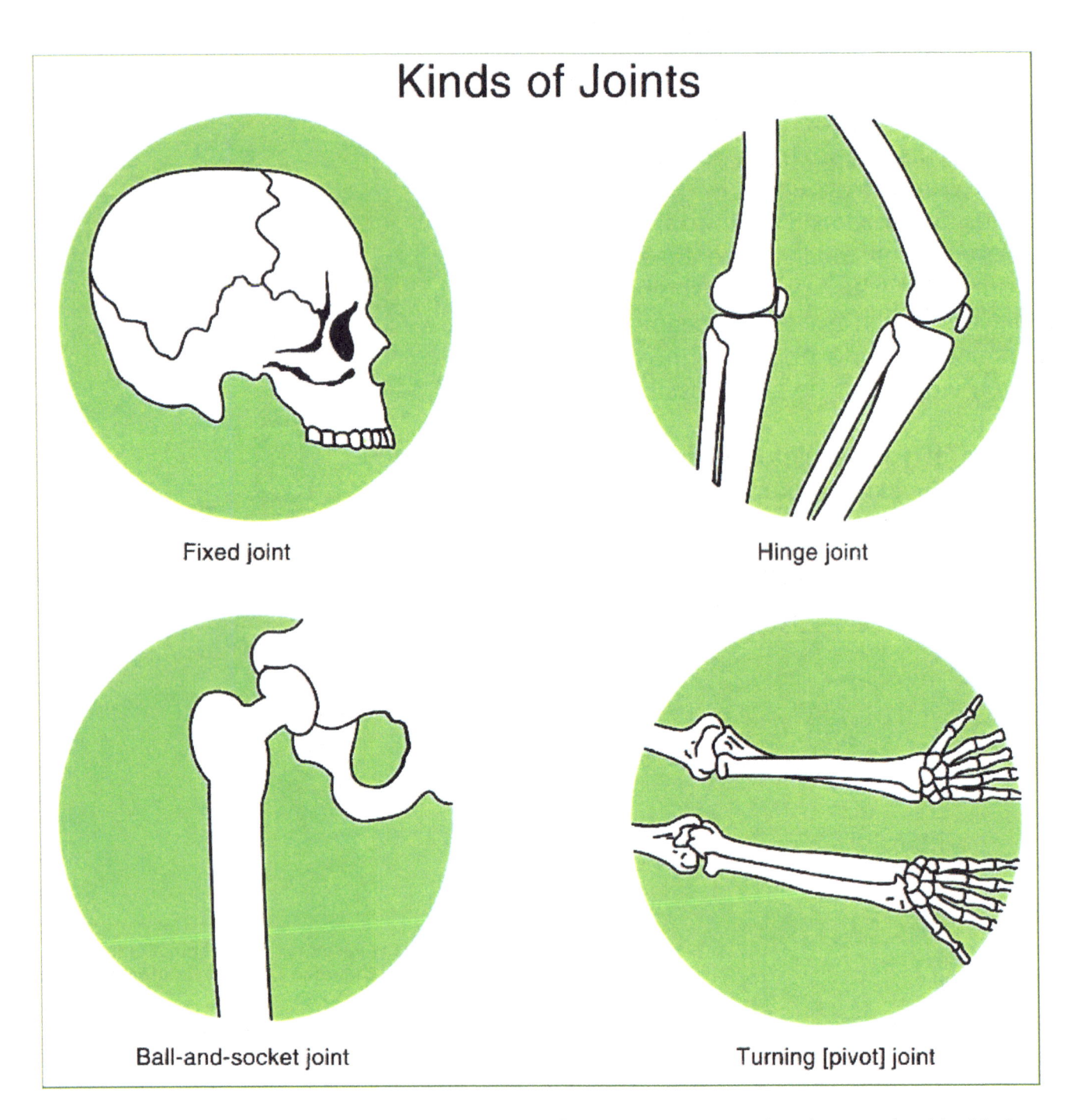

Turning [pivot] joint

Some parts of your skeleton are not made of bone. They are made of **cartilage** (KAR-tuh-lij). Cartilage is somewhat firm, but softer than bone. The outside of your ears and the tip of your nose are made of cartilage.

JOINTS. A place where two or more bones meet is a **joint** (JOYNT). Those joints that you can move are tied together by tough bands called **ligaments** (LIG-uh-munts).

You have several kinds of joints. The joint at your shoulder is a **ball-and-socket joint**. Your upper-arm bone has a ball at its end. This ball fits into a cup, or socket, in your shoulder blade. Your hip joint is also a ball-and-socket joint. At the shoulder and the hip, you can make twisting movements. You also can swing an arm or a leg around in a full circle.

A **hinge joint** moves back and forth like a door on its hinge. Your knees and fingers have hinge joints.

Your elbow has two joints. Its hinge joint works when you raise and lower your forearm. Its **turning joint** works when you flip over your hand. Try it!

Some joints are called **fixed joints** because they do not move. The "cracks" in your skull are fixed joints. When you were born, your skull bones were small. You had a soft spot on your head where the bones did not come together. As you grew, your skull bones grew too - until they finally joined.

---------------------------------- **REVIEW** ---------------------------------- U-5 L-4

I. In each blank, write the word that fits best. Choose from the words below.

vertebrae   skeleton   supports   ligament   protects   shape   move

The body's inside frame is its _____. The skeleton gives your

body its _____, and holds it up, or _____ it.

Muscles attached to bones allow the body to _____. The bones of

the backbone are called _____.

II. If the statement below is true, write **T**. If it is false, first write **F** and then **correct** the underlined words (to make the statement true).

A. _____ New blood is made in the spinal cord.

B. _____ The shoulder joint can be moved in a full circle.

C. _____ The vertebrae enclose and protect the heart and lungs.

D. _____ The tough bands that tie bones together are called muscles.

E. _____ The tip of your nose is made of bone.

F. _____ The cracks in your skull are fixed joints.

G. _____ The ligament carries messages from the brain to the rest of the body.

H. _____ The knee is an example of a hinge joint.

III. Your wrist contains a type of joint called a **sliding joint**. Where else in the body are sliding joints found? Describe their movement.

_____

_____

IV. The root word "**arth**" means joint. The suffix "**itis**" means inflammation or not normal. What word would a doctor use to describe a person with joint pain?

_____

Scientist-artist prediction of a Neanderthal

How Do Your Muscles Work?

U-5 L5

Exploring Science / Historical Steps

The Amazing Bo. Bo Jackson was always an awesome batter. Hitting home runs seemed like second nature. He even broke his bat once and still sent the ball flying into the waterfalls at Royals Stadium in Kansas City. On the football field, he was clocked running 40 yards in 4.2 seconds. He set a record by running over 90 yards for a touchdown twice in his career.

Bo seemed unstoppable. He had an enormous amount of muscle power packed into one body. He was the best in not just one, but two professional sports.

Unfortunately, in January of 1991, when playing football for Los Angeles, Bo fractured his left hip. Even worse was that the top of Bo's thighbone and hip joint were decaying. He was through with football. Maybe he was through with baseball, too.

Bo continued to exercise to keep his leg muscles in shape. He rode a bike and swam, but the pain was too great.

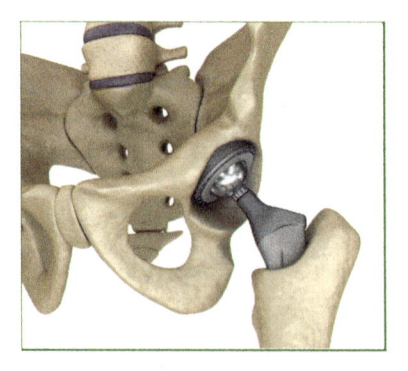

Artificial hip

Bo decided to have hip replacement surgery to correct the damage in his joint. Doctors cut off the top of his upper thigh bone (the "ball"). They put in a new metal "ball." In his hip they inserted a plastic and metal socket. In other words, his ball-and-socket joint was completely replaced.

With his new joint, Bo started to work out again. He got his muscles back in shape. In his very next game, he hit a home run.

➤ To Do Yourself (After Reading This Lesson)   See Muscles and Tendons!

You will need:

Boiled chicken leg including a thigh, hand lens, dissecting needle or probe, scissors

1. Skin the chicken leg and thigh.
2. With the dissecting needle or probe, pull apart the muscle layers. Look for tendons.
3. Pull on the tendons. What happened?
4. Now find the muscle in your upper leg. Straighten your knee as you feel the muscles work. Bend your knee and find the muscles that make this happen.

Questions

1. To what structure are the muscles attached? _____

2. What structure attaches the muscles to the bones? _____

3. Which muscle bends the knee? Which straightens it? _____ / _____

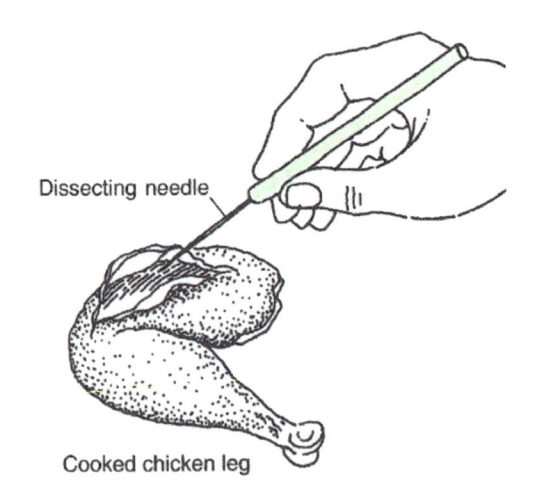

134

Your Body's Movers

When you throw a ball, you move your bones. But bones cannot move by themselves. Your body's movers are your muscles.

There are more than 600 muscles in your body. Many muscles are attached to bones by tendons. A **tendon** is a strong band of tissue. When a muscle shortens, or **contracts** (CON-trakts), it pulls on the tendon. At the same time, the other end of the tendon pulls on the bone. This causes the bone to move.

Muscles work in pairs. The two muscles of a pair cause opposite movements. In your upper arm, the **biceps** (BY-seps) and **triceps** (TRY-seps) muscles work as a pair. Prove this to yourself. Feel the biceps on one upper arm. It is on the top side of the arm. Now feel the triceps on the underside of the arm. Now bend your elbow. Feel how your biceps gets firm as it contracts. At the same time, you can feel how the triceps gets soft (as it relaxes). Now straighten the elbow. The triceps contracts. You can feel how it gets firm. How does the biceps feel as it relaxes?

You have control over your arm muscles. In fact, you can control all of the muscles that are attached to bones. These muscles are **voluntary** (VOL-un-tare-ee). The muscles in your arms, hands, legs, feet, face, neck, and trunk are voluntary. Since these muscles move the skeleton, they are called **skeletal muscles.** And the type of tissue they are made of is called skeletal muscle tissue.

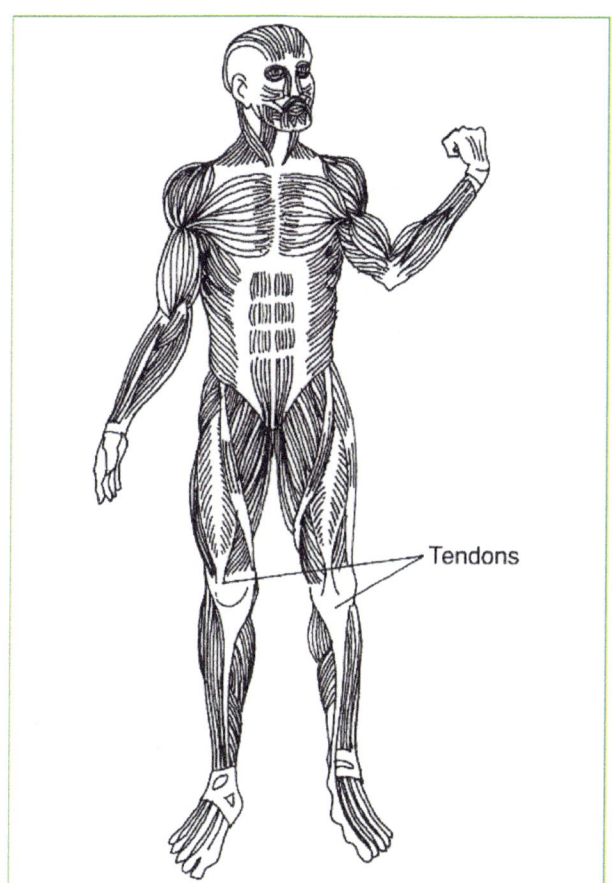

A tendon connects a bone to a muscle. What is the best known tendon in the human body, and where is it?

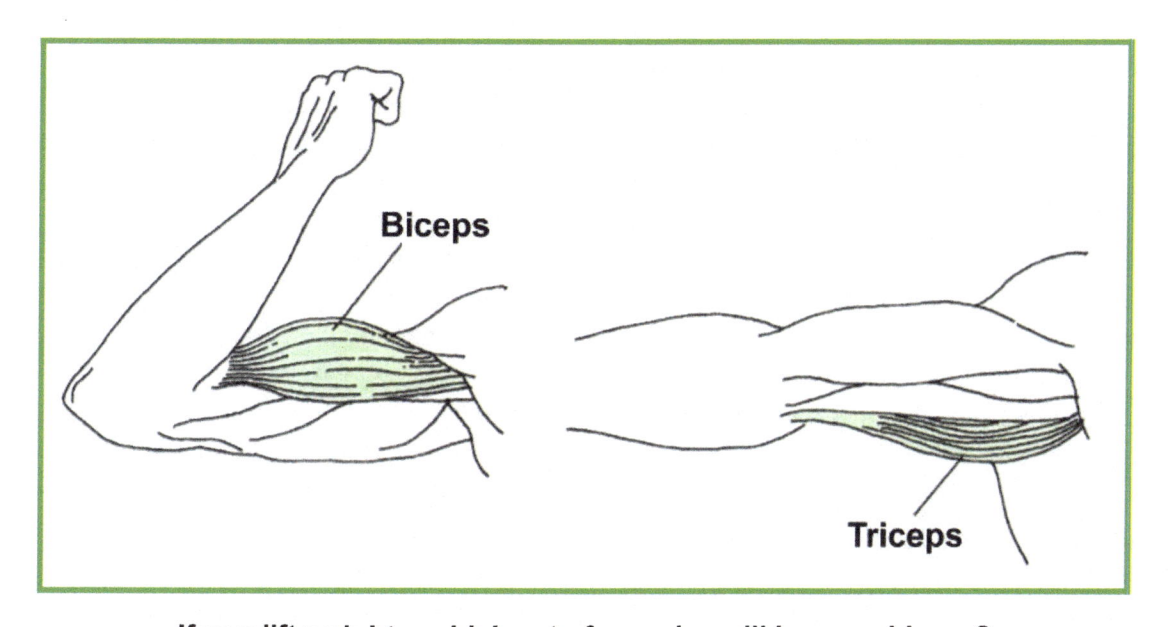

Biceps

Triceps

If you lift weights, which set of muscles will become bigger?

135

Some muscles are *not* under your control. These are called **involuntary muscles**. Certain involuntary muscles, such as those in your stomach and blood vessels, are made of a kind of tissue called **smooth muscle**.

A different kind of involuntary muscle tissue, **cardiac** (KAR-dee-ak) **muscle**, makes up the heart. "Cardiac" means "of the heart." Cardiac muscle works harder than any other kind of muscle. Even when you sleep, it works! Among other things, it must pump blood containing food and oxygen to your arm muscles so that they are strong enough to throw a ball.

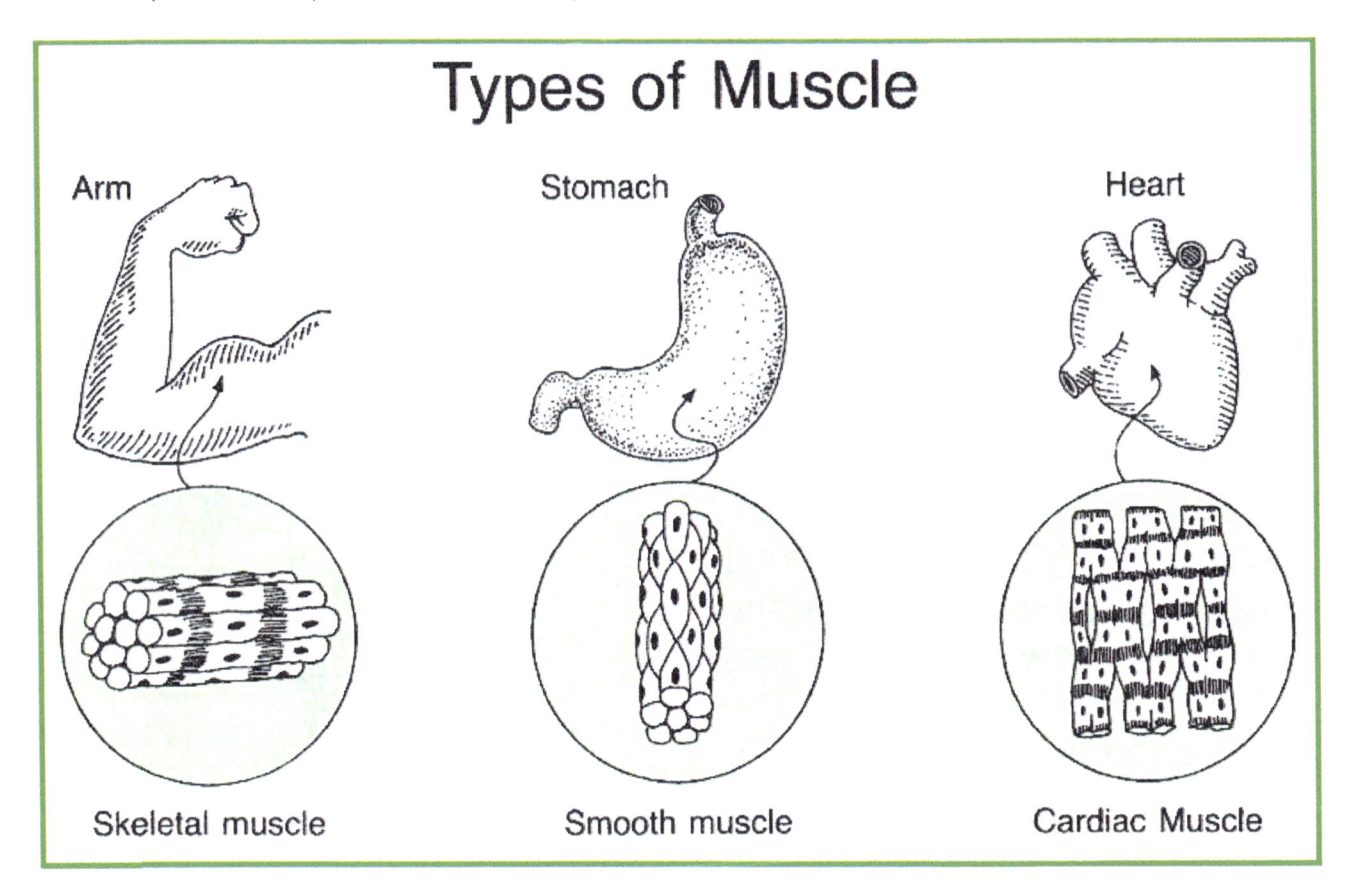

Types of Muscle

Arm — Skeletal muscle

Stomach — Smooth muscle

Heart — Cardiac Muscle

Use them or lose them - - and chase away the blues! Voluntary and cardiac muscles stay strong only if they are used. Unfortunately, they also get weaker if they are not used. If, over time, you do very little movement, your skeletal muscles will actually shrink. You have probably noticed weak ('flabby') arms in older people. For years, these people did too little exercise.

You have likely heard the words "cardio workout." This term describes exercise that works your heart (keeping it strong). It is wise to find an enjoyable way to raise your heart rate each day. You might like brisk walks, or bicycle rides, or the use of various exercise machines. You might enjoy playing vigorous sports such as tennis or soccer.

As a bonus, people who move have less trouble staying at a healthy weight. Strong muscles burn a great deal of fuel - especially if they are often used! A strong heart will let you continue to move (and burn fuel) for longer periods of time!

As a second bonus, moving actually improves a "down" mood. The brain releases "feel good" molecules while you move. Most people notice that if they *force* themselves to move, a sad mood improves faster. Feeling sad? Even if you don't feel like it, get up and go for a walk, or shoot a basketball, or throw a frisbee -- anything that uses skeletal muscles.

Finally, here is one more way to stretch your years of happiness. *S - t - r - e - t - c - h!* Your tendons are somewhat like strong rubber bands. As you age, these strong bands become stiffer - unless they are stretched often. A gentle stretching routine seems best. Most physical therapists recommend that you first "warm up" before stretching - perhaps with a short stroll.

REVIEW U-5 L-5

I. In each blank, write the word that fits best. Choose from the words below.

involuntary   biceps   contracts   triceps
ligaments   relaxes   tendons   voluntary

Muscles are attached by _____ to bones. When a muscle _____, it shortens. A pair of muscles in your upper arm are your _____ and _____. Muscles under your control are _____. Muscles not under your control are _____.

II. Match the **letter** of the muscle tissue to the numbered body parts.

a. cardiac muscle   **1.** ___ stomach   **4.** ___ blood vessel

b. smooth muscle   **2.** ___ heart   **5.** ___ finger

c. skeletal muscle   **3.** ___ leg

III. Circle the word or phrase (between the brackets) that makes each statement correct.

  A. When the biceps muscle contracts, the triceps muscle [contracts / relaxes].

  B. The muscles of your leg are [voluntary / involuntary].

IV.   Observe the diagram below. The large muscles in front of your thigh are the "**quad**" muscles. Behind the thigh are large muscles called **hamstring** muscles. The tendons that attach hamstring muscles to bones are also called hamstrings.
   Sometimes an athlete, while running or stopping, will feel a sharp pain and quickly grab the back of their leg. "Pulling a hamstring" simply means that these muscles or tendons have been damaged. This hurts (and may take weeks or even months to heal). If treated properly and given time to rest, healing will occur. Without adequate care and rest, another pull is common.
   Describe the action of the quad and hamstring muscles as you bend your knee.

_____

_____

_____

_____

_____

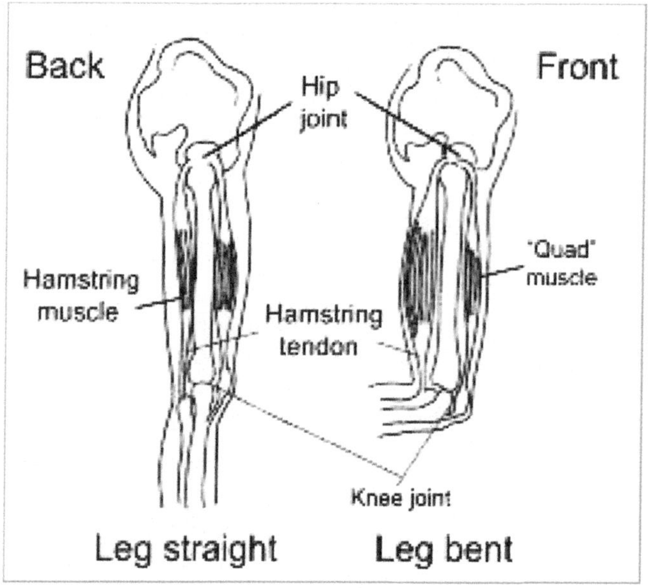

137

Unit 5 -- Review What You Know --

A. Unscramble the groups of letters to make science words. Write the words in the blanks.

1. PYRRATESOIR / _____ = system that moves air
2. NIRUE / _____ = produced in the kidneys
3. KNLESEOT / _____ = the body's framework
4. CDAAICR / _____ = "of the heart"
5. XITEENOCR / _____ = getting rid of wastes
6. GERTLIAC / _____ = found in the outer ear

B. <u>Write</u> the ending that best completes each statement.

1. Oxygen enters the blood through
   a. pores   b. cilia   c. air sacs (alveoli)   1. _____

2. A joint in the elbow is called
   a. fixed   b. hinge   c. ball-and-socket   2. _____

3. The biceps is made of muscle that is
   a. skeletal   b. smooth   c. cardiac   3. _____

4. Bones are tied together at joints by
   a. tendons   b. ligaments   c. triceps   4. _____

5. A waste excreted through sweat glands is
   a. carbon dioxide   b. mucus   c. urea   5. _____

6. A tube that carries urine from the kidney to the bladder
   is a   a. ureter   b. bronchus   c. capillary   6. _____

7. An example of a voluntary muscle is one in the
   a. leg   b. stomach   c. heart   7. _____

8. Carbon dioxide is excreted when we
   a. inhale   b. exhale   c. sweat   8. _____

9. New blood cells are made in
   a. cartilage   b. hard bone   c. marrow   9. _____

10. Salt is excreted through the
    a. kidneys   b. trachea (windpipe)   c. joints   10. _____

11. The muscle in a blood vessel is
    a. cardiac   b. skeletal   c. smooth   11. _____

12. A muscle that bends the elbow when it contracts is the
    a. triceps   b. diaphragm   c. biceps   12. _____

13. A waste produced from cell respiration is
    a. urea   b. carbon dioxide   c. salt   13. _____

C. Apply What You Know

1. Below is a drawing of some parts of the body. Write labels for each numbered part. Choose from the labels below.

ureters	trachea (windpipe)	diaphragm	nose	bronchus
liver	urethra	lung	bladder	

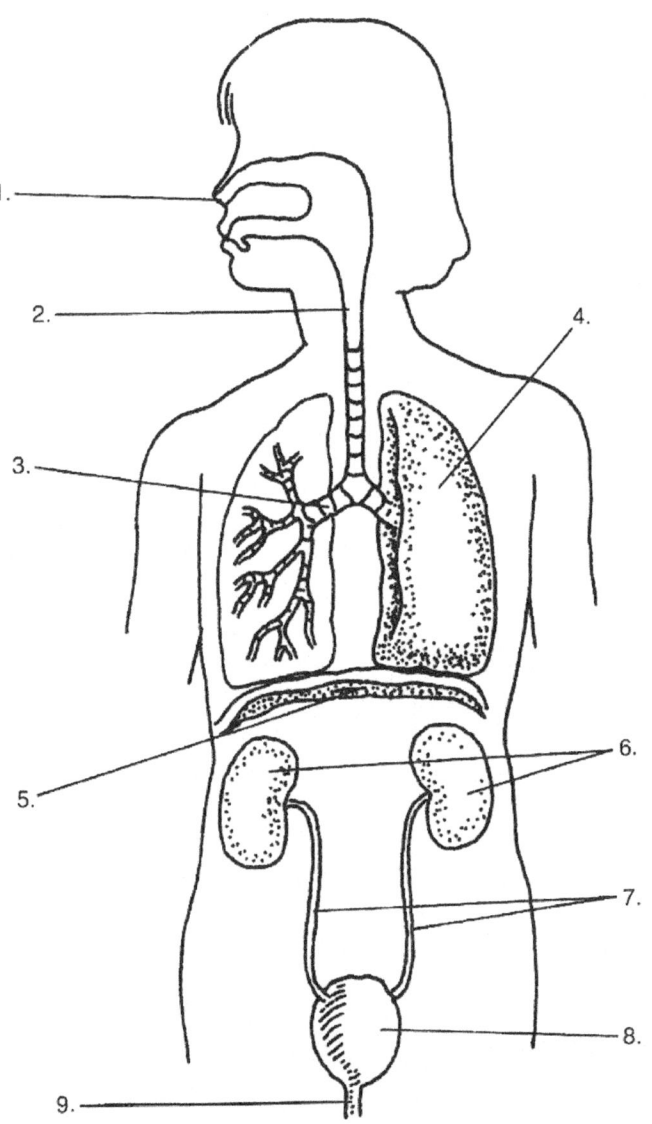

2. On each blank below, write the name of the correct part. Choose from the labels that you used on the drawing in question 1.

   a. Becomes more curved when air leaves the body. _____

   b. Made up of many tiny filters. _____

   c. Carries air from trachea (windpipe) into lungs. _____

   d. Passes urine out of the body. _____

   e. Made up of many tiny alveoli (air sacs). _____

139

3. Study the drawing of some of the parts of a human leg. Write labels for each numbered part. Choose from the labels below.

    tendon   lower leg bones   knee joint   hip   ligaments   calf muscle

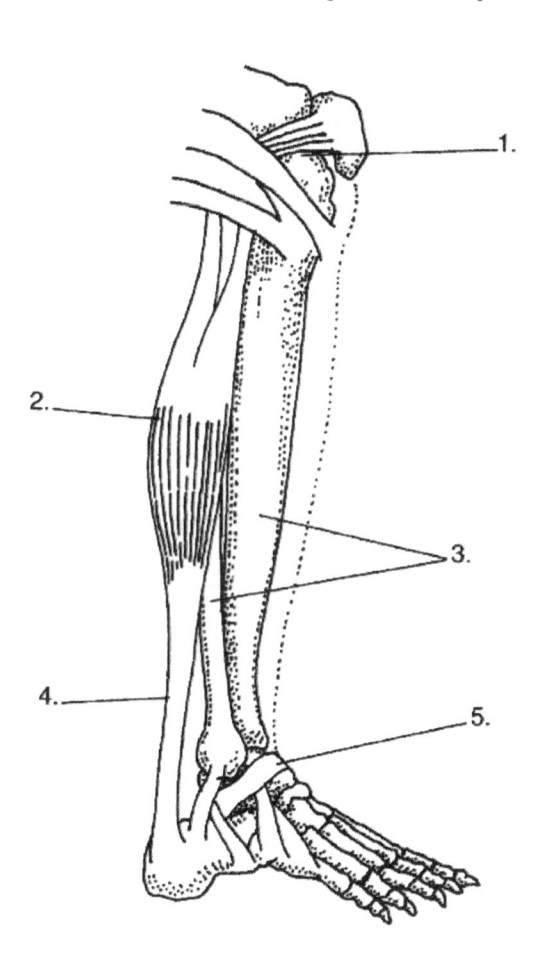

4. Write the number of the part from the illustration that matches each statement below.

    a. ____ Attaches a muscle to a bone.

    b. ____ Has an action like a hinge.

    c. ____ Are joined by ligaments to the thigh bone.

    d. ____ Tie lower leg bones to ankle bones.

    e. ____ Contracts when you stand on tiptoe.

D. Find out more.

1. Calcium and other minerals make bones hard. What would a bone be without minerals? Try this with two chicken leg bones. Place one bone in a jar of vinegar. Keep the other one in the air. After a week, try to bend both bones. What happens? What can you do to be sure that there is enough calcium in your bones?

2. Part of a physical checkup is to have your urine tested. Visit a hospital laboratory to see how tests are completed. What can doctors find out by testing urine?

3. Research the causes of **acne**. Talk to a dermatologist - a doctor who specializes in skin diseases. How can acne be cured - or at least managed? Report your findings.

SUMMING UP:
Review What You Have Learned So Far

A. Study the drawings. Each item is shown as it looks under a microscope.

1. On the lines below each drawing, write its name. Choose from these names.

| cork | villi | diatoms | air sacs (alveoli) | capillaries |
| green leaf | blood | polio viruses | sweat gland | skeletal muscles |

1. _____ 2. _____ 3. _____

4. _____ 5. _____ 6. _____

7. _____ 8. _____ 9. _____ 10. _____

B. Each statement below refers to one of the items shown above. The number of the statement is the same as the number of the item. Circle the word (or words) that makes each statement correct.

1. The gas that goes into the blood through their "walls" is [oxygen / carbon dioxide].
2. Scientists do not all agree as to whether they [are living things / can reproduce].
3. These are golden [fungi / algae] with glassy shells.
4. This tissue is part of muscles that are [involuntary / voluntary].
5. The name of the scientist who called these "cells" is [Leeuwenhoek / Hooke].
6. Digested food passes through their "walls" from the [large / small] intestine.
7. These cells float in a liquid called [plasma / mucus].
8. When the waste that it excretes evaporates, the body is [warmed / cooled].
9. The cytoplasm of these cells has parts that contain [cellulose / chlorophyll].
10. They connect veins with [arteries / atria].

UNIT SIX
BEHAVIOR AND CONTROL

Why Do Living Things Act As They Do?   U-6 L1

Exploring Science / Historical Steps

"Mother" Konrad Lorenz. The 1973 Nobel Prize winning scientist **Konrad Lorenz** is known as the "father of **ethology.**" Ethology is the study of animal behavior.

One day in the late 1930s, while walking near his home in Austria, he crouched down and began to waddle. He flapped his arms. He called out "gork-gork." After a while he stood up. Over the fence, a row of people stared at him. Why was a grown man acting like a duck?

Actually, Lorenz made an important discovery about ducks. For a short time after hatching, the first moving object that a baby duck sees becomes its "mother." The term for this kind of animal behavior is **imprinting**.

A group of baby ducks had imprinted Lorenz as their mother. They became attached to him and followed him everywhere. By quacking, they let him know that they needed him to act like a duck. If he didn't, they became afraid and upset. That day in the garden, he wanted the baby ducks to feel happy. So he acted like a good mother to his "family." Tall grass, however, hid the baby ducks from view. The people watching could only see Lorenz. No wonder they stared!

➤ Before a baby duck hatches it can hear, but not see, through the egg shell. How might a duck become attached to its mother even before "birth?"

An adult black swan (in Australia) has been imprinted by the three cygnets (baby swans).

An adult Canada goose has been imprinted by the five goslings (baby geese).

The Behavior of Living Things

All of the actions of a duck, a person, or any living thing make up its **behavior** (bih-HAYV-yur). What makes living things act, or **behave**, as they do? The behavior of living things helps them survive. It helps them get food, protection, and other things that they need.

Like all organisms, ducks receive many messages from their environment. These messages are called **stimuli** (STIM-yuh-lie). For Lorenz's ducks, the sight and sound of him were stimuli. (One message is a **stimulus**). A reaction to a stimulus is a **response** (rih-SPONS). The ducks' response to "Mother" Lorenz was to follow him everywhere. How is this behavior useful to baby ducks in the wild?

Some responses are under your control. These are **voluntary responses**. If you are standing at home plate during a softball game, an incoming pitch is a stimulus; you decide whether or not to respond (by swinging or not swinging your bat).

When you are thirsty, your dry mouth is a stimulus. You then decide whether to take a drink of liquid. Obviously, this response is voluntary. Can you name some other voluntary responses?

On the other hand, some of your responses are not under your control. These are **involuntary responses**. If you enter a dark room, the low light is a stimulus; the openings of your eyes - called **pupils** - get larger. You have

143

no control over this response. If you step from this dark room into a well lit area, the pupil size rapidly shrinks involuntarily (meaning, without your control).

If you touch a hot stove, you quickly jerk your hand away. The painful heat is a stimulus. Pulling your hand away is a response. This action prevents your hand from being badly burned. Did you *decide* to pull your hand away - or did it happen without you making a decision?

Here is one more example. When food that you just swallowed enters your stomach, do you control your stomach's release of gastric juice?

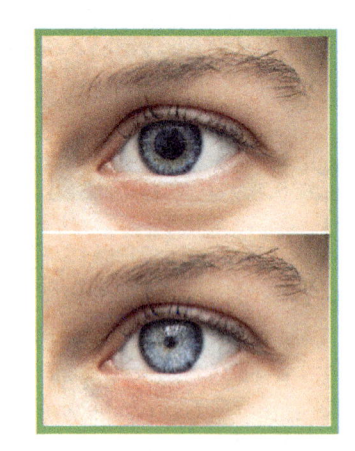

Pupil differences: top = dim light
bottom = bright light

➤ To Do Yourself — How Do Some Animals Respond to Stimuli?

You will need: earthworm, shoebox lid, pencil, board, damp paper towels, flashlight, aluminum foil

1. Observe the earthworm as it responds to the touch of your hand.
2. There are nerve endings in each segment of the earthworm. Gently touch the top of the earthworm. Observe its response.
3. Place the earthworm on a wood board covered with a damp paper towel. Tap the board with a pencil. Observe the earthworm's response to the sound.
4. Place the earthworm on a damp paper towel in the shoebox lid. Cover half of the lid with aluminum foil. Shine a flashlight on the earthworm. Observe its responses.

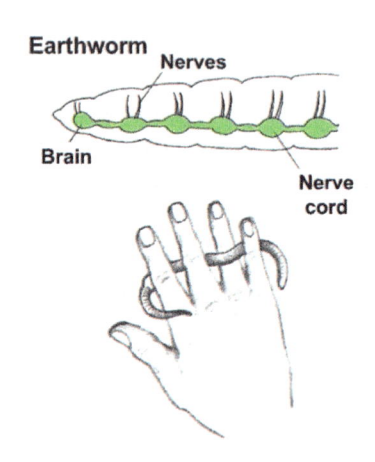

Questions

1. What stimuli did the earthworm respond to? How did it respond? _____
_____

2. How does the earthworm's body receive the messages of the different stimuli? _____

Plants also respond to stimuli. You might be surprised to learn that plants even respond to gravity. Note the illustration at the right.

If a seed is planted *upside down,* the roots will sprout and begin to grow upward, but then they turn and grow downward. On the other hand, the stem will sprout and begin to grow downward, and then it turns upward. This response to gravity is called **geotropism**.
[Note: "Geo" means earth; "tropism" means turning in response to a stimulus.]

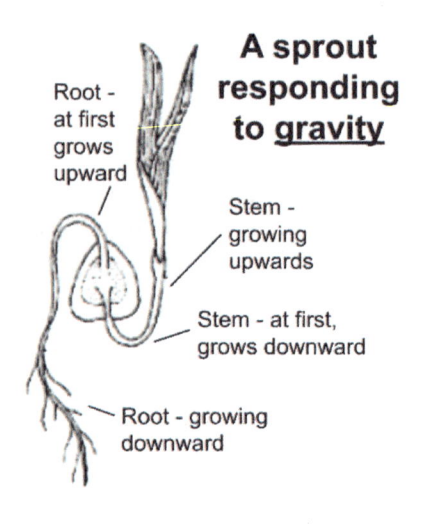

A sprout responding to gravity

Root - at first grows upward
Stem - growing upwards
Stem - at first, grows downward
Root - growing downward

Have you ever grown a plant on a windowsill? If so, you have noticed that a plant's stem and leaves grow toward light. Light is a stimulus. The plant's slow movement toward the light is a response. How is this behavior useful to the plant? (Hint: What do plants' leaves need in order to carry out photosynthesis?). Can you guess what scientists call this behavior? [Answer: See the caption of the *photo* below].

At left: Sunflowers - with sun at upper right

[Do you see how these flowers received their name?]

Above: Sunflowers - with sun behind photographer
[Phototropism = turning in response to light]

-- **REVIEW** ---------------------------------- U-6 L-1

I. In each blank, write the word that fits best. Choose from the words below.

| survive | stimuli | behave | involuntary |
| response | stimulus | behavior | voluntary |

The _____ of a living thing includes all of its actions. A

message, such as heat or light, is a _____. A reaction, such

as a movement, is a _____. The responses of living things help

them to _____ . Hitting a softball is a(n) _____

response. The flow of gastric juice is a(n) _____ response.

II. To the right of the number, place the letter of the response from column **B** that would result from the stimulus in column **A**. (One item in **B** is *not* used)

A (stimulus)
1. ____ dog barks
2. ____ odor of food
3. ____ bright light
4. ____ whistle
5. ____ pepper in nose

B (response)
(a) dog's ears lift
(b) pupils get smaller
(c) person sneezes
(d) plant's leaves turn away
(e) cat's hair stands on end
(f) saliva flows in mouth

III. The roots of a willow tree grow toward a nearby river. Why do you think they do this?

_____

145

How Does Your Body Receive Messages? U-6 L2

Exploring Science / Historical Steps

While You Were Out. Your life today is filled with messages. Does this imaginary student sound like you? When Ming got home from school there was a handwritten note on the refrigerator. It told Ming where her mother was, and what was available for a snack. Ming checked her voicemail and found that a friend had left a message for her to call soon. A few minutes later, the doorbell rang. It was a delivery person bringing an overnight package containing her family's new laptop! The package came from a factory halfway around the world. She next received a text message from her father reminding her to feed the hungry cat.

That afternoon, Ming and her brother competed in an online, interactive video game with their cousins - who live in another State. Finally, Ming searched online for information that she needed for her history paper.

➤ What form of energy makes most kinds of long-distance communication possible?

Receiving Messages

People and animals receive stimuli, or messages, through **sense organs**. These include the eyes, skin, tongue, nose, and ears. Each sense organ is sensitive to certain kinds of stimuli. Let's look at each of these senses:

SEEING. When you look at a cat, what causes you to see the cat? Light bouncing off of the cat passes through the **cornea** (a tough, clear covering), then enters your eye through the **pupil**. The size of this opening is controlled by the thin sphincter muscle called the **iris**. Next, the light goes through the **lens**. The lens focuses light on the **retina** (RET-in-nuh) at the back of the eye. The retina is sensitive to light. A "picture" of the cat forms on the retina. From the retina, messages pass along the **optic nerve** to the brain. Next, the brain interprets the messages. You see the cat!

If you have brown eyes, the iris of the eye is brown.

TOUCH. Your skin is a sense organ that covers your whole body. Suppose that you gently touch your skin with a wire. Under each point where you feel the wire, there is a **nerve ending** sensitive to touch. Your skin has special nerve endings that receive five kinds of stimuli: touch, pressure, heat, cold, and pain.

Messages from your skin go through nerves to your spinal cord. The spinal cord relays messages from the skin (and other parts of the body) to the brain. Next, your brain interprets those messages. Only then, when the messages reach your brain, can you feel what you touch.

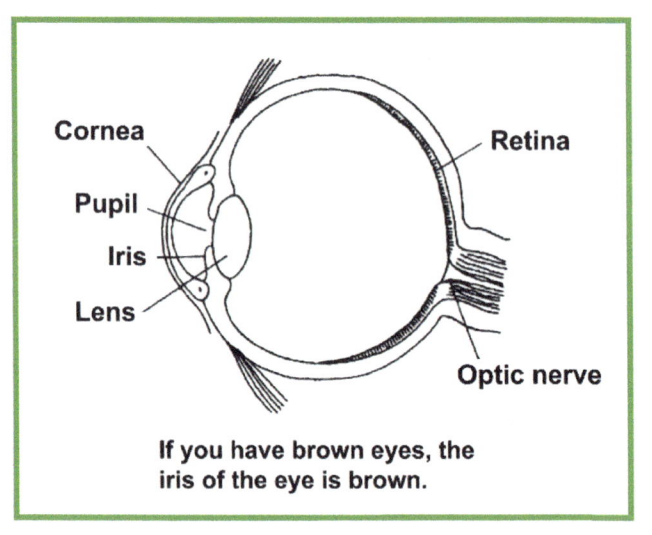

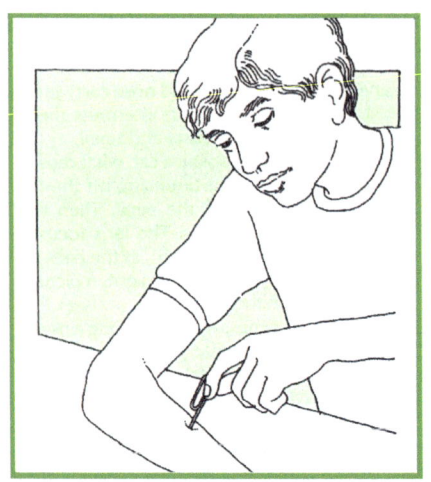

TASTING AND SMELLING. On the tongue, we have long known that there are four major kinds of taste-sensitive nerve endings. These four tastes are: sweet, sour, salty, and bitter.

In addition, scientists have discovered two other types of nerve endings in the tongue. One can sense whether we are eating a high protein food. The other lets us know if we put "spoiled" fat into our mouth.

Suppose that you close your eyes, hold your nose, and have someone place a slice of apple or pear in your mouth. Your sense of taste will tell you if the fruit is sweet or sour. But you may have trouble telling whether it is an apple or a pear unless you let go of your nose. Why is this?

Part of what we call taste is actually a food's odor (smell). The nose senses odors. Odors are actually molecules. We can identify many different food molecules (flavors) as well as other smells. Each time we eat, nerves from our tongue *and* nerves from our nose carry messages to our brain. The brain interprets these messages, and we identify specific tastes and odors.

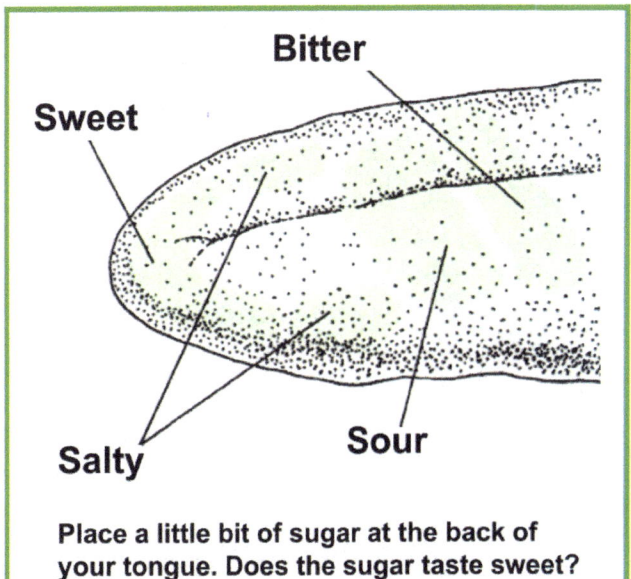

Place a little bit of sugar at the back of your tongue. Does the sugar taste sweet?

Your nose helps you taste food.

HEARING. When a friend speaks to you, how do you actually hear what they said? He or she makes sound waves when speaking. Sound waves that enter your **outer ear** strike your **eardrum** and cause it to vibrate.

Attached to the eardrum are three tiny bones in the **middle ear**. When the eardrum vibrates, so do the three bones. The vibrations then pass to the **inner ear**. From the inner ear, nerves carry messages to your brain. There, the messages are interpreted so that you hear the sound of your friend's voice.

Fun fact: The three tiny bones of your middle ear are named after their shape. From left to right in the illustration to the right (and the one on the next page), they are the <u>hammer</u>, the <u>anvil</u>, and the <u>stirrup</u>.

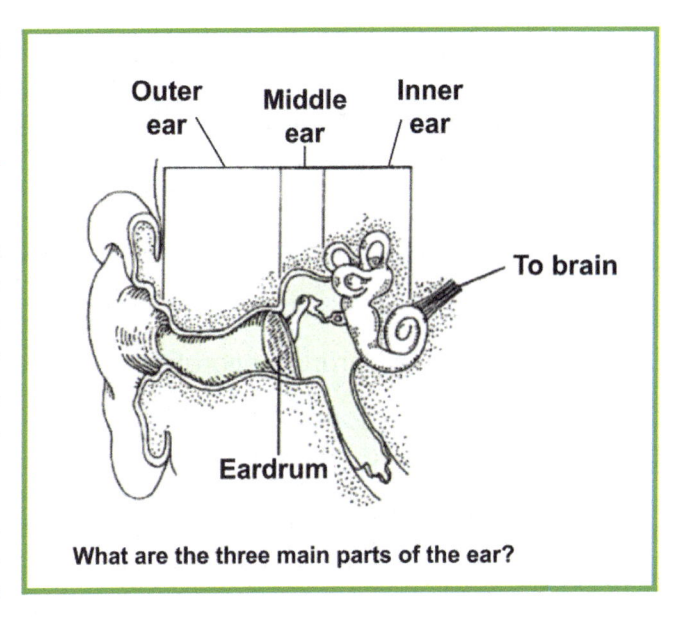

What are the three main parts of the ear?

147

BALANCE. Three tubes in the inner ear are filled with fluid. These tubes, called **semicircular canals**, help you to keep your balance. Anything that upsets the workings of these tubes can cause dizziness or motion sickness.

Balancing an apple on her head

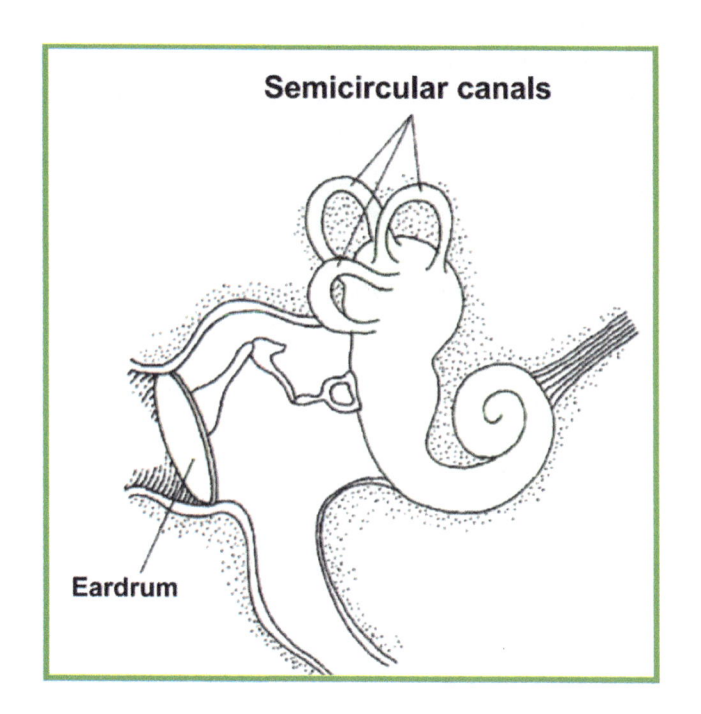

Semicircular canals

Eardrum

REVIEW — U-6 L-2

I. In each blank, write the word that fits best. Choose from the words below.

brain   stimuli   spinal cord   nerves   sense organs   responses

Through the work of your _____ you <u>receive</u> messages

from around you. From each sense organ, _____ <u>carry</u> messages.

Messages from the skin are passed to the brain by the _____.

In the _____, the messages from each sense organ are interpreted.

II. Name of the five kinds of stimuli received by the skin. _____

III. What are the four *major* kinds of taste buds? _____

IV. Circle the word or phrase (between the brackets) that makes each statement correct.

    A. In the eye, a "picture" of what you see is formed on the [lens / retina].

    B. From the eardrum, vibrations pass to three tiny bones in the [middle / inner] ear.

    C. When you touch something hot, you feel the heat when messages reach the [spinal cord / brain].

V. Why might it be dangerous if you lost your sense of smell?

_____

_____

148

What Does Your Nervous System Do? U-6 L3

Exploring Science / Historical Steps

<u>Take Care of Body and Mind</u>. This common phrase is puzzling. It seems to say that our mind is something separate from our body. In fact, our mind is part of our body!

Like *all* parts of our body, we need to take care of our brain! In other words, mental health <u>is</u> health!

Look at the illustration of a nerve cell at the top of the next page. You have likely heard that nerve cells are much like wires. In fact, electric pulses do move along the thin, wire-like part of a nerve cell. But when one nerve cell sends a signal to the *next* nerve cell (across the tiny gap between the two), special **molecules** are involved.

The amount of these molecules is carefully controlled. Unfortunately, in some people these molecules get out of balance.

In rare cases, this imbalance might begin at birth. It can also happen after a person has had an upsetting event in their lives - such as the loss of someone they love. Witnessing or experiencing violence may also cause an imbalance. Finally, using illegal drugs can lead to a serious, sometimes deadly, imbalance.

The good news is that proper balance can often be restored. Sometimes, as time passes, the brain gradually restores itself. However, in many cases, help is needed. Help is particularly important if a person feels trapped in sadness.

The most common help-professionals are counselors and psychologists. At the end of this unit, other mental health workers are described.

A mental health professional helps a young woman.

Researchers have learned a great deal about the molecules made by nerve cells. As a result, we now have medicines that help people recover from stressful experiences. **Psychiatrists**, doctors with advanced training regarding mental issues, determine the type and amount of these medicines for each patient.

Since the brain is so complex, science has much to learn about mental health. A particularly challenging issue is the decline that happens to so many people as they age. The term for this decline is **dementia** (duh-MEN-shuh).

➤ A research challenge: Seek the difference between a type of dementia called Alzeimer's disease and one due to Parkinson's disease.

Your Nervous System

The centers of control in your nervous system are your **brain** and **spinal cord**. Together, these are called the **central nervous system**.

Cells that are part of the nervous system are called **neurons**, or simply **nerve cells**.

Your nervous *system* is built to send and receive messages, but <u>each</u> nerve cell is a one-way street. Nerve cells carry messages in only one direction. A long extension of each neuron, called the **axon**, always carries the message <u>away</u> from the cell body. At the end of the axon, the message is shared with the shorter extensions that are near the cell body of other nerve cells.

149

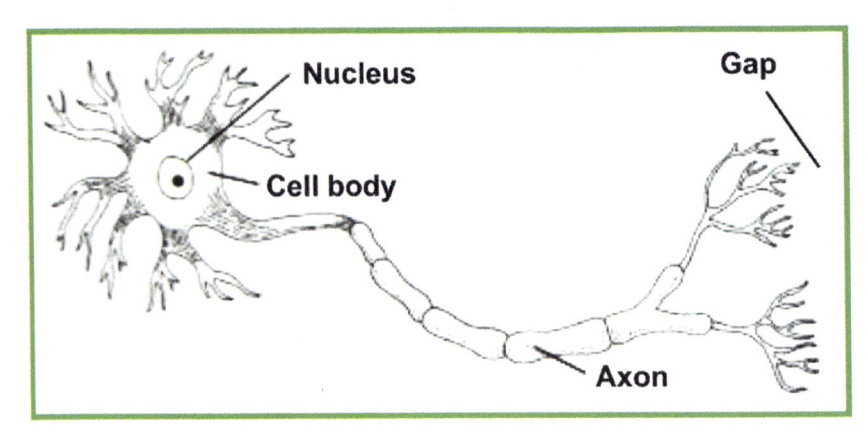

Across the "gap," special molecules send information to the next nerve cell.

A group of axons that work together is called a **nerve fiber**, or sometimes simply a **nerve**. There are two basic types of nerves - sensory nerves and motor nerves.

Sensory nerves carry messages from the sense organs *to* the spinal cord and the brain. They tell the spinal cord and brain what is happening to your body.

Motor nerves carry messages *from* the brain and spinal cord to other parts of the body. These messages tell the body parts what to do. Some motor nerves carry messages to muscles. They tell the muscles to move. Other motor nerves carry messages to glands, telling them to make or to release their chemicals (see Lesson 6).

Sensory nerves and motor nerves have many branches. These branches spread their messages to many parts of the body.

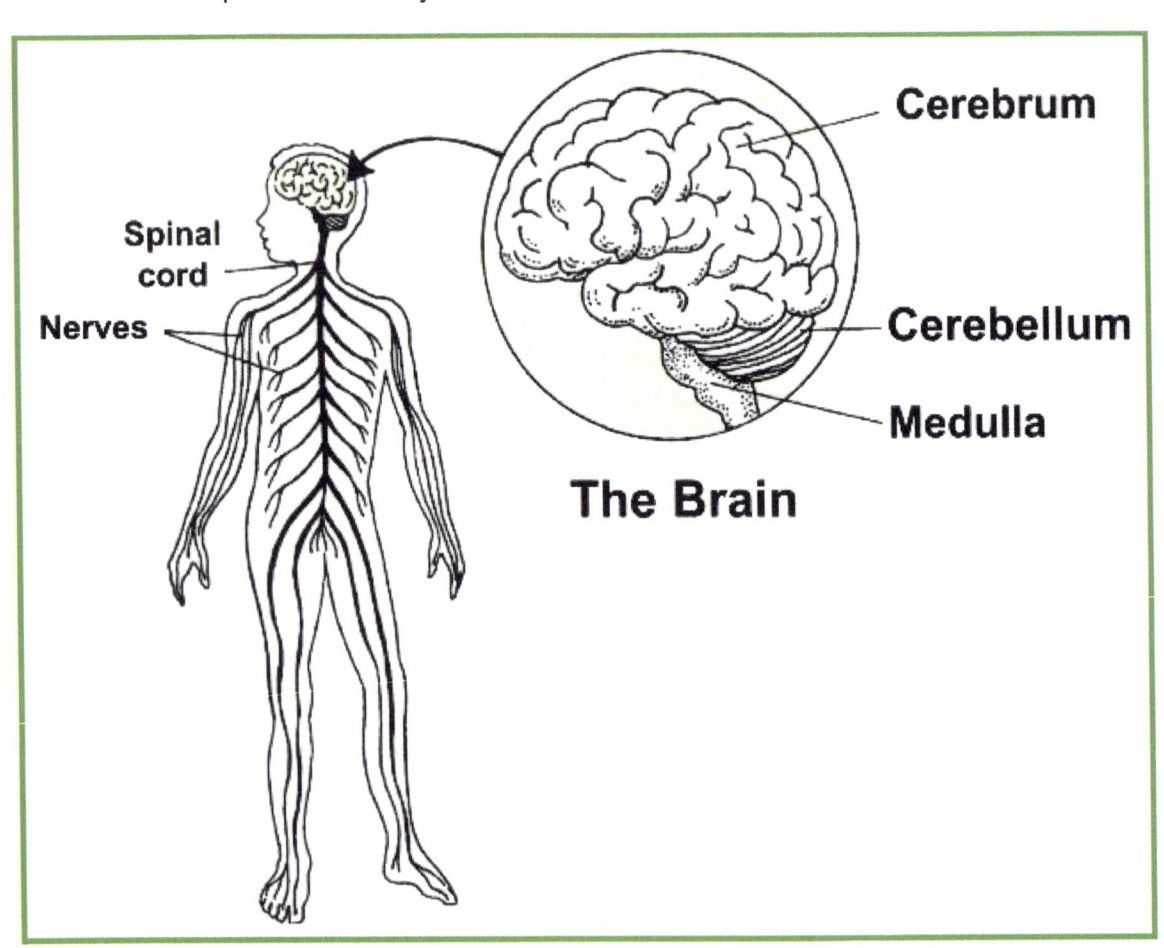

Nerves connect the brain and spinal cord to all parts of the body.

150

Suppose that you stub your toe on a sharp rock. What happens? Pain-sensitive nerve endings in the toe are stimulated. They send a message to your spinal cord. The spinal cord immediately passes the message to motor nerves. These motor nerves carry an urgent message to your muscles - "Pull away!" Your muscles move your foot away from the rock. But this is just your nervous system's first response to the stimulus.

Less than a second later, the spinal cord sends another message - this one to your brain. Your brain quickly interprets the message. Only then do you feel pain (and say - "Ouch")!

Pulling your foot away is an involuntary action. You do it before you even know what you have done. This kind of involuntary action is called a **reflex** (REE-fleks).

Can you name some other reflexes?

A doctor tests a patient's involuntary reflex.

➤ To Do Yourself   What Are Some Parts of a Fish's Nervous System?

You will need:

Fresh fish which has been cut in half, paper plate, paper towels, large pins, hand lens

1. Place a fresh fish on a paper towel. Use the drawing to help you locate the backbone. Then locate the dark spinal cord, below and inside of the backbone. Trace it from the tail to the head of the fish. Use the pins to mark its location.

2. Follow the spinal cord to the place where it enters the head. With a pin, locate the brain at the end of the spinal cord. Since the fish was cut in half, only part of the brain may be present. Use the hand lens and the drawing to help you observe the brain.

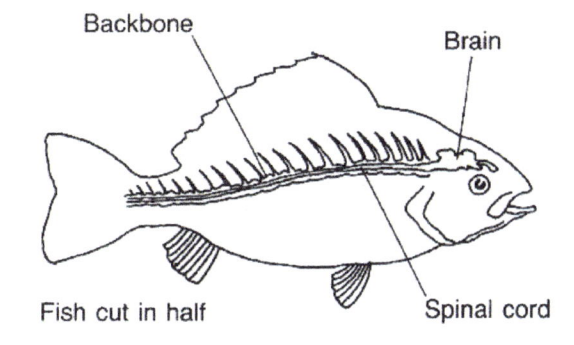

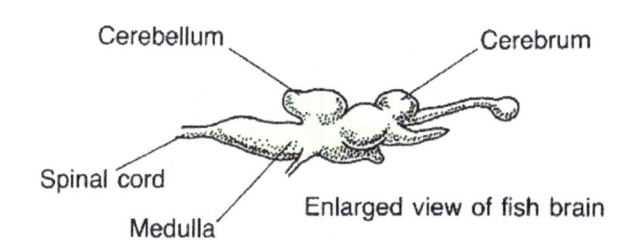

Questions

1. Why is the spinal cord located inside of the backbone? _____

2. How do messages get from body surfaces to the spinal cord? _____

Your **spinal cord** is the center of control for many <u>involuntary</u> actions. However, the brain does control some involuntary actions, such as "mouth watering" when you smell food that you enjoy.

Of course, the brain controls all of your <u>voluntary</u> actions. Deciding to stand or walk is a voluntary action. So is watching out for sharp rocks!

--- **REVIEW** --- <u>U-6 L-3</u>

I. In each blank, write the word that fits best. Choose from the words below. Words may be used more than once or not at all.

| sensory | nervous | brain | spinal cord |
| reflex | motor | voluntary | stimulus |

Receiving and sending messages is the job of your _____ system.

Your _____ and _____ are centers of control.

Messages *from* sense organs go through _____ nerves. Messages *to* muscles or *to* glands go through _____ nerves. A _____ is an involuntary action.

II. If you put your hand into water that is too hot, a <u>reflex</u> action takes place. Use numbers 1 through 6 to indicate in what order the following things happen.

A. ____ the spinal cord relays the message to the brain

B. ____ you feel the hot water

C. ____ hot water touches the skin

D. ____ the message goes through a motor nerve

E. ____ your hand pulls away from the hot water

F. ____ the message goes through a sensory nerve

III. Some motor nerves carry messages to glands, such as the salivary glands. (These are the glands that make your mouth water). Has the smell of food ever made your mouth water? This is a <u>reflex</u> centered in the brain. Describe what happens in your nervous system during this reflex.

_____

_____

_____

_____

What Are The Jobs of Your Brain? U-6 L4

Exploring Science / Historical Steps

Down Memory Lane on the Operating Table. In the 1930s, **Dr. Wilder Penfiled** did his now famous work to "map the brain"

As a **neurosurgeon** (a surgeon who is an expert on neurons (nerve cells) he sometimes operated on the brain to help people with seizures. Fortunately, the brain itself has no nerve endings for pain. While patients were awake, he touched exposed parts with an electric probe. This helped him find any areas that were not working properly.

Penfield was amazed to discover that touching certain parts caused patients to "hear" or even "see" events from their past! The memory of the experiences were somehow stored in their brains.

He discovered that each of these experiences seemed to be stored in an exact spot in the brain.

➤ Do you think that some sort of change takes place in brain cells when you learn something? Explain.

Your Brain

The work of Dr. Penfield helped scientists draw more accurate maps of the human brain. Different parts of your brain have different jobs.

The brain's largest part is the **cerebrum** (ser-EE-brum). The top of your skull covers and protects the cerebrum. You use most of your cerebrum to think, reason, and remember. Like a computer, your cerebrum sorts and stores information.

Messages from each sense organ go to certain areas of the cerebrum. There the messages are interpreted. You can find these control centers on the brain map.

You do not "see" anything until nerve messages reach your brain's center for sight. The same is true for your senses of hearing, smell, taste, and touch. You can locate these areas on the brain map.

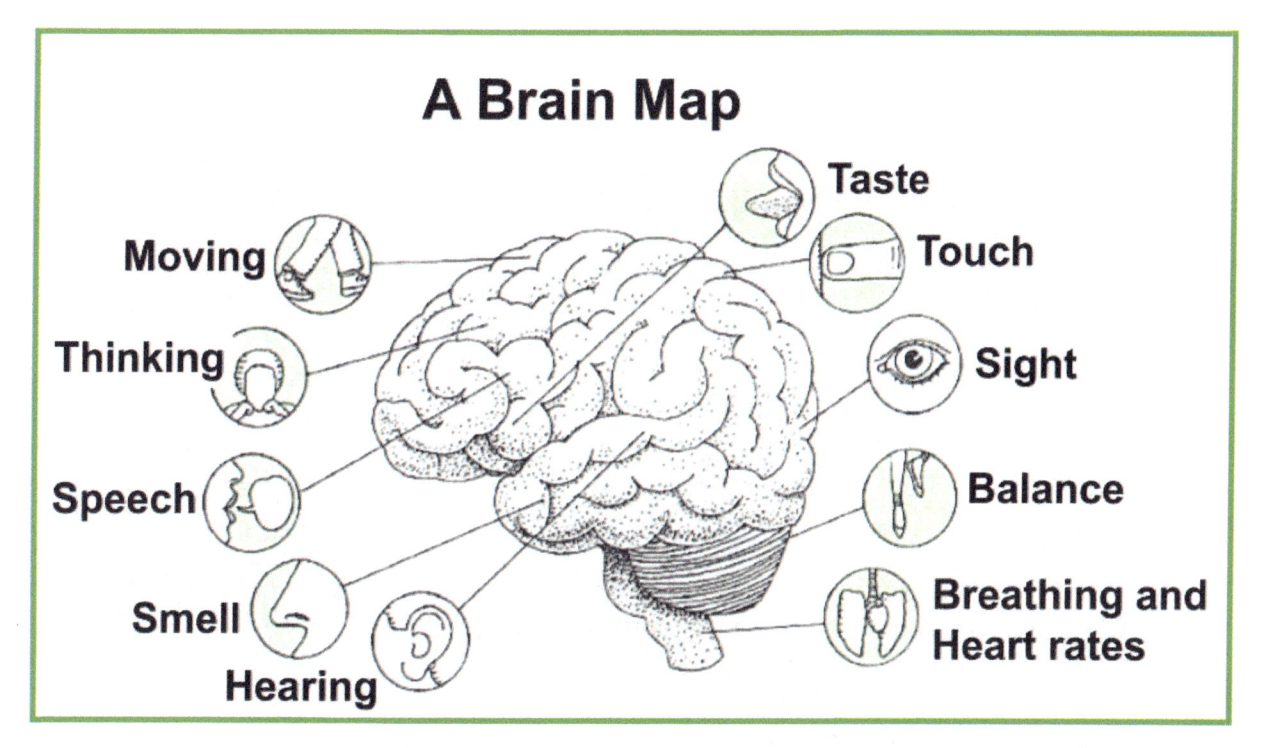

A Brain Map

- Moving
- Thinking
- Speech
- Smell
- Hearing
- Taste
- Touch
- Sight
- Balance
- Breathing and Heart rates

Your brain controls most of your activities.

153

Other areas of the cerebrum control your voluntary muscles. (Remember that you are able to move these muscles at will).

The **cerebellum** (ser-uh-BEL-um) is the part of the brain located below the cerebrum, at the back of your head. The cerebellum helps your muscles work together. It also helps you keep your balance.

The **medulla** (mih-DUL-uh) connects the brain to the spinal cord. Your heart rate and breathing are controlled by the medulla. So are movements of involuntary muscles, such as those in your digestive system. Some reflex actions, such as blinking and yawning, are centered in the medulla.

➤ To Do Yourself   What Are The Parts of the Vertebrate Brain?

Sheep's Brain (Top View)

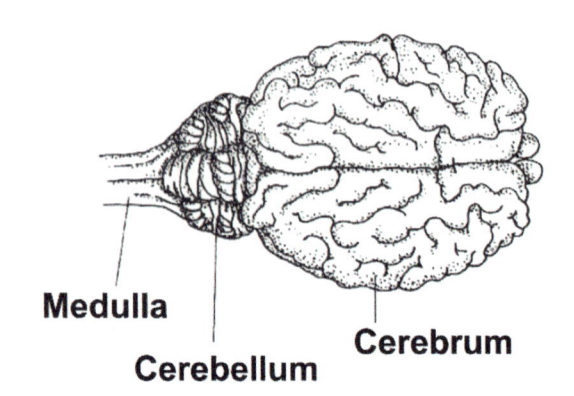

You will need:

Sheep brain from a butcher or a supply company, tray, large pins, paper, drawings of the brain of a sheep and a human

1. Place the sheep brain in the tray. Use the drawing of the human brain to locate the different parts.
2. Locate the cerebrum of the sheep's brain. How does it compare to that of a human?
3. Locate the cerebellum of the sheep's brain. Compare it to that of the human.
4. Locate and compare the medulla of the sheep's brain to that of the human.

Questions

1. How are the sheep's brain and human brain alike and different? _____

_____

2. What part of the brain connects it to the rest of the body? _____

3. How does the sheep brain compare with the fish brain (from the last lesson)? _____

_____

---------- **REVIEW** ---------- U-6 L-4

I. Write the letter of the part of the brain that controls each of the following functions:

    a. cerebrum   **b.** cerebellum   **c.** medulla

1. ____ balance
2. ____ reason
3. ____ heart rate
4. ____ memory
5. ____ hearing
6. ____ muscle coordination
7. ____ stomach contractions
8. ____ voluntary movement

II. To swallow, the action starts as a voluntary action, but continues as an involuntary action. What parts of the brain control … **(a)** the start of swallowing? **(b)** the rest of the action?

    a) _____   b) _____

III. A person may "see stars" after a blow to the back of the head. Check the brain map, and then infer why this happens. [A reminder: Inferring is guessing based on knowledge.]

How Do You Learn New Behavior? U-6 L5

Exploring Science / Historical Steps

<u>Goodbye to Cold Hands</u>. People usually have a hand temperature of about 32⁰ Celsius. But one woman who came to **Dr. Keith Sedlacek** had a hand temperature of only 22⁰ Celcius. She wore heavy gloves, but her hands were still cold and very painful.

The woman had **Raynaud's disease**. This can be made worse by stress. When a person with this condition is tense or upset, nerves cause blood vessels in the hand to narrow. Not enough warm blood can get into the hands.

Dr. Sedlacek taught the woman some special mental exercises to help her relax. As she did the exercises, her hands were connected to a machine that recorded their temperature. Bandage-like sensors attached to the outside of her head recorded her brain waves.

As shown in the illustration, the brain produces certain kinds of brain waves when we are relaxed, and other kinds when we are tense. While the woman carried out Dr. Sedlacek's mental exercises, she was getting *feedback*; she could immediately view her brain waves as well as her hand temperature. While she did the mental exercises, the blood vessels in her hands expanded and her hands warmed up.

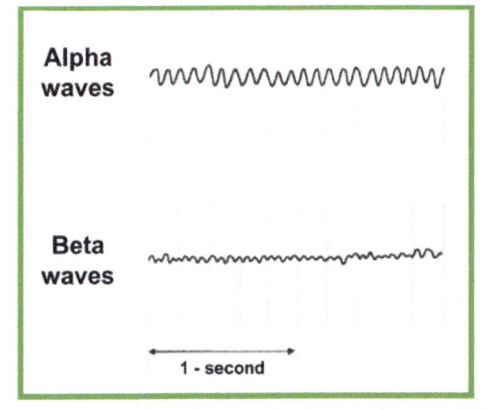

When you are relaxed, your brain produces alpha waves. They are slow and high. When you are excited, your brain produces beta waves. They are fast and flat.

Eventually, the woman learned to raise her hand temperature without the machine. This method of providing patients with feedback while their *thinking* improves their condition became known as **biofeedback** (by-oh-FEED-bak).

Over his long career from the <u>1970</u>s through the <u>1990</u>s, Dr. Sedlacek became an admired expert in the use of biofeedback.

➤ Some people with high blood pressure and others with migraine headaches have been successfully treated with biofeedback. What is one probable cause of these ailments?

Learned and Inborn Behavior

Biofeedback exercises are a type of **learned behavior**. One of the first (and now one of the most famous) people to study learned behavior was the Russian scientist **Ivan Pavlov**.

Pause now, look ahead to the next page, and try to find the key differences between the three images - top to bottom. Then, continue reading.

<u>CONDITIONING</u>. In the <u>1890</u>s, Pavlov observed that a dog's mouth waters when the dog first smells food. Pavlov tried a simple, but now widely known experiment. He rang a bell each time he fed a dog. He repeated this many times. Pavlov always rang the bell when he gave food to the dog.

Then Pavlov rang the bell *without* giving the dog any food. Still the dog's mouth watered. The dog learned to connect the sound of the bell with being fed. Before the experiment, the dog's response (mouth watering) was made to one stimulus (the smell of food). The dog now made the same response (mouth watering) to a new stimulus (a bell).

The dog had <u>learned</u> to respond to a new stimulus. A response (or behavior) learned in this way is called a **conditioned** (kun-DISH-und) **response**. This kind of learning is called **conditioning**. Can you give an example of how you have been conditioned to respond to a bell or other sound?

Now, look once more at the illustration at the right. Did you find all of the differences? Make certain that you clearly understand this illustration. "Pavlov's dog" is the most famous experiment in the field of **psychology**, the study of the brain and behavior. Experts in psychology are called **psychologists.**

TRIAL AND ERROR. Another way of learning is by **trial and error**. In trial-and-error learning, you learn something by trying different ways of doing it. You learn many skills - such as riding a bicycle - by trial and error. The more you practice, the fewer mistakes you make.

Animals also use trial and error. A rat was put in a maze similar to the one shown on the next page. A food reward was placed at the end of the maze. Each time the rat ran the maze, it made fewer mistakes. In time, the rat found its way to the food without making any mistakes. It had learned the correct pathway.

REASONING. **Reasoning** (REE-zuh-ning) is learned behavior in which you think things through. Your ability to solve a math problem depends on reasoning. So does doing a puzzle or finding a lost object.

Many animals also use reasoning. At the bottom of the next page is a good example. A chimpanzee was put in a room with a bunch of bananas that were hanging too high to reach. Three boxes were also in the room. The chimp looked over the situation for some time. Then the chimp piled the boxes on top of each other. Climbing on the boxes, the chimp was able to reach the bananas.

INBORN BEHAVIOR. Behavior that does not have to be learned is **inborn behavior**. At this moment, your heart is beating. Your digestive system is breaking down food. You are breathing. You may also be coughing, sneezing, or yawning. Every few seconds, your eyes blink. All of these behaviors can be considered types of reflexes. Remember that a reflex happens outside of your control. Reflexes are examples of inborn behaviors.

"Pavlov's dog" was trained to salivate when a bell rang, even though no food was present.

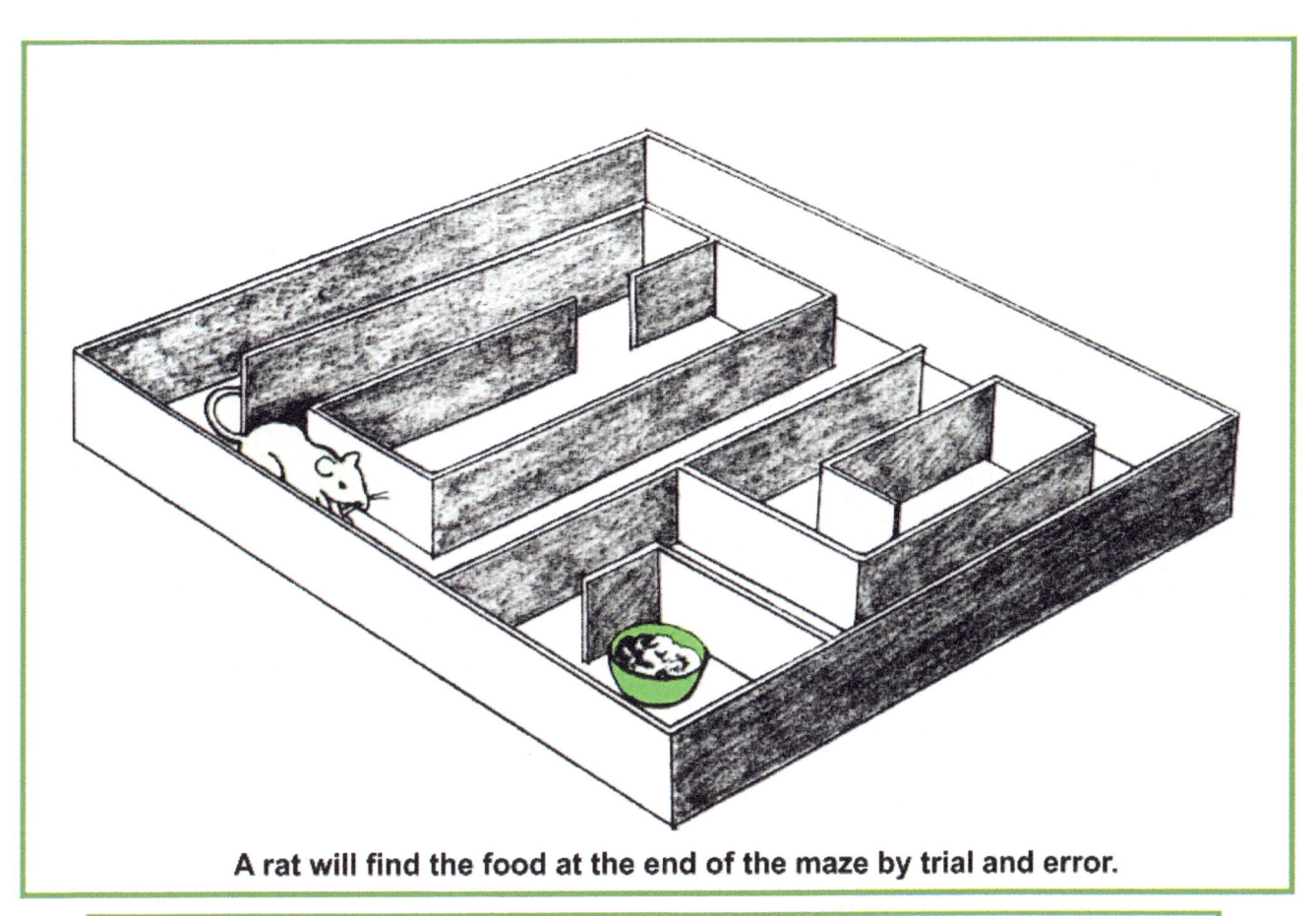

A rat will find the food at the end of the maze by trial and error.

The chimpanzee uses reasoning to get to the bananas.

➤ To Do Yourself   Can Animals Such As Fish Learn?

You will need:

A fish tank, goldfish, fish food, pencil

1. Set up a simple fish tank with several fish of one kind.
2. Before you feed the fish, tap the side of the tank several times with the pencil. Do this each time that you feed the fish. Feed them at the same time each day. Also, feed them at the same place in the tank.
3. After a week or so, tap the side of the aquarium, but do not feed the fish. What happens?

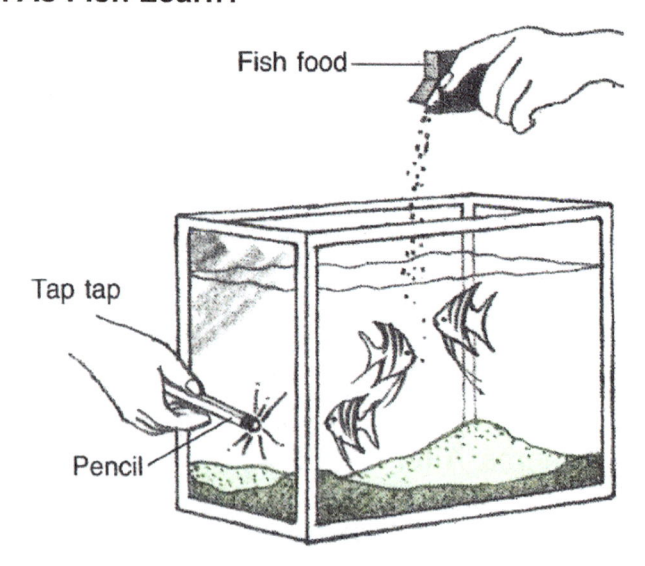

Questions

1. What was the stimulus in the fish experiment? What was the response? _____
   _____
2. How long did it take for your fish to learn the behavior? _____
3. This is an example of what type of response? _____

REVIEW — U-6 L-5

I. In each blank, write the word that fits best. Choose from the words below.

| trial and error | response | learned | stimulus |
| reasoning | inborn | conditioning | |

A reflex is _____ behavior. A conditioned response is

_____ behavior. Pavlov's dog's mouth still watered when the

_____ was changed. An animal learns a maze by

_____. Thinking things through is _____.

II. For each statement, write the kind of learning described. Choose from this list:

conditioning   reasoning   trial and error

A. You figure out where to find a lost sweater. _____

B. Your dog comes when you call its name. _____

C. At first, a newly hatched chick pecks at both the ground and at its food (called "feed"). Later, it pecks only at the feed. _____

158

What Are Chemical Messegers?

U-6 L6

Exploring Science

No Joking Around. When you visit sick friends, you try to cheer them up. Perhaps you tell them funny stories or jokes. Almost everyone has heard that making people laugh can help them get well. In fact, scientists have data that indicates why!

In one hospital, a comedian often performs for the patients. After laughing for 45 minutes, patients report that their pain has gone away or become less. How does this happen?

When you laugh, the body's chemical messenger system gets going. Chemicals released by this system have many good effects. One group of these chemicals are called **endorphins** (en-DOOR-finz). They are the body's own painkillers. They kill pain like drugs do, but without side effects.

So: no joking around - laughter really is good for you!

➤ Besides the pleasure that makes you laugh, what other feelings might cause the release of chemical messengers? Explain.

A girl laughing - and producing endorphins!

Chemical Messengers

The endorphins that are released when you laugh are a kind of hormone. **Hormones** (HORE-mohns) are chemical messengers. They control many activities in your body. Hormones are used not only by animals, but by plants!

➤ To Do Yourself   How Are Hormones Helpful?

You will need: Two plastic bags that may be sealed easily, two green bananas, one orange

1. Add a green banana to each bag.
2. Add an orange to only one bag.
3. Seal both bags and place them where they will not be disturbed. Fruits produce a hormone (in the form of a gas) that causes them to ripen.
4. Examine the bags each day. After the second day, take the bananas from the bags and compare them to see which one has ripened the most.

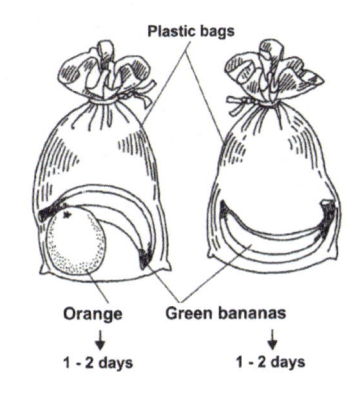

Plastic bags

Orange → 1 - 2 days   Green bananas → 1 - 2 days

Questions

1. Why do you think that the banana in the bag with the orange ripened first? _____

_____

2. What must the orange have produced? _____
3. How is the hormone in this activity different from hormones in the human body? [see next page.]

_____

ENDOCRINE GLANDS & THEIR HORMONES (See the illustration on the next page).

Pituitary gland. How tall you will be depends on **growth hormone**. This hormone is made by your **pituitary** (pih-TOO-ih-teh-ree) **gland**. A giant has too much growth hormone. A midget has too little. The pituitary also makes other hormones. These control many body activities; some even control other glands. For this reason, the pituitary is called the "master gland."

Thyroid gland. The **thyroid** (THY-roid) **gland** produces the hormone **thyroxin** (thy-ROK-sin). Thyroxin controls the rate at which your body produces energy. An adult with too little thyroxin may be tired and gain weight. Too much thyroxin can make a person nervous and cause weight loss.

Adrenal glands. On top of each of your kidneys you have an **adrenal** (uh-DREEN-ul) **gland**. These make a hormone called **adrenalin** (uh-DREN-ul-in). When you feel angry, excited, or afraid, your adrenal glands send adrenalin into your bloodstream. Adrenalin makes you breathe faster, helping you take in extra oxygen. Adrenalin also causes more sugar to go into the blood so your muscle cells have plenty of fuel to burn. Finally, adrenalin makes your heart beat faster, helping oxygen and sugar get to your cells faster. All of these changes give you an extra burst of energy. In an emergency, this energy can help you do things that you would not be able to do normally. For example, if you were in real danger, you could run faster than normal. Not surprisingly, adrenalin is sometimes called the "fight or flight" hormone.

Pancreas. You may recall that your pancreas makes digestive juice. In addition, the pancreas makes **insulin** (IN-suh-lin). Insulin is a hormone that controls the use of sugar by the body cells. If the pancreas makes too little insulin, the person has **diabetes** (dy-uh-BEE-tis). In this disease, extra sugar builds up in the blood. The kidneys remove some of this sugar; as a result the urine has some sugar in it. It is easy for your doctor to check whether your urine contains sugar.

Diabetes can sometimes be controlled by proper diet and exercise, but many people with diabetes must also take insulin shots.

Ovaries and Testes. If you are a girl, you have two **ovaries** (OH-vuh-reez). If you are a boy, you have two **testes** (TES-teez). The ovaries and the testes are part of the reproductive system, (which is covered in much more detail in Unit 8, Lesson 5).

The ovaries produce the female sex hormones that cause many changes during adolescence. For example, as a girl becomes a woman, her breasts develop and her hips widen.

Testes produce male hormones. These also have many effects during adolescence. As a boy becomes a man, his beard grows, his voice deepens, and his muscles grow larger and stronger.

All of these glands are part of your **endocrine** (EN-duh-krin) **system**. Endocrine glands produce hormones that pass directly from the glands into the bloodstream (via capillaries). The blood then carries the hormones to the parts of the body where they are needed.

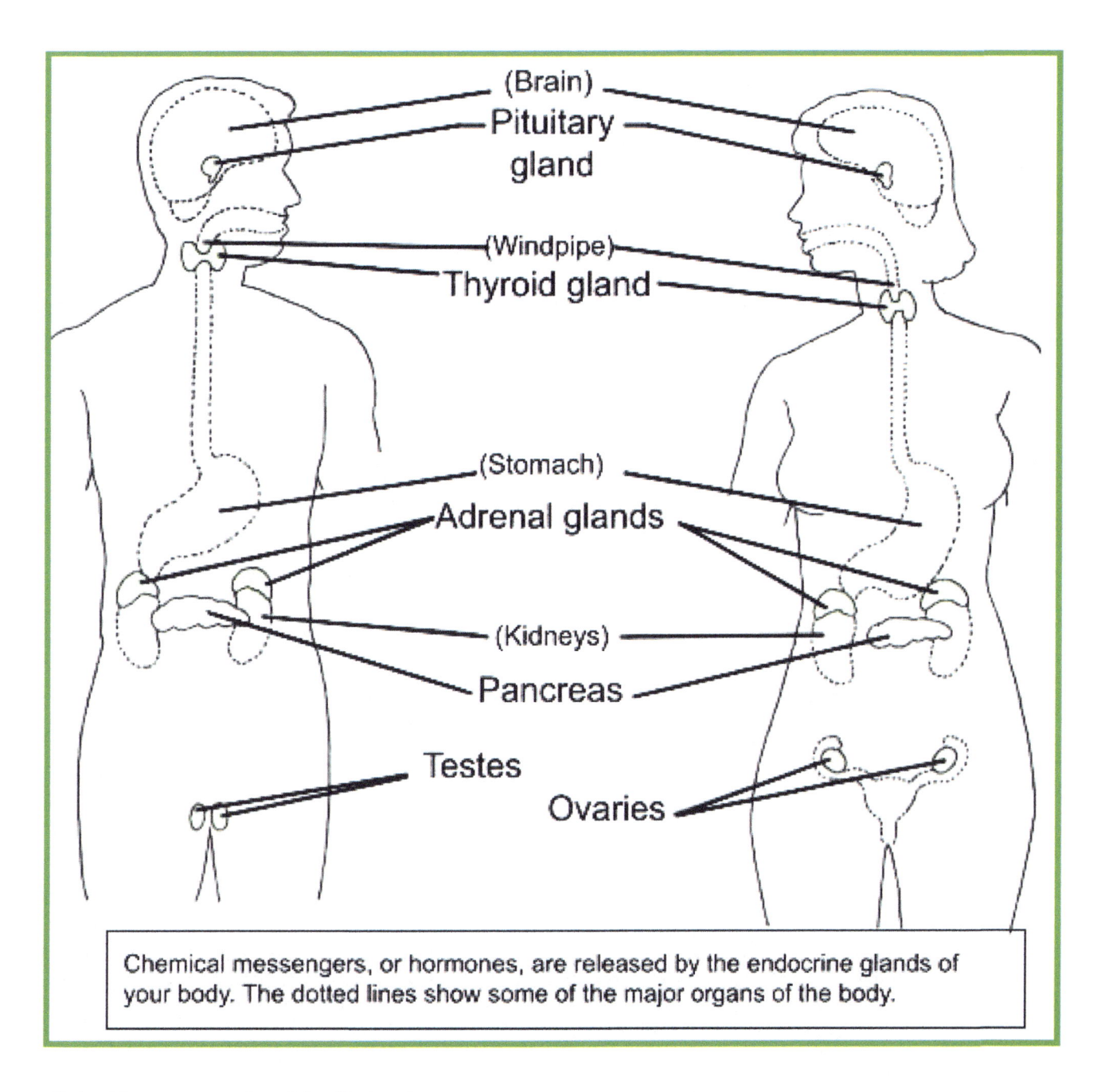

[Diagram of male and female figures showing endocrine glands: Pituitary gland (Brain), Thyroid gland (Windpipe), Adrenal glands (Stomach), Pancreas (Kidneys), Testes, Ovaries]

Chemical messengers, or hormones, are released by the endocrine glands of your body. The dotted lines show some of the major organs of the body.

Note: The pancreas is a "dual" gland.

Some tissues in the pancreas produce the hormone <u>insulin</u>. Insulin is released by these cells directly into the capillaries nearby. Therefore, the pancreas is an endocrine gland.

Other tissues in the pancreas produce digestive juices that contain enzymes. These juices are carried by a small tube - from the pancreas to the small intestine. Therefore, the pancreas is also a digestive gland.

------------------------------ **REVIEW** ------------------------------ U-6 L-6

I. In each blank, write the word that fits best. Choose from the words below.

thyroxin   **insulin**   **endocrine**   **adrenalin**
testes   **ovaries**   **hormones**   **pituitary**

The body's _____ system is made up of glands that produce

_____. The "master gland" is the _____.

The hormone that helps you to meet an emergency is _____.

(This hormone may also help an athlete perform better during a game than during practice).

Diabetes is caused by not enough _____. Female body changes are

caused by hormones from the _____.

II. If the statement is true, write **T**. If it is false, write **F,** and then <u>correct</u> the underlined word.

A. ____ Too little hormone from the <u>pancreas</u> can cause a person to become a midget.

B. ____ A hormone made by the <u>adrenal</u> glands speeds up the heart.

C. ____ Too much <u>insulin</u> can make a person nervous and cause them to lose weight.

III. <u>Name</u> the numbered endocrine glands.

1. _____

2. _____

3. _____

4. _____

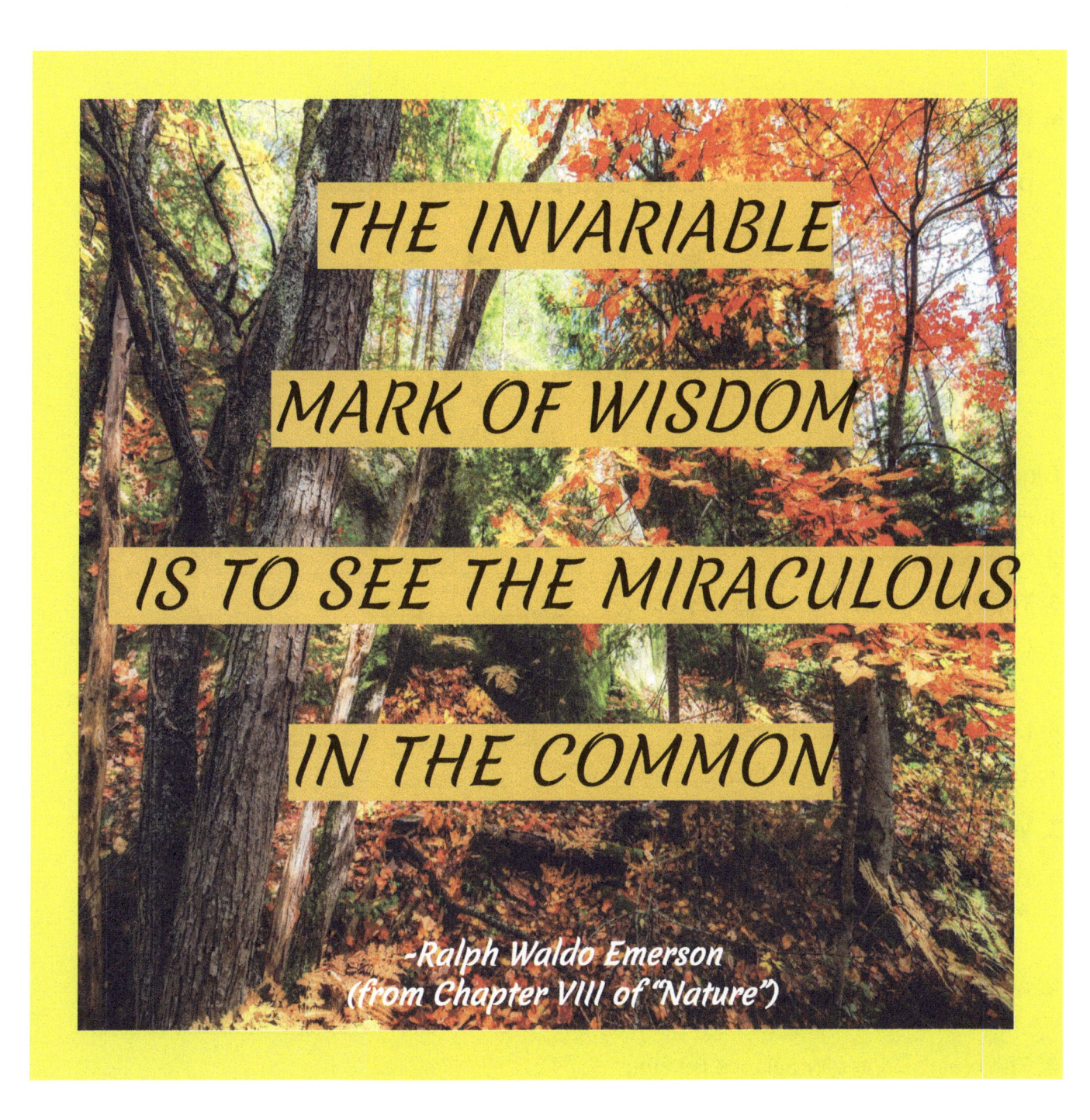

> *THE INVARIABLE MARK OF WISDOM IS TO SEE THE MIRACULOUS IN THE COMMON*
>
> —Ralph Waldo Emerson
> (from Chapter VIII of "Nature")

Most species are able to respond to stimuli.
.... but ....
- *How many understand how this happens?*
- *How many can appreciate nature's beauty?*
- *How many can act to protect it?*

Unit 6 -- Review What You Know ---

A. Hidden in the puzzle below are the names of seven nutrients. Use the clues to help you find the names. Circle each name you find in the puzzle. Then write each name on the line next to its clue.

```
M  O  T  O  R  O  D  S
Z  B  H  F  E  V  H  G
I  J  Y  L  N  A  P  P
C  E  R  E  B  R  U  M
N  Q  O  M  O  I  P  R
O  T  I  X  Q  E  I  S
S  W  D  A  V  S  L  K
E  M  E  D  U  L  L  A
```

Clues:

1. Has nerves sensitive to odors. _____
2. Where light enters the eye. _____
3. The brain's center for involuntary acts. _____
4. Controls the rate at which the body uses energy. _____
5. The type of nerve that carries messages to muscles. _____
6. Make female sex hormones. _____
7. The brain's center for memory. _____

B. <u>Write</u> the word (or words) that best completes each statement.

1. When a bell rings near your ear, you hear the sound when messages reach your
      a. eardrum   **b.** inner ear   **c.** brain   1. _____

2. An example of inborn behavior is
      a. blinking   **b.** trial-and-error   **c.** reasoning   2. _____

3. The brain's center for balance is in the
      a. medulla   **b.** cerebrum   **c.** cerebellum   3. _____

4. Light that enters the eye is focused as it passes through the
      a. retina   **b.** lens   **c.** nerve   4. _____

5. When you touch a hot object, you pull your hand away when messages reach your
      a. sensory nerve   **b.** spinal cord   **c.** brain   5. _____

6. An example of learned behavior is
      a. yawning   **b.** conditioning   **c.** sneezing   6. _____

7. A hormone that can speed up your breathing rate is
   a. adrenalin   **b.** insulin   **c.** thyroxin   7. _____

8. Male sex hormones are produced in the
   a. thyroid   **b.** pancreas   **c.** testes   8. _____

9. Diabetes is caused by too little
   a. insulin   **b.** thyroxin   **c.** adrenalin   9. _____

10. The body's master gland is the
   a. thyroid   **b.** adrenal   **c.** pituitary   10. _____

C. Apply What You Know

Study the drawings of three steps in teaching a puppy its name. In a few sentences, explain what is happening. Use these words in your answer:

stimulus   **response**   **brain**   **conditioned**

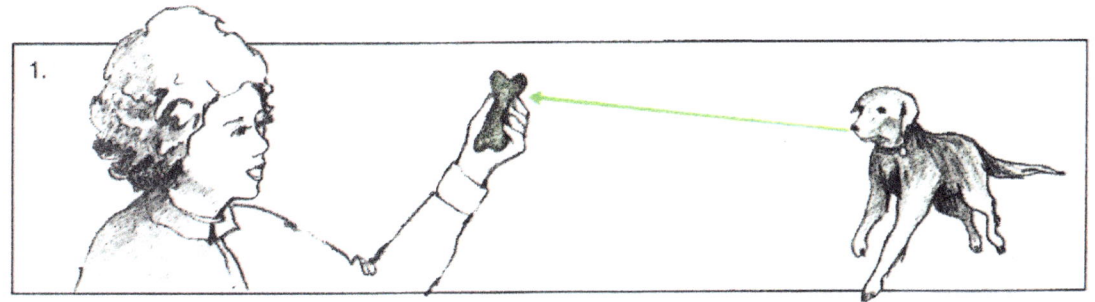

1. Girl holds food; puppy comes.

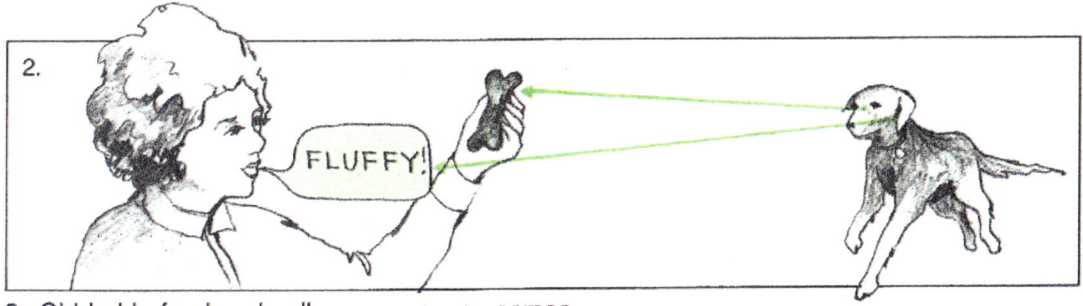

2. Girl holds food and calls name; puppy comes.

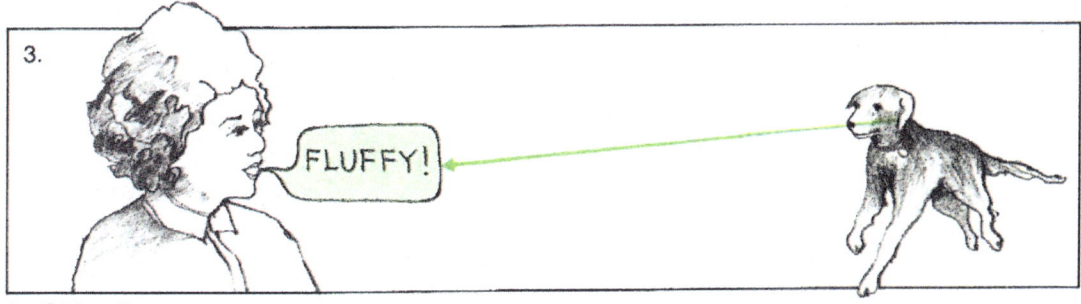

3. Girl calls name; puppy comes.

_____

_____

D. Find out more.

1. Do you know that you have a **blind spot** in each of your eyes? Try this. Close your left eye. Look at the square with your right eye. Slowly bring the page close. Stop when you cannot see the circle. You have found the blind spot in your right eye. Now close your right eye and look at the circle. Bring the page closer until you cannot see the square. Research why humans have a blind spot in each eye.

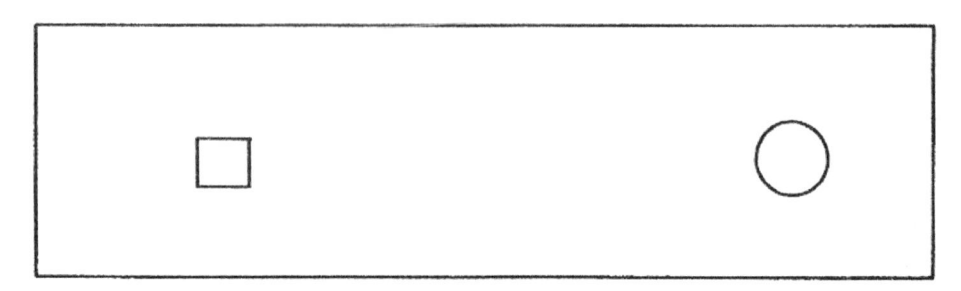

2. Plan how you would train a pet to do a trick. Carry out your plan.

3. To find nerve endings for hot and cold in the skin, try this. You will need a partner. Use a washable marker to mark off a 2-centimeter square on the back of your partner's wrist. Make 16 evenly-spaced dots within the square. Place one iron nail in hot water. Place another nail in ice water. Blindfold your partner. Touch 8 scattered dots lightly with the hot nail. Then touch the other 8 dots with the cold nail. Have your partner tell you what she or he feels with each touch. Make a map of the nerve endings for hot and cold; use red pencil for hot and blue for cold. Trade places with your partner and repeat the investigation.

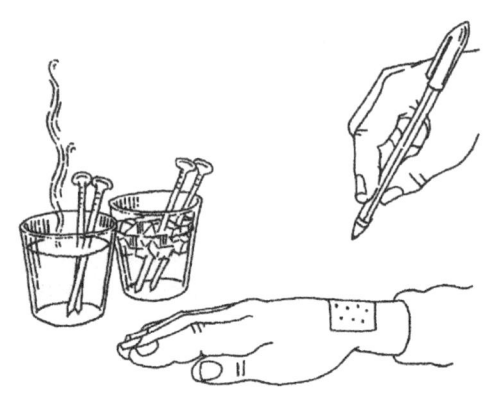

4. **Blinking** is a reflex that can sometimes be hard to control. Work with a partner to test your blinking reflex. Put a clear sheet of plastic in front of your eyes. (If you don't have such a sheet, goggles are adequate). Have your partner throw wads of paper (or cotton balls if you are using goggles) so that they hit the eye protection. Can you keep from blinking each time an object hits the protection? Trade places with your partner. Can they avoid blinking?

5. Posters are often used as stimuli to encourage people to respond with a change in their behavior. Observe the "poster" on the page prior to this unit review. Discuss with a classmate or family member what you think it means.

    Is there a behavior change that you'd like to see happen? Consider creating your own poster to help this occur!

Careers in Life Science

A Healthy Mind in a Healthy Body. "Health is wealth." "A healthy mind in a healthy body." These slogans remind us that health is not only physical well-being. Health is also emotional well-being. People with emotional problems often go to mental health professionals for help.

Psychology is the study of the mind and behavior. Are you interested in what makes people tick? Are you able to tune in to other people's feelings? Are you in touch with your own inner self? Then you may want a career related to psychology and mental health.

Mental Health Technician. A mental health technician (also called a psychiatric technician) is part of the team of caregivers in a mental health center. As a mental health technician, you might be the first person seen by new patients. Your job would include listening to them and observing their behavior.

Psychologist supports an injured woman

The mental health technician's relationship with a patient is the start of that patient's treatment, or therapy. The therapy continues as the patient is seen by other members of the team. A two-year training program after high school can prepare you for a career as a mental health technician.

Social worker assists a family under stress

Clinical Social Worker. The main therapy for mental or emotional problems may be conducted by a clinical, or psychiatric, social worker. This worker helps patients gain a thorough understanding of themselves, and helps them learn of helpful resources in their community. Patients use these resources as well as their new self-knowledge to cope better with problems.

Four years of college can prepare you to be a licensed social worker, but two more years of education are needed to become a clinical social worker.

Psychologists, Psychiatrists, and Psychiatric Nurse Practitioners

The education and training for these careers takes many years. These professionals devote most of their time to counseling patients, but psychiatrists also prescribe medications. Psychiatrists are medical doctors who have at least four years of training and education beyond medical school. They are often assisted by psychiatric nurse practitioners, who have first earned a college degree in nursing and then completed at least 2 years of additional training.

All mental health careers demand that you give a great deal of yourself. But the rewards of helping people during challenging times of their lives are great.

UNIT SEVEN

REPRODUCTION IN SIMPLE ORGANISMS AND PLANTS

How Do Living Things Reproduce?

U-7 L-1

Exploring Science / Historical Steps

Kelly's Unlikely Baby. Which adult could be the parent of the young animal to the right? Horses can give birth only to horses - right? If you let nature take its course, the answer is "right." But when scientists take a hand in the matter, the answer can be "wrong." In 1984, Kelly was a horse at the Louisville Zoo in Kentucky. She gave birth to a baby zebra. How did this happen?

The baby zebra's real parents were, of course, zebras. The baby began its life in the body of its real mother. As an **embryo** (a very early stage of growth) the zebra embryo was moved from its mother into Kelly's body. For 12 months, the baby zebra grew in its substitute mother's body. Then it was born.

Kelly was the first horse to have a zebra baby. But, thanks to zoos, she is just one of a number of animal mothers with unlikely babies. For example, a cow gave birth to a wild ox, an endangered animal similar to a cow. At the Cincinnati Zoo in Ohio, endangered antelopes have been born to more common animals. Even a rare desert kitten was born with a common domestic cat for its mother. Using substitute mothers is one way of increasing the population of endangered animals.

➤ Why do you think scientists chose a horse instead of a cow (or other animal) as the substitute mother for a zebra?

Young zebra

Adult female zebra

Adult female horse

169

How Living Things Reproduce

When zebras and other living things reproduce, they make more of their own kind. The young are called **offspring**. The offspring of zebras are zebras. Those of humans are humans. Those of oak trees are oak trees.

Like the zebra in the experiment, you came from two parents. One parent, the mother, is female. The other parent, the father, is male. A dog, a cat, a fish - nearly any living thing that you can name - comes from two parents. Reproduction from two parents is called **sexual reproduction**.

Some organisms reproduce from only one parent. Reproduction from one parent is called **asexual** (ay-SEK-shoo-ul) **reproduction**. Protozoans and many other one-celled organisms can reproduce asexually.

The simplest kind of asexual reproduction is called binary fission, or often simply **fission** (FISH-un). These terms mean "splitting in two."

Remember that a cell has a center of control called the nucleus. During fission, the nucleus splits into two equal parts. So does the cytoplasm - the part of the cell around the nucleus. Each half of the parent cell becomes a new cell, enclosed by its own cell membrane.

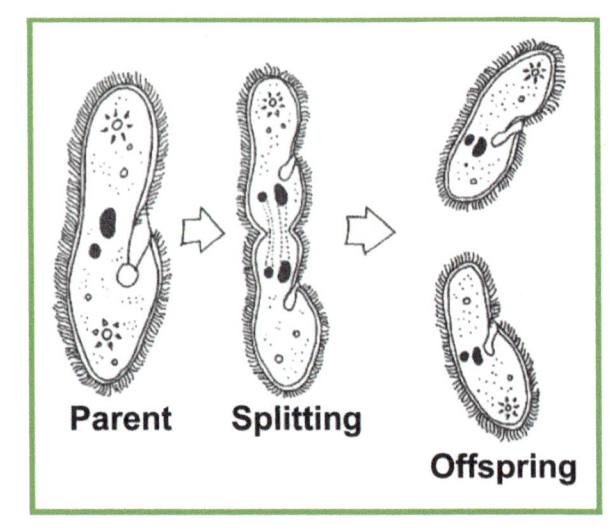

Parent   Splitting

Offspring

A paramecium undergoes binary fission.

➤ To Do Yourself   How Do Yeast Cells Reproduce?

You will need:

Sugar, 1 package of dry yeast, 3 glass jars, a medicine dropper, warm water, cold water

1. In one jar place ½ cup warm water, ⅓ package dry yeast, and ½ spoonful of sugar. Label the jar "warm."

2. Set up the second jar the same way, but use cold water. Label the jar "cold."

3. Set up the third jar the same as the first, but omit the sugar. Label the jar "0 sugar."

4. Stir each jar and allow them to stand.

5. After 5 -10 minutes, examine the jars. When yeast reproduce, they make alcohol and carbon dioxide. Smell each jar to decide whether yeast reproduced.

Questions

1. In which jar did the yeast grow best? _____

2. What does a yeast cell need in order to reproduce? _____

3. How could you tell that your yeast was reproducing? _____

The two new cells are called **daughter cells**. Each daughter cell is just like the parent cell, but half its size. Each daughter cell grows until it reaches full size. Then it may also reproduce by fission. Two organisms that can reproduce this way are amebas and paramecia.

Another way that some living things can reproduce asexually is by **budding**. Yeasts, the one-celled organisms used in making bread, can reproduce by budding.

When a yeast cell buds, a part of its cell wall pushes out. This is the start of the bud. The cell nucleus moves toward the bud and then splits into two equal parts. One part goes into the bud. The other part stays in the parent cell.

The yeast bud grows larger while attached to the parent cell. When the bud becomes large enough, a new cell wall grows between the parent cell and the bud. Then the bud may break away. The newly formed yeast cells can also reproduce by budding.

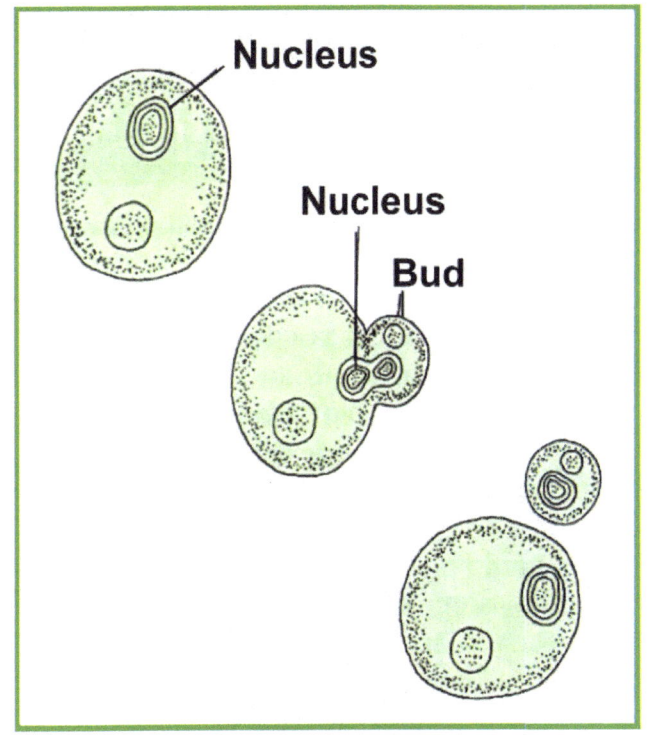

A yeast cell forming a bud

REVIEW — U-7 L-1

I. In each blank, write the word that fits best. Choose from the words below.

| daughter | parent | budding | fission | cytoplasm |
| sexual | bud | nucleus | offspring | asexual |

The young that result from reproduction are the _____ of the parents. Two parents are needed for _____ reproduction. Only one parent is needed for _____ reproduction. During _____, the whole cell divides into two equal parts. During _____, a small part of the parent cell becomes a new cell. Two _____ cells form as a result of fission. A _____ forms as a result of budding. In both fission and budding the cell _____ divides equally.

II. An <u>alga</u> is one cell of a particular protist. (The plural for alga is <u>algae</u>. The adjective for alga is <u>algal</u>). Imagine one algal cell of the genus *Protococcus,* commonly called "tree green." It grows on the bark of trees. It reproduces by fission.
 • Circle the word (between the brackets) that makes each statement true.

   A. The new algal cells that form are [equal / unequal] in size.

   B. Each new algal cell is called a [bud / daughter] cell.

   C. The form of reproduction of the alga called *Protococcus* is [sexual / asexual].

III. Bacteria reproduce by fission. If a single bacterium reproduces once every 20 minutes, how many bacteria would be formed at the end of 1 hour? 2 hours? 3 hours?

How Do Molds Reproduce? U-7 L-2

Exploring Science / Historical Steps

Dr. Fleming - Lucky? In September of 1928, **Dr. Alexander Fleming** had the windows open in his London lab. He was growing bacteria in flat glass dishes with lids, called **petri dishes**. As he examined a dish with its lid off, something that he did not see blew into the open dish.

Sir Alexander Fleming

A few days later, Fleming examined the dish again. The bacteria were still growing in the dish, but so was a green mold! The mold had "spoiled" the dish. Fleming's first thought was to throw the petri dish away.

Then something unusual caught his eye. The bacteria *very near* the mold had died. Why? Fleming discovered that this type of mold makes a substance that kills deadly bacteria.

Later it was found that the substance destroys cell walls. Fortunately for us, bacteria *have* cell walls, but animal cells *do not*. The substance that kills bacteria does not harm us!

Since the mold's name was *Penicillium* (pen-ih-SIL-ee-um), Dr. Fleming named the substance **penicillin** (pen-ih-SIL-in). It was the first **antibiotic** (a medicine that kills bacteria that have invaded the human body). Penicillin has saved countless lives. In 1945, Fleming was awarded a Nobel Prize.

Was Fleming's discovery a "lucky" find? A famous quote by **Louis Pasteur** is, "Chance favors the prepared mind." Lucky for humanity, Fleming had a prepared mind!

➤ What do you think blew in Fleming's window? Below is a strong hint!

Reproducing by Spores

Have you ever seen green mold growing on an orange or a lemon? It is the same mold that grew in Fleming's dish of bacteria. The first penicillin was made from this common mold. How did the mold get into Fleming's laboratory?

If you used a microscope, you would see that *Penicillium* has root-like parts. These are attached to the food on which it grows. The mold takes in food through these "roots." The mold also has stem-like parts that grow up into the air. At the ends of the "stems" are bead-like strings made up of extremely tiny reproductive cells, called **spores**.

The spores are made in enormous numbers. They are so tiny and light that the slightest breeze carries them away from the parent mold. Each spore can grow into *Penicillium* mold. Since spores come from only one parent, this is a kind of **asexual reproduction**.

Mold spores are blown everywhere and fall on everything. When inhaled, the spores of some molds cause allergies in many unlucky people. On the other hand, we are all lucky that at least one *Penicillium* spore landed in Fleming's famous petri dish.

Another kind of mold grows on old bread. Without magnification, the mold looks like white fuzz filled with tiny dots.

Penicillium mold

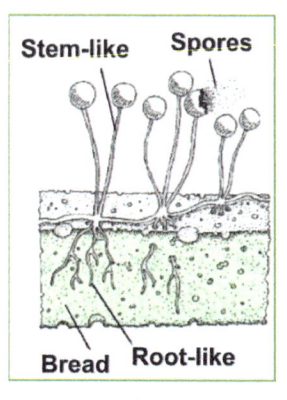

Bread mold

Under a microscope, you can see that the fuzz on the bread is the mold itself. You can also see that the dots are spheres at the tips of the bread mold's "stems." Each sphere breaks open and releases thousands of microscopic spores. Each spore can grow into another bread mold.

➤ To Do Yourself   How Can We Slow the Growth of Mold?

You will need: Potato, apple-coring tool, cotton balls, household disinfectant, soapy water, hand lens, 3 test tubes

1. Use the apple-coring tool to take 3 cylinders (cores) from a potato.
2. Roll each potato core on a dusty table.
3. Spray one potato core with disinfectant. Wash one with soapy water. Leave the third one alone.
4. Place a wad of moistened cotton in the bottom of each tube. Put a potato core inside of each and label the tubes.
5. Mold grows best in a dark, warm place. Place your labeled tubes where the mold can grow for several days. Observe the potato cores each day with a hand lens. Record your observations.

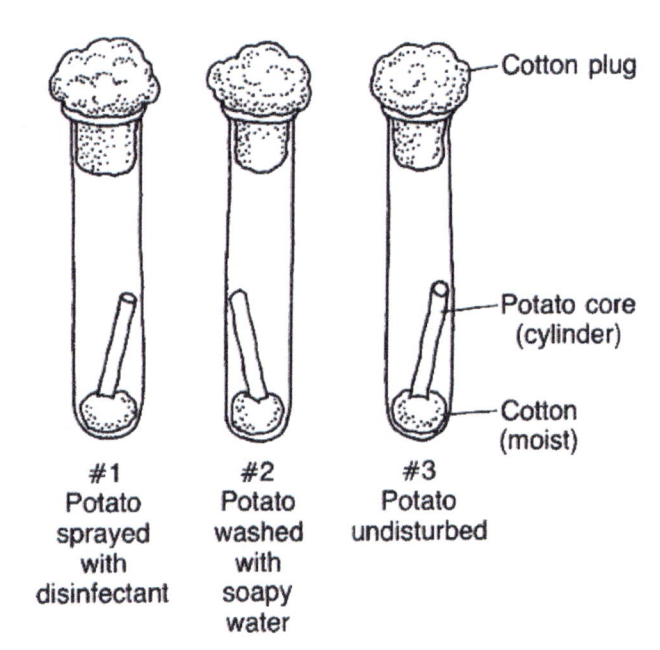

#1 Potato sprayed with disinfectant
#2 Potato washed with soapy water
#3 Potato undisturbed

Cotton plug
Potato core (cylinder)
Cotton (moist)

Questions

1. In which test tube did you first notice mold growing? _____
2. Which mold grew the most? The least? _____
3. What might people do to slow down the growth of mold? _____

REVIEW — U-7 L-2

I. In each blank, write the word that fits best. Choose from the words below.

   "stems"   *Penicillium*   "roots"   fission   spores   mold   buds

   A mold that grows on oranges and lemons is called _____.

   A mold reproduces by _____. A mold takes in food through its

   _____. Spores form on a mold's _____.

   One spore can grow into a whole new _____.

II. Explain how making spores is like fission and budding. Explain how it is different from fission and budding.

   _____

III. A doctor refuses to prescribe an antibiotic for a patient who has a bad cold. Can you infer why the doctor made this decision?

How Do Animals Grow Back Lost Parts? U-7 L-3

Exploring Science

<u>A Tale of a Lizard's Tail</u>. What good is a tail? For many lizards, a tail can be used for self-defense. When a snake or a bird attacks a lizard, the lizard's tail break's off. While the predator pays attention to the tail, the lizard tries to escape. The predator may not catch on to what's happened until the lizard is far away. As for the lizard, it simply grows itself a new tail.

This defense works better for some lizards than others. Scientists in Texas observed how snakes react to the broken-off tails of two types of lizards: anoles (un-NOH-lees) and skinks (SKINGKS). After an anole's tail comes off, the tail doesn't move very much. But a skink's broken-off tail keeps thrashing about. The more the broken-off tail moves, the longer the snake "thinks" it has caught the whole lizard.

In an experiment, snakes caught many more anoles than skinks. For fooling snakes, it is good for a lizard to have a tail that can both come off and keep moving.

➤ Which kind of lizards - skinks or anoles - do scientists think have a special way to store energy in their tails? Explain.

A skink escapes (and survives) while a snake goes after the skink's thrashing tail.

Growing Back Lost Parts

Like lizards, some other animals can grow back a lost part. This power to replace lost parts is called **regeneration** (rih-jen-uh-RAY-shun). A lizard whose tail breaks off can regenerate, or grow back, a new tail. Lobsters and crabs can regenerate lost claws.

The starfish has an unusual ability to grow back lost parts. If a starfish loses an arm, it can grow a new arm. This is regeneration. But its cut-off arm can also grow into a whole new starfish. This is a kind of asexual reproduction.

The planarian (pluh-NAIR-ee-un) is a type of flatworm. It also can reproduce asexually from a cut-off part. Cut a planarian into two, three, or even four pieces - each becomes a whole worm!

A. Arm regeneration

B. Broken-off arm *with* part of center → Whole new starfish (asexual reproduction)

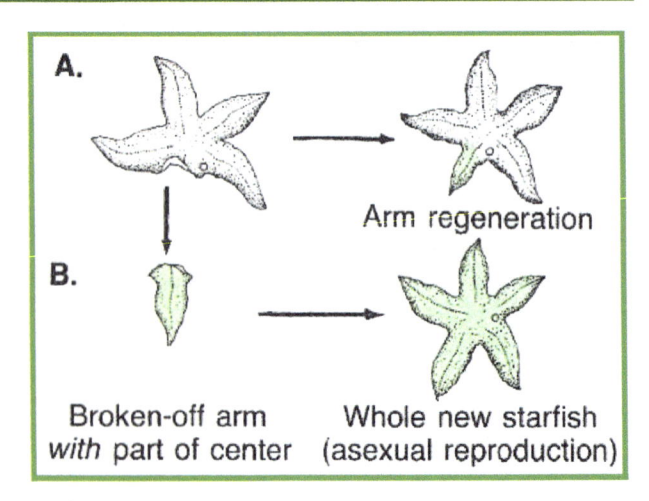

A. A starfish regenerates a lost arm.
B. An entire starfish grows from one arm and a small part of the center.

What about our body? We can repair cuts and heal broken bones, but not much else. However, like other organisms, our body has **stem** cells. These special cells are able to grow into a _variety_ of tissues that are near them.

The day may come when scientists are able to "trick" our stem cells into regenerating entire organs - such as failing hearts or kidneys! Who knows, someday humans may even be more like lizards or starfish in our ability to regenerate!

➤ To Do Yourself   Is Regeneration a Form of Reproduction?

You will need:

An adult to supervise the cutting, planaria, a box-cutter, petri dish and lid, hand lens or microscope, medicine dropper, dechlorinated water or creek water

1. Use the medicine dropper to transfer a planarian to the petri dish.

2. With an adult nearby, use the box cutter to cut the planaria into one of the patterns shown.

3. Add the dechlorinated (or creek) water to the dish. Cover it and add a label.

4. Observe your planarian each day with the hand lens. Record any regeneration that you observe.

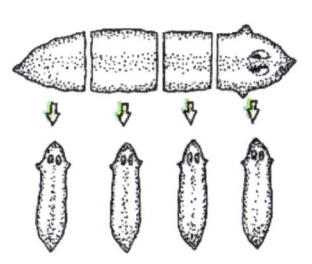

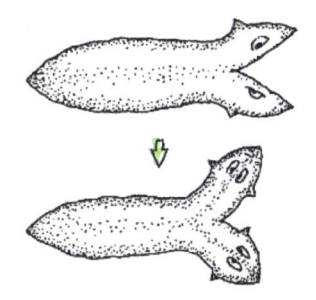

Questions

1. When is regeneration also a type of reproduction? _____

2. Is this type of reproduction sexual or asexual? _____

REVIEW — U-7 L-3

I. For each statement, write **A** for regeneration or **B** for (asexual) reproduction.

1. _____ Your body makes 1 billion new blood cells each day.

2. _____ A snake grows a new skin under its old skin, then sheds the old skin.

3. _____ A sponge animal cut into 3 parts grows into three new sponge animals.

4. _____ A deer sheds its antlers, then later grows new antlers.

5. _____ A small piece of a ribbon worm grows into a whole new ribbon worm.

6. _____ Stem cells in a liver divide to replace liver tissue damaged during a car accident.

II. Starfish like to eat oysters. In the past, people who gathered oysters would cut up starfish and throw the pieces back into the water. They did this to reduce the number of starfish. What do you think actually happened?

III. In 1996, an entire sheep (called Dolly) was copied (cloned) using its stem cells. Do you think that animals should be cloned?

How Do Plants Reproduce Asexually? U-7 L-4

Exploring Science / Historical Steps

Humans can "copy" plants.

You heard in the last lesson that, in 1996, a sheep called (Dolly) was **cloned** (copied) from stem cells of another individual. In fact, decades earlier, other animals (like fish and frogs) were cloned.

What about plants? In the 1930s, a method called **plant tissue culture** was developed to produce adult plants from very tiny pieces of plants. An expert in this field, **Dr. Valeria Pence**, has long directed the Cincinnati Botanical Garden. Here, endangered plants from all over the world are being cloned using tissue culture. Tiny pieces of plant tissue are used to make many exact copies of plants that are in danger of becoming extinct.

This method requires special plant nutrients, including **hormones** (chemical messengers). The plant hormones stimulate plant tissue to grow roots, stems, and leaves.

In Thailand and India, plant tissue culture is used to grow (and then sell) beautiful orchids (like the one to the left). Thousands of identical flowers can be made from a single beautiful plant.

Perhaps most importantly, plant tissue culture is used to make copies of selected individual food-making plants - such as tomatoes and potatoes. The tastiest and the most disease-resistant plants are "multiplied" and grown.

And humans can make "mixed" plants.

Since 1994, scientists have been "improving" and then copying plants. To do this, they take a "good" trait from a different organism and add it to a plant. Plants made in this way are called **GMOs** (genetically modified organisms).

Let's look at an example. American Chestnut trees, which once grew in the billions, have been nearly wiped out by an invasive fungus. But, this fungus can't hurt a type of wheat plant.

In 2013, scientists put this particular wheat trait into cells taken from Chestnut trees. They then used tissue culture methods to grow these cells into adult trees. These modified (changed) Chestnut trees survive attacks from the fungus.

This same method has been used to make improved soybean, corn and other crops that insect pests can't hurt. Farmers are able to grow these crops with less insect-killing chemicals.

While there are obvious benefits from GMOs, many people feel strongly that science should not be changing or mixing organisms' traits. In fact, GMO foods are limited and even banned in many nations across the globe - including France and Germany. A key concern is that we need to learn more about how GMOs could affect other organisms in their ecosystems.

➤ Want more? Research the term **transgenic organisms**. ("trans" means across; "genic" refers to genes - which will be covered in Unit 9).

Asexual Reproduction in Plants

Growing plants by tissue culture takes great skill. Many plants, however, can be reproduced quite easily from their parts (stems, leaves, or roots). Reproducing plants from growing parts is called **vegetative propagation** (VEJ-uh-TAYT-ive prahp-uh-GAY-shun). Like plant tissue culture, this is a kind of asexual reproduction. Do you see why?

Vegetative propagation can happen in nature. Two ways that plants propagate in nature are by bulbs and tubers (TOO-burs). People also use bulbs and tubers to grow new plants.

BULBS. A **bulb** is a short, thick underground stem. It has thick underground leaves on it that store food. These unusual, underground leaves have no chlorophyll; they are not green.

Onions are bulbs. The next time that you see a cut piece of onion, notice that it is made of layers. You can peel these layers off, one by one. The layers are actually leaves that have become "fleshy." Instead of *making* food (like most leaves do), these *store* food.

Other bulbs that you may have seen are tulips and garlic. When conditions are right, a bulb will grow into a new plant that looks like the parent. It will send up green leaves and send down new roots. The food stored in the bulb is used by the young plant until its leaves begin to make food by photosynthesis.

TUBERS. A **tuber** is another kind of underground plant stem. It has a fleshy part that looks like a root. A white potato (shown below) is a tuber. The fleshy part of the potato is stored food. The "eyes" are actually buds on the plant's stem. Each eye can grow into a new potato plant.

Farmers use cut-up potatoes to grow new crops of potato plants. First, the potatoes are cut into pieces. Each piece has an **eye** (bud) on it. When the pieces are planted, a new potato plant grows from each eye. The fleshy part of the potato provides food for the developing young plants.

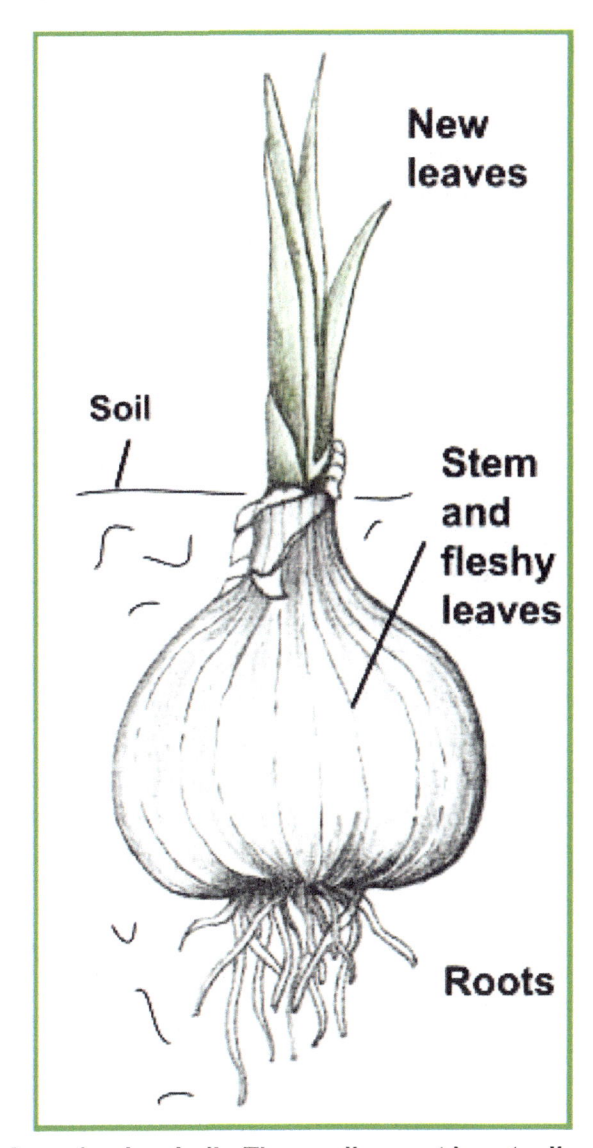

An onion is a bulb. The swollen part is actually an underground stem with fleshy leaves. From the bulb can grow new *green* leaves and new roots.

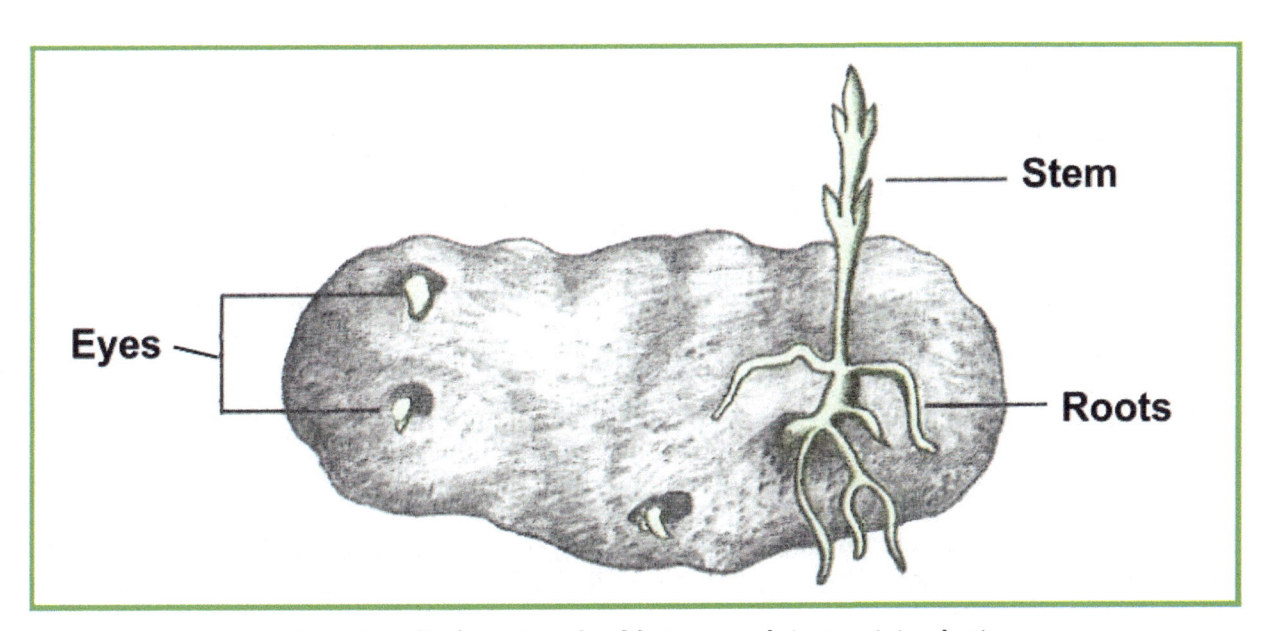

Each "eye" of a potato is able to grow into a potato plant.

177

Two other ways that people propagate plants are by cutting and grafting.

CUTTING. Some plants can be reproduced from a cut-off piece of stem with a few leaves on it. The piece of stem is called a cutting. Ivy and geraniums are plants that you can propagate from stem cuttings.

GRAFTING. You may recall this term from Unit 2 Lesson 6. Plant growers use grafting to produce large numbers of plants with certain traits. In grafting, a stem cutting with leaves on it is taken from one plant. The cutting is joined to the stem of a related plant.

By grafting, the good features of both plants are combined. The first plant may produce beautiful flowers. The second plant may have strong roots.

So grafting is a way to reproduce large numbers of plants with selected traits more quickly than by planting seeds.

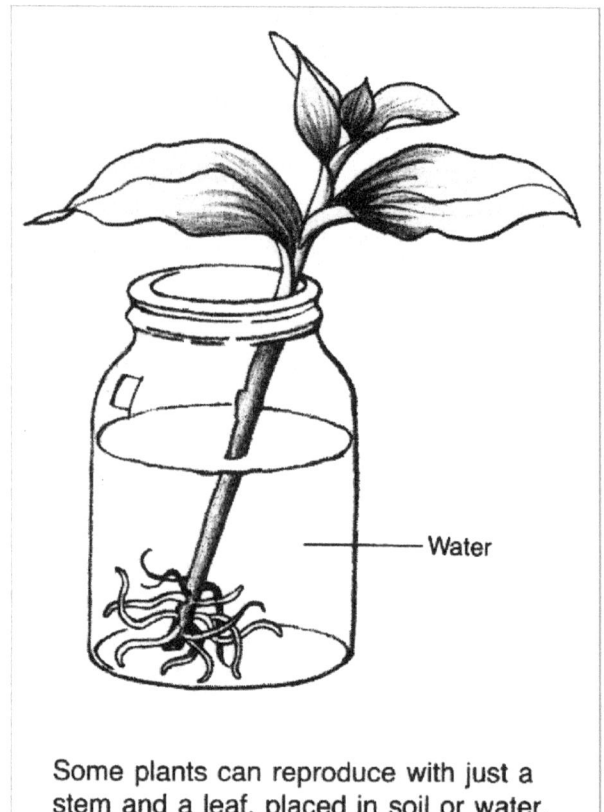

Some plants can reproduce with just a stem and a leaf, placed in soil or water.

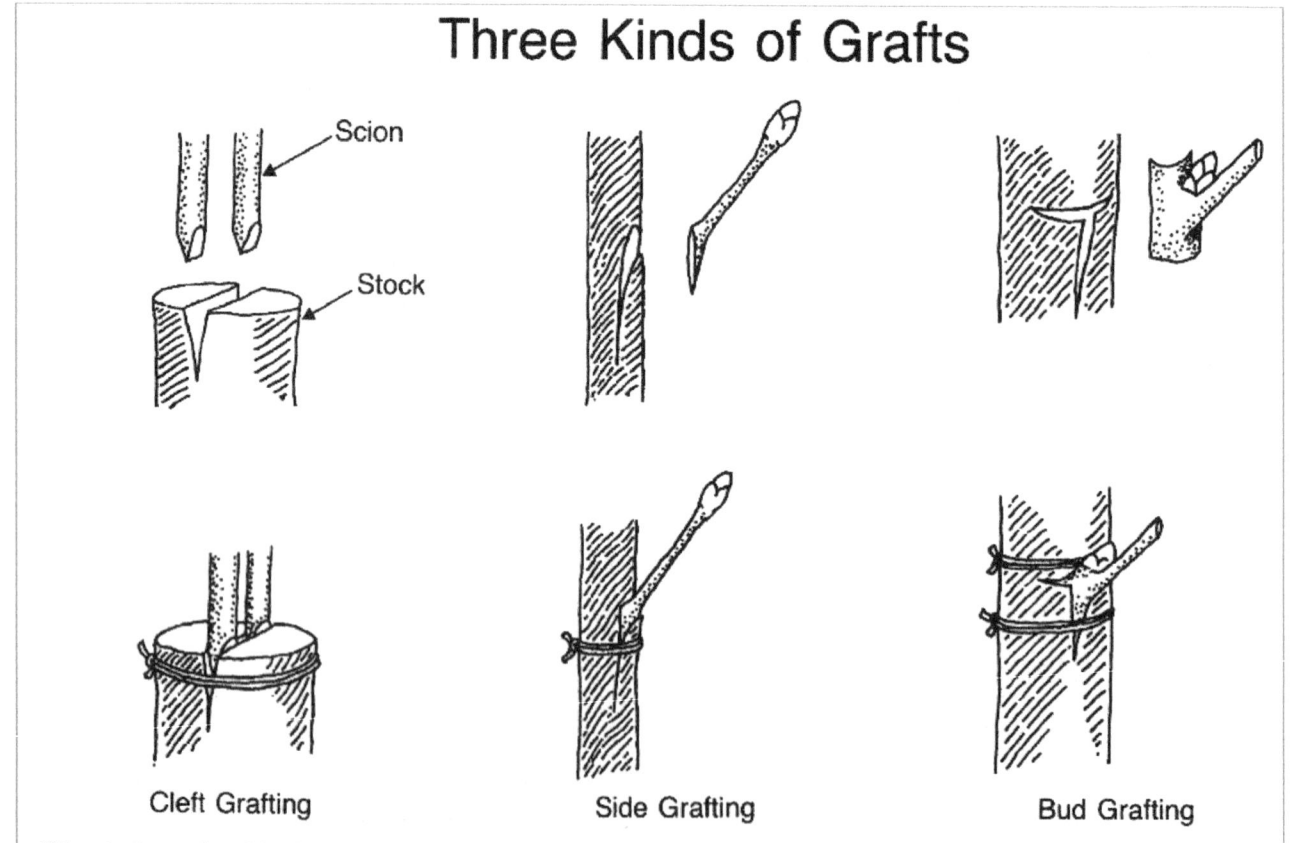

Three Kinds of Grafts

Cleft Grafting

Side Grafting

Bud Grafting

The twig or bud being grafted is called the *scion* (SEYE-un). The rooted plant is called the stock. After the graft is made, the scion is tied to the stock. Soft wax is then placed over the cut to prevent infection.

> **To Do Yourself**   How Can Tubers Be Used To Reproduce Plants?

You will need:

Plastic knife; toothpicks, 2 white potatoes, water, paper cups

1. Place toothpicks around the middle of one of the white potatoes. Place it in a cup of water.
2. Cut the second potato into sections.
3. Place toothpicks into each section.
4. Suspend each section in a cup of water.
5. Label each cup and place it in a dark, warm place to grow.
6. After a week or two, observe the whole potato and each section of potato. Record the number of "eyes" that have changed on each potato.

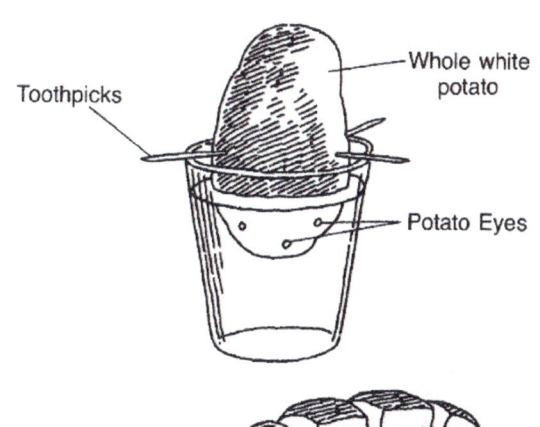

Questions

1. On which potato sections did a new potato plant begin to grow? _____
2. Why would a farmer cut potatoes into sections and plant them? _____

REVIEW — U-7 L-4

I. In each blank, write the word that fits best. Choose from the words below.

   cutting   asexual   sexual   grafting   tuber   bulb   spore

Vegetative propagation is a kind of _____ reproduction. An underground stem with fleshy leaves is a _____. An underground stem with buds is a _____. A piece of stem or leaf may be a _____. In _____ a branch of one plant is joined to the stem of another.

II. Circle the word (between the brackets) that makes each statement correct.

  A. Potatoes can be reproduced from [tubers / bulbs].

  B. A stem cutting is a way to propagate [onions / ivy].

  C. Joining an orange tree branch to a lemon tree stem is a kind of [cutting / grafting].

  D. A bulb is a way to reproduce [onions / geraniums].

III. The thick part of a carrot is its main root. If planted, it can grow into a new carrot plant. Is this a kind of vegetative propagation? Explain.

What Is the Work of a Flower? U-7 L-5

Exploring Science / Historical Steps

A Risky Move. Around 1900, a big change came to New Zealand (an island nation near Australia). Without knowing it, some New Zealanders risked seriously damaging their nation's ecosystem.

A bumblebee visits a red clover flower

Red clover makes very good hay, but there was no red clover in New Zealand. Sheep need large amounts of good hay for their feeding. So sheep ranchers imported red clover seeds from Europe and planted them.

The seeds grew into healthy clover plants with sweet-smelling purple-red flowers. The clover crop was a big success. However, the second year, there was no red clover. In other parts of the world, red clover formed plenty of seeds. These seeds drop to the ground and grow into next year's red clover crop. Why didn't this happen in New Zealand?

The ranchers soon solved the mystery. There were no bumblebees in New Zealand! The ranchers imported and planted more seeds, but they also imported and released 100 bumblebees. With help from the bees, the clover plants made lots of seeds. The clover grew richly, year after year. The bees also reproduced. Soon there were thousands of bumblebees.

What do bees have to do with seed-making in clover? When a bee visits a clover flower, the bee picks up tiny grains of **pollen** (POL-un). The bee also drops off pollen that it picked up from other clover flowers. This transfer of pollen from flower to flower must happen before clover seeds can form.

Losing the bet. In New Zealand, the ecosystem seems to have adjusted with little damage. But it is always a big gamble to introduce a new organism to an area. Today, around the globe, there are many cases where the results of similar decisions (or accidents) have been disastrous.

A row of huge Asian honeysuckle bushes at the base of trees in Louisville, KY

A key example in the U.S. is the **Asian bush honeysuckle** (of the genus *Lonicera*). These tall bushes were first sold in America to decorate lawns and to control erosion. Today they are found in much of eastern and central U.S.

Since they prefer well lit areas, they often take over the edge of woody areas. They aggressively crowd out many **native plants** (plants that grow naturally in an area). As they spread, they use up more and more of the nutrients in the soil. Eventually even nearby trees are weakened.

➤ The "burning bush" (*Euonymus alatus*), also called "winged euonymus," has leaves that turn a beautiful red in the fall. It has been widely planted in cities and suburbs across the United States. How do you think it "moves" out of suburbs into nearby woods?

Burning bushes that have escaped from lawns to woods

180

The Work of a Flower

Like the red clover, most plants with flowers reproduce by seeds. Reproduction by seeds is a kind of sexual reproduction. For a seed to be made, a female sex cell and a male sex cell must join. A female sex cell is called an **egg**. A male sex cell is called a **sperm**. The joining together of a sperm cell and egg cell is called **fertilization**. The sperm and eggs in a plant join together in a flower. Note the diagram below as you read parts A, B, and C below.

[A] THE PARTS OF A FLOWER

- **sepals:** Green leaf-like parts are called sepals (SEE-puls). They cover and protect the inner flower parts before it opens. After the flower bud opens, you may see the sepals where the flower is attached to the stem.

- **petals:** The colored parts of the flower are the petals. Like sepals, they help protect the inner parts of the flower. But petals also attract bumblebees and other pollinators.

- **stamens:** Inside the petals are the flower's male parts, called stamens (STAY-muns). Each stamen is a thin stalk (the **filament**) with a knob (the **anther**) at the top. The anthers produce the fine yellowish powder, called **pollen**. Pollen is made of many barely visible pollen grains. Pollen grains contain sperm.

- **pistil:** At the center of the flower is the pistil. This is the flower's female part. The pistil has a sticky top called a **stigma**. It helps trap pollen. At the lower part of the pistil is a bulge called the **ovary** (OH-vuh-ree). Inside of it are the eggs.

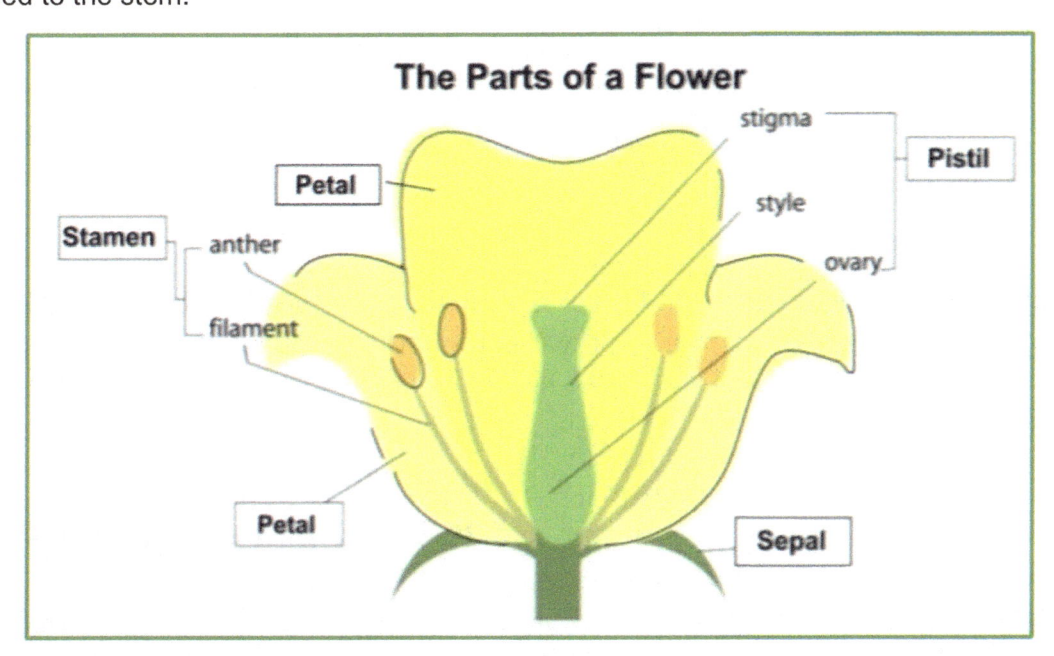

The Parts of a Flower

(Diagram labels: Petal, Stamen, anther, filament, stigma, style, ovary, Pistil, Petal, Sepal)

➤ To Do Yourself   What Are the Parts of a Flower?

You will need: A flower such as a tulip, scissors, hand lens, drawing paper, pencil.

1. Carefully examine your flower. Compare it to the drawing above.
2. Use the hand lens to observe the stamens and find the pollen.
3. Use the scissors to cut off some of the petals. Locate the pistil. Then carefully cut through the ovary at the base of the pistil. Observe the developing seeds.
4. Make a drawing of your flower and label the parts that you observe.

Questions

1. What are the male parts of your flower? _____
2. What are the female parts of your flower? _____
3. What part of your flower attracts insects for pollination? _____

[B] POLLINATION Remember the bees and the clover? Bees, other insects, birds, or the wind may carry pollen from flower to flower. The way that pollen is carried depends on the kind of plant. The transfer of pollen from the male (anther) to the female (stigma) is called **pollination** (pol-uh-NAY-shun).

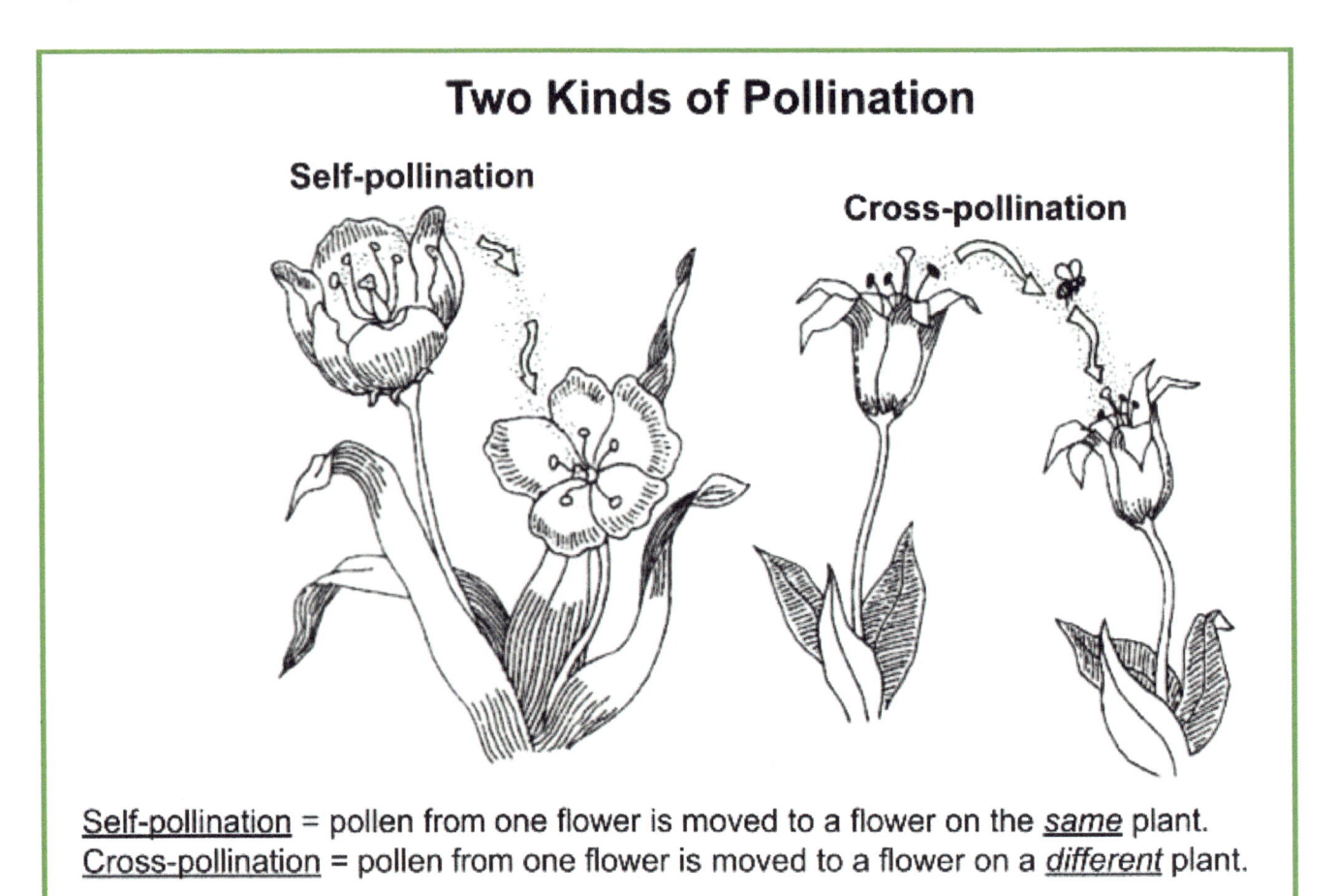

Two Kinds of Pollination

Self-pollination

Cross-pollination

<u>Self-pollination</u> = pollen from one flower is moved to a flower on the <u>same</u> plant.
<u>Cross-pollination</u> = pollen from one flower is moved to a flower on a <u>different</u> plant.

[C] FERTILIZATION DETAILS. When a pollen grain lands on the stigma, the grain starts to grow. It grows a long tube down through the pistil. The tube grows until it reaches the ovary. Finally, a sperm that was inside the pollen tube joins with an egg inside the ovary. In other words, **fertilization** occurs.

After fertilization, the flower can form seeds. Soon, young plants start to form inside of the seeds.

What about plants that are GMOs, such as the genetically modified American Chestnut tree mentioned in the last lesson? Do these plants make their own seeds? The answer is yes. And, of course, if these plants are allowed to produce flowers, and then seeds, they will likely spread in the wild.

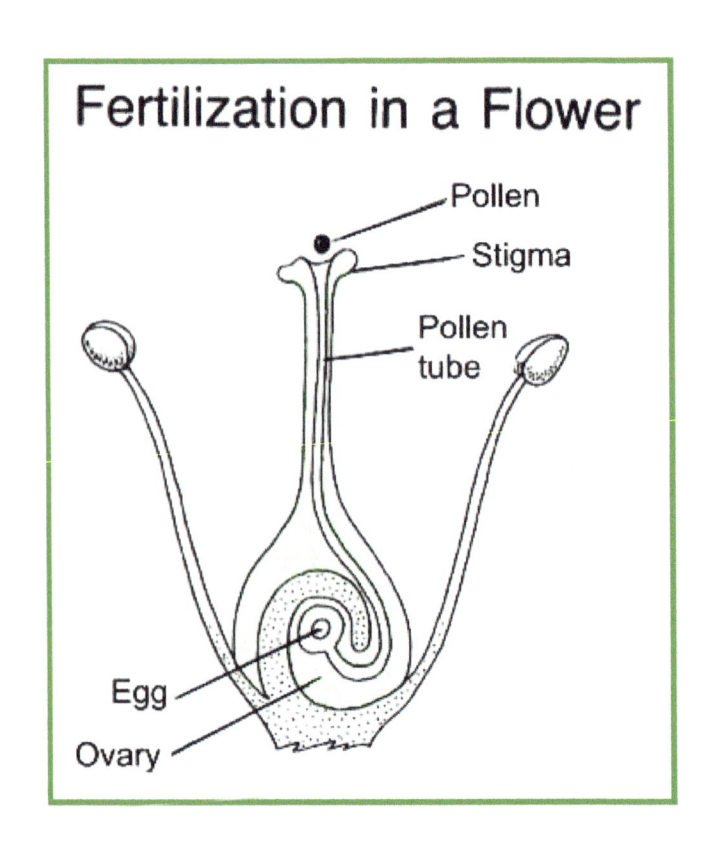

Fertilization in a Flower

- Pollen
- Stigma
- Pollen tube
- Egg
- Ovary

REVIEW — U-7 L-5

I. Match the flower part in column **A** with its description in column **B**.

A	B
1. ____ stamen	a. they attract insects and help protect the inner parts
2. ____ pistil	b. its top portion (anther) produces pollen
3. ____ petals	c. it has an ovary at its lower end and a stigma at the top
4. ____ sepals	d. they cover the petals before they open

II. Circle the word (between the brackets) that makes each statement correct.

A. Egg cells are found inside the [pollen / ovary].

B. Moving pollen from a stamen to a pistil is [propagation / pollination].

C. Male sex cells are called [sperm / egg] cells.

D. The joining of a male cell and a female cell is called [fertilization / propagation].

III. Corn flowers are pollinated by the wind rather than by insects. Which parts - petals or ovaries - are missing in the female flowers of corn?

_____

IV. In corn plants the male flowers are often called tassels, and the female flowers are often called silks. Tassels must therefore have what parts? Silks must have what parts?

Tassels have _____

Silks have _____

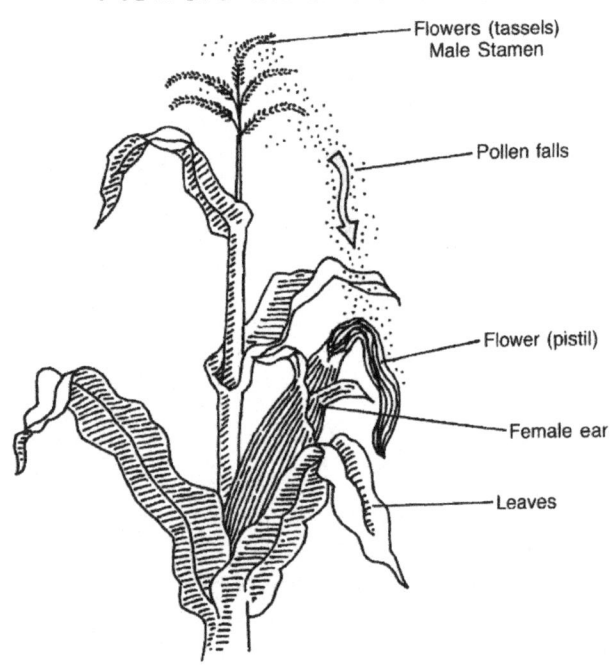

Flowers on a Corn Plant

- Flowers (tassels) Male Stamen
- Pollen falls
- Flower (pistil)
- Female ear
- Leaves

183

What Are Seeds and Fruits? U-7 L-6

Exploring Science/ Historical Steps

Alive After 1,000 Years? A thousand years ago in China, seeds from the sacred lotus were buried in mud. The mud dried. Without water, the seeds did not sprout. These seeds had coats that were rock-hard. They stayed buried for 1,000 years. Then they were found by scientists. The scientists used a special test to find out how old the seeds were. Later, they put the ancient lotus seeds in a museum.

Then the scientists had an idea. They knew that the seeds of some plants stay alive only for days or weeks, but others survive for many years. Could the ancient seeds with the "rock" coats still be alive? They decided to find out.

First, the scientists softened the hard seed coats with an acid. Then they planted the seeds. The seeds sprouted! Even more amazing, the new lotus plants bloomed. On July 29, 1952, they made beautiful pink flowers!

Each year since 1952, the lotuses have continued to bloom. Now, each July, visitors to the Kenilworth Aquatic Gardens in Washington, D.C. get a special treat. They see the blossoms of lotuses whose seeds lived for 1,000 years.

The lotus flower is from a seed that was buried for a thousand years.

➤ Wheat seeds have much softer seed coats than lotus seeds. Decide which claim below is more likely to be true, and explain your answer:

A. Lotus seeds from Japan that were 3,000 years old were sprouted.
B. Wheat seeds from Egypt that were 3,000 years old were sprouted.

Seeds and Fruits

The story of the 1,000-year-old seeds starts with the lotus flowers that produced them. After the flowers were pollinated, they lost their petals. Inside of each flower's ovary, a number of eggs were fertilized by sperm that were in the pollen. Then the ovary began to grow. It grew until it became a large seed pod. Each section of the pod contained a seed.

Just what is a seed? After a flower's egg is fertilized, it develops into a tiny young plant called an **embryo** (EM-bree-oh). At the same time, other cells within the ovary form a food supply for the embryo. Around the seed, a **seed coat** is formed. The seed coat protects the embryo and its food during its resting stage. Together, the embryo, food, and seed coat are called a **seed**.

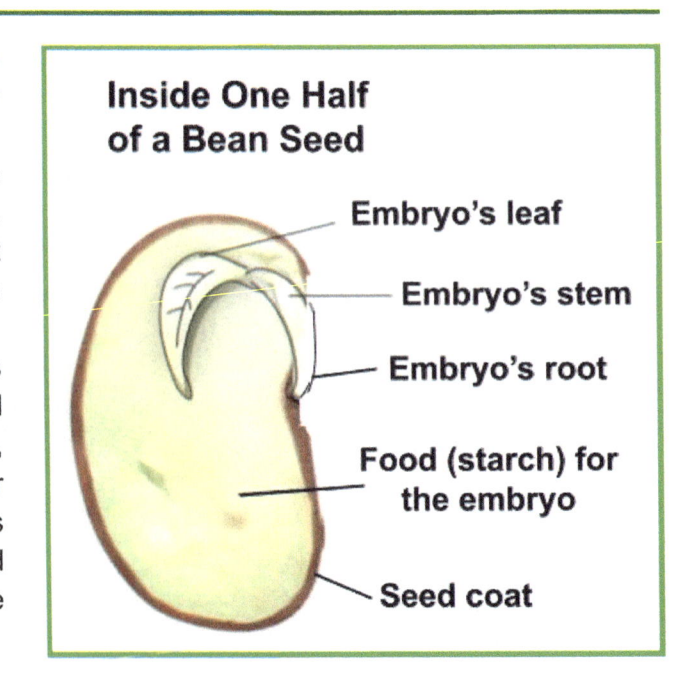

Inside One Half of a Bean Seed
- Embryo's leaf
- Embryo's stem
- Embryo's root
- Food (starch) for the embryo
- Seed coat

As you just read, the resting stage of a seed may last for a short time, or a very long time. This depends on the kind of seed it is. It also depends on whether conditions are right for the embryo to start growing.

Remember the seed pod of the lotus? It may come as a surprise to you that a **seed pod** is the plant's fruit. A **fruit** is simply a ripened, enlarged ovary that contains seeds. You have eaten many kinds of fruits. Think about what you find when you bite into a fruit such as a peach or an apple. There is a tasty fleshy part, and there are one or more seeds.

What about vegetables? Foods that people call vegetables or nuts are - to a scientist - simply fruits. Some fruits are fleshy, and some are dry. Tomatoes, cucumbers, and squash are fleshy fruits. Walnuts and acorns are dry fruits.

All fruits, fleshy or dry, contain seeds. Seeds usually have a better chance of growing if they are some distance away from the parent plant. Can you guess why?

The fruit not only protects, but helps to scatter the seeds. How? Animals (like us) eat the fleshy parts of fruits and throw away the seeds. Many animals eat fruits "seeds and all." These seeds often survive the animal's digestive system, and then pass out with the feces. A bird may release seeds far from the parent plant. Look carefully at some bird droppings and you will often see small seeds!

The wind also carries some seeds away from their source. The dry fruits often help this happen. Maple trees have two seeds, each surrounded by a wing; you likely call these "helicopters." The dry fruits of dandelions are carried by fluffy "parachutes." Other dry fruits, such as the burdock, have hooks that stick to fur or clothing. Have you ever helped to scatter seeds that stuck to your clothes?

Plants make many more seeds than they need. Most will never grow into new plants. Many will die and decay. Many will be eaten, providing nutrients to animals. But making and scattering *many* seeds increases the odds that some of them will land where conditions for growth are good.

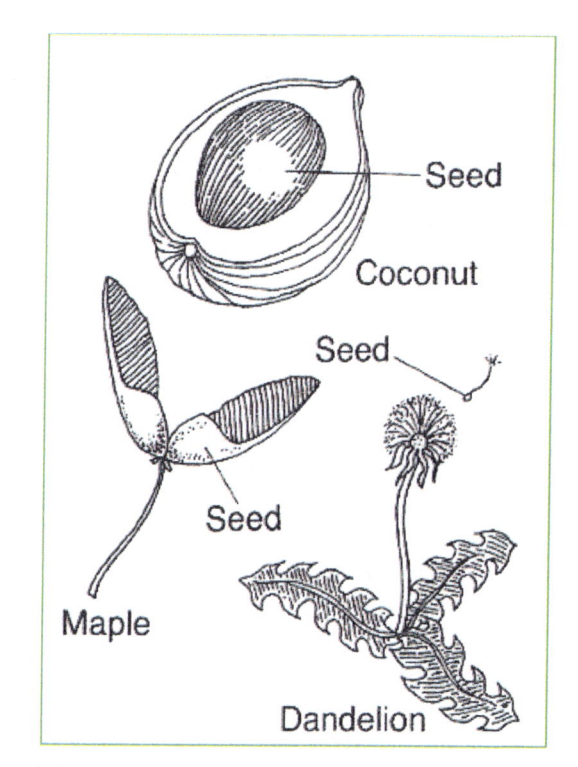

The dry fruits of maple trees and dandelions help to carry their seeds in the wind. Coconuts float, and are carried by water; the outer case rots and releases the coconut seed.

If oxygen, warmth and moisture are present, the seed can sprout (**germinate**). Moisture softens the seed coat, and the embryo begins to grow. Its root will emerge first to anchor the plant and to draw in water and minerals. Next, the young stem will rise; often it brings with it the food supply from the seed. Finally, the first leaves will open. [Note: Seeds do NOT need light or soil to sprout, but the seedling will soon need both in order to survive.]

When the new plant produces its own flowers, the cycle of reproduction and growth starts over.

How a Flower Becomes a Seed

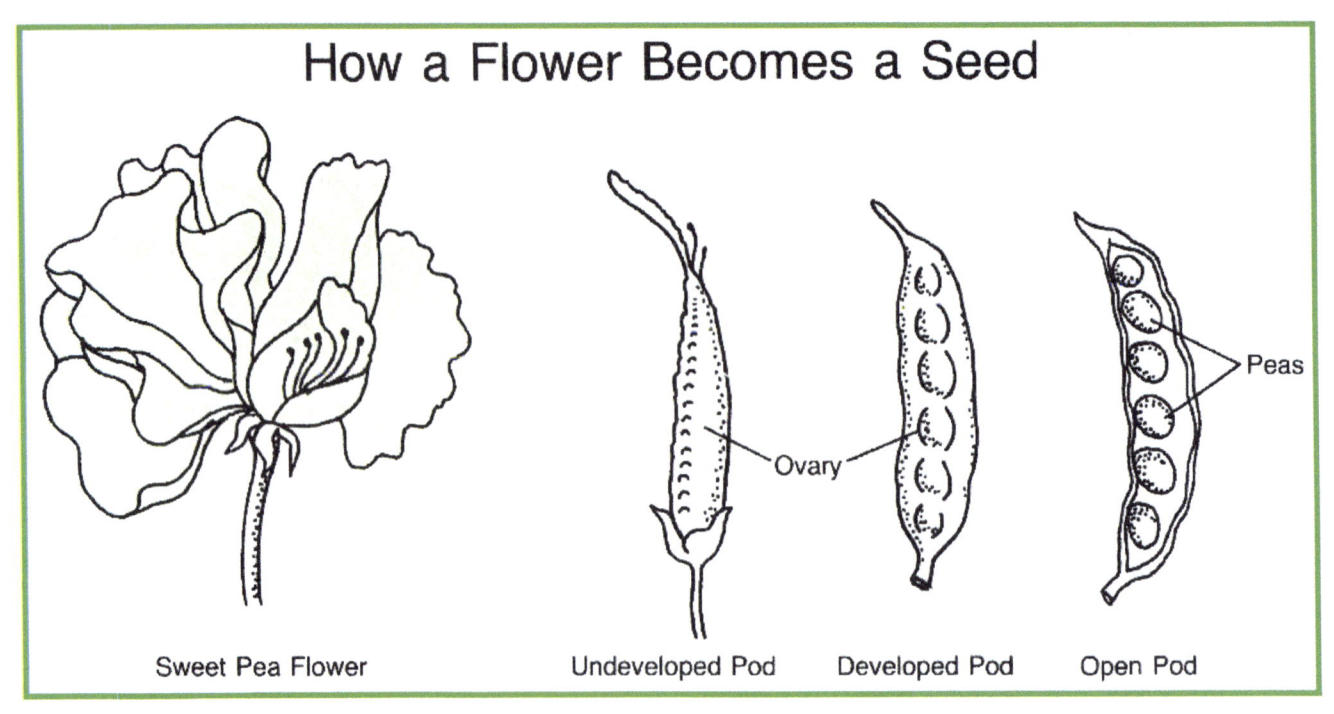

Sweet Pea Flower — Undeveloped Pod — Developed Pod — Open Pod — Ovary — Peas

After pollination and fertilization, the ovary becomes the pod (fruit). The mature pod contains seeds.

➤ To Do Yourself   How Can You Find the Seeds of Plants?

You will need:

Dry fruits (such as lima beans, peas, peanuts in the shell); fleshy fruits (such as tomatoes, oranges, apples, grapefruits, grapes, peppers); water; hand lens; iodine; opened paper clip.

1. Soak the lima bean and pea in water overnight. Carefully remove the seed coats and open the two halves of these seeds. Use a hand lens to examine and identify the parts. For help, look at the bottom right illustration on the first page of this lesson.
2. With the paper clip, scratch the flat surfaces of these seeds.
3. Carefully add a drop or two of iodine solution to each scratched area. Record what you observe. Draw it.
4. Open the peanut shell and observe it. Compare it to the bean and pea. Draw these.
5. Open the fleshy fruits. Locate and collect the seeds. Observe and draw them.

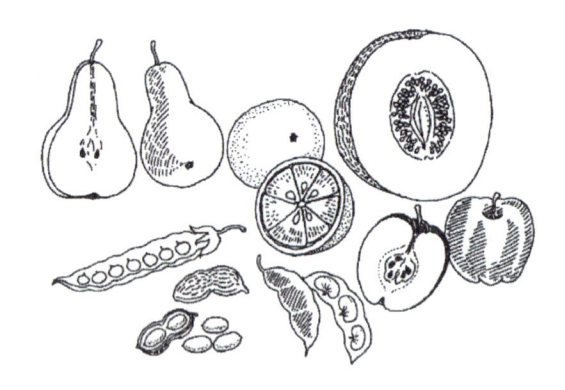

Questions

1. When iodine touches starch, a very dark color appears. Where in a seed was starch located?

_____

2. What part of the seed *uses* this starch? _____

3. What do all of the fruits have in common? _____

186

REVIEW — U-7 L-6

I. In each blank, write the word that fits best. Choose from the words below.

ovary	fruits	animals	wind
eggs	stamens	embryo	seed coat

A fertilized egg develops into an _____. Pea pods and tomatoes are kinds of _____. An embryo and its food supply are covered by a _____. A fruit is a ripened _____.

Seeds can be scattered by _____ or by _____.

II. Circle the word (between the brackets) that makes each statement true.

  A. An example of a fleshy fruit is [squash / peanut].
  B. A tree that makes seeds that travel by the wind is the [maple / cherry].
  C. After a flower is pollinated, it loses its [ovary / petals].

III. Seed coats are often so tough that an animal's digestive system cannot break them down. How does this help some plants to scatter seeds?

_____

IV. Pines and other conifers have cones instead of flowers. Some cones are male (and make pollen) while others are female (and make eggs - and eventually seeds). Male cones are small and usually grow in clusters. Female cones are larger than male cones, and usually grow alone. In the drawings below, write "male" or "female" under the matching drawing.

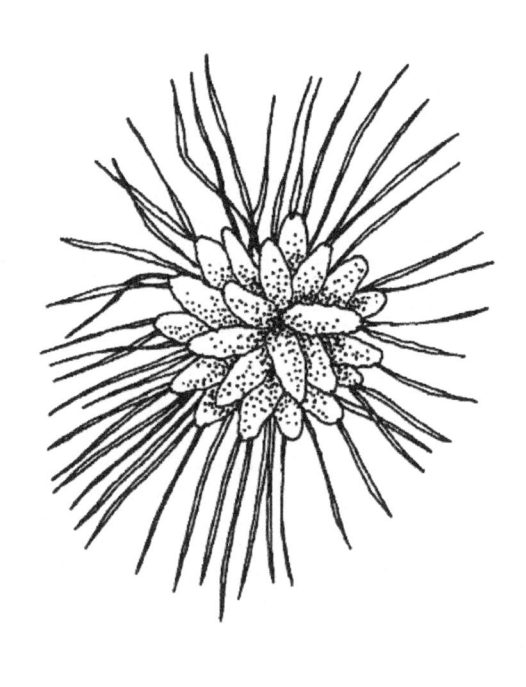

_____   _____

Unit 7 -- Review What You Know ---

A. Use the clues below to complete the crossword.

Across
1. A sperm _____ an egg.
4. A tulip or garlic
5. Tiny cell of a mold
7. Can grow into a whole potato
9. Found in a fruit
11. Root-like stem
12. Reproduction from one parent

Down
1. How one paramecium becomes two
2. Tiny plant inside seed
3. To transfer pollen
6. Sex cells found in ovaries
8. Reproduction from two parents
10. A sperm is a male sex _____

B. Write the word (or words) that best completes each statement.

1. Two daughter cells of the same size result from
   a. budding   b. fission   c. grafting   1. _____

2. The black dots in bread mold are its
   a. stems   b. roots   c. spore cases   2. _____

3. The regrowth of a lizard's tail is
   a. reproduction   b. regeneration   c. propagation   3. _____

4. Onions reproduce in nature from
   a. bulbs   b. stems   c. tubers   4. _____

5. Pollen grains contain cells called
   a. eggs   b. buds   c. sperm   5. _____

6. A male flower part is a
   a. stamen   b. pistil   c. sepal   6. _____

7. Vegetative propagation can be used to reproduce
   a. worms   b. starfish   c. geraniums   7. _____

8. An example of a fruit is a
   a. mushroom   b. tomato   c. potato   8. _____

9. A flower's female part is a
   a. petal   b. pistil   c. stamen   9. _____

188

10. A seed contains a food supply and a(n)
   a. egg   **b.** embryo   **c.** sperm   10. _____

11. Asexual reproduction can result from
   a. pollination   **b.** regeneration   **c.** fertilization   11. _____

12. A mold reproduces by
   a. buds   **b.** spores   **c.** seeds   12. _____

13. Potatoes reproduce in nature from
   a. tubers   **b.** grafts   **c.** bulbs   13. _____

C. Apply What You Know

1. On the line below each picture, write the term that best describes the picture. Choose from these terms:

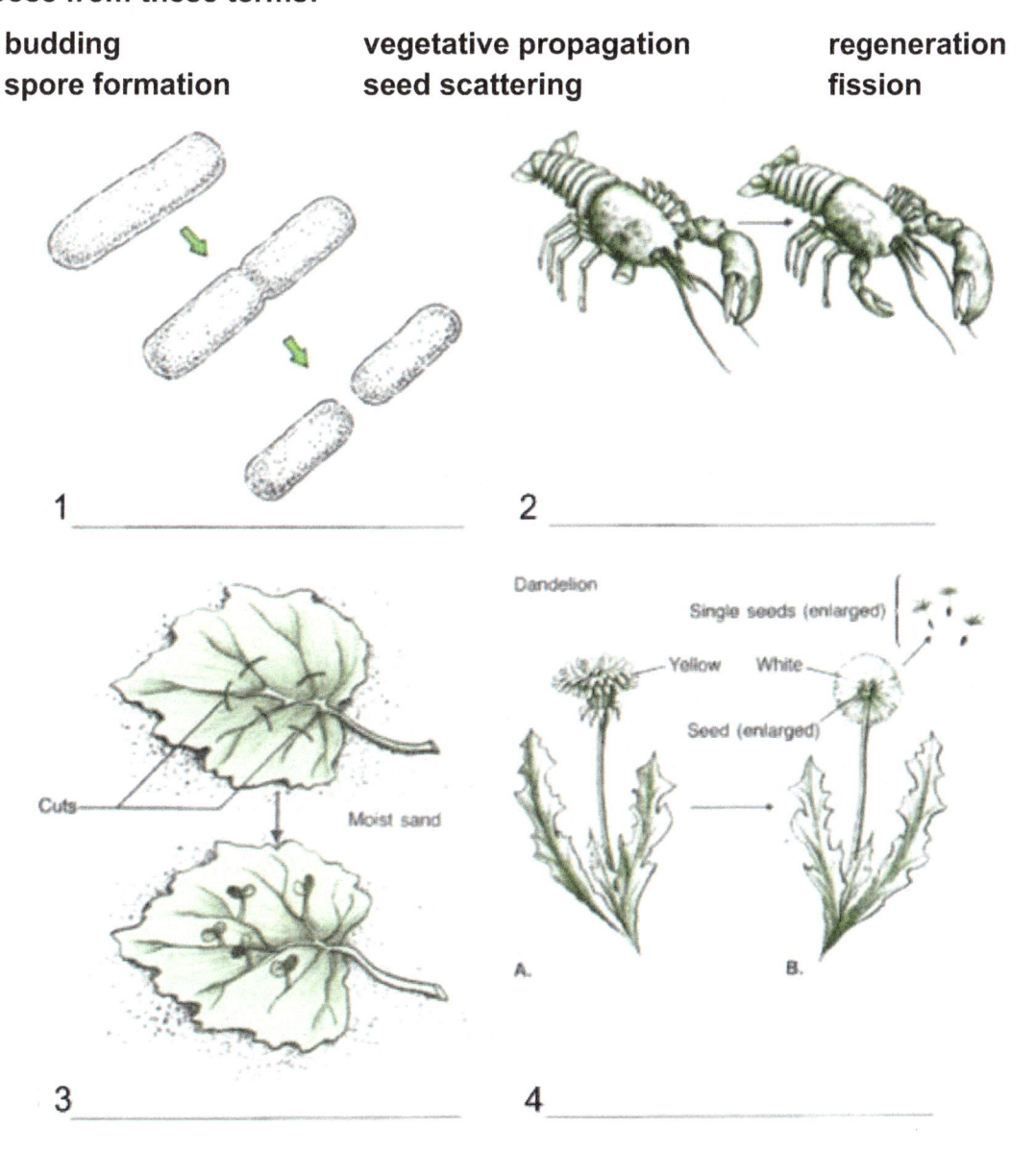

   budding   vegetative propagation   regeneration
   spore formation   seed scattering   fission

1 _____   2 _____

3 _____   4 _____

189

2. Each statement below refers to one of the pictures on the previous page. The number of the statement is the same as the number of the picture. Circle the word (or words) from within the brackets that makes each statement true.

    a. The offspring of a bacterium are called [buds / daughter cells].

    b. a broken-off claw of a lobster [can / cannot] grow into a whole new lobster.

    c. Young begonias grow from [cuttings / tubers].

    d. When a dandelion turns from yellow to white, it loses its [petals / embryos].

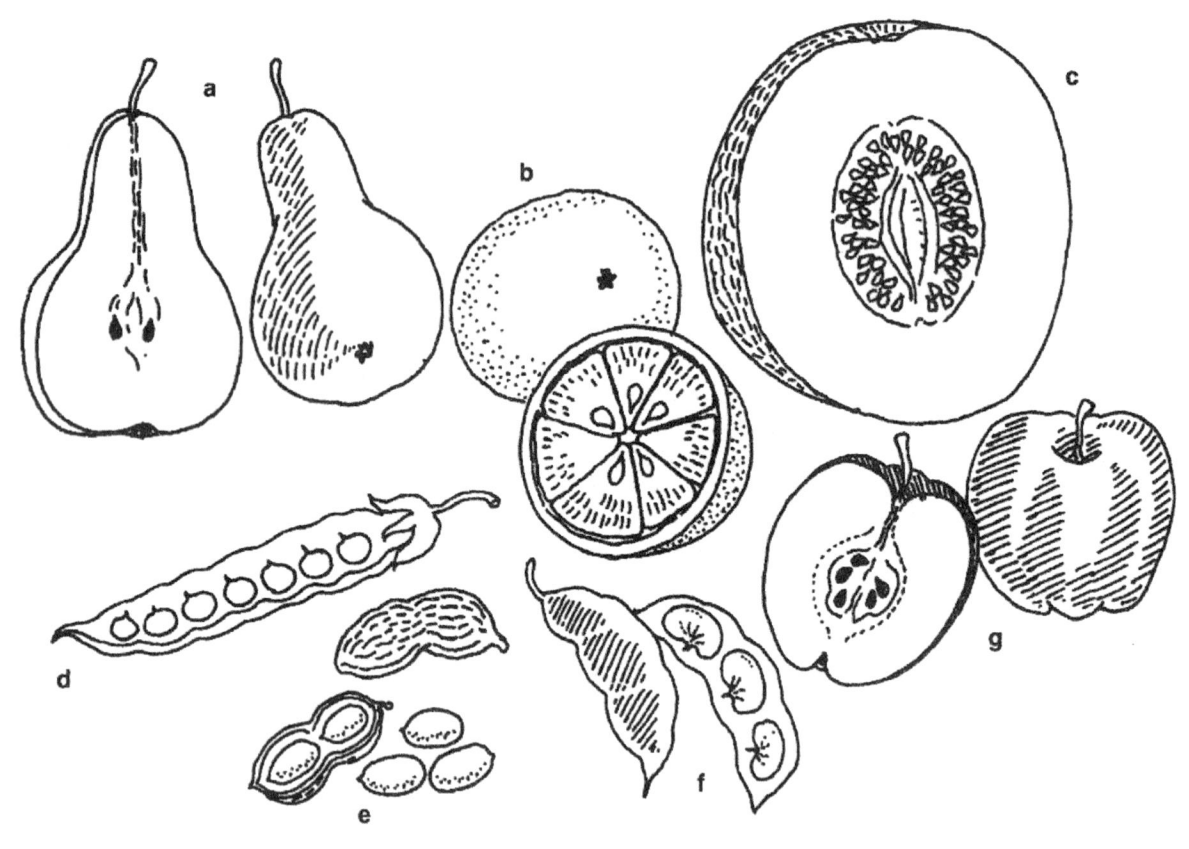

3. Identify the fruits above.

    a. _____   e. _____

    b. _____   f. _____

    c. _____   g. _____

    d. _____

D. Find out more.

1. Make a seed collection. Get real seeds from nature or from foods. You can use pictures of some. Make a display of your collection. Label each seed to tell how it is scattered. (Don't forget coconuts).

2. Botanists - scientists who study plants - consider all "vegetables" that contain seeds to be fruits. But on grocery store shelves, not all fruits contain seeds! Find out how seedless oranges and grapes first came to be. How are they propagated?

SUMMING UP:
Review What You Have Learned So Far

A. On the drawing of a plant, write labels for each numbered part. Choose from these terms:

stem   root   leaf   flower

Each statement below refers to one of the plant parts shown. The number of the statement is the same as the number of the part. Circle the word (or words) between the brackets that makes each statement true.

1. After fertilization, the [ovary / stamen] ripens into a fruit.
2. Plant cells that make food give off [carbon dioxide / oxygen] during the daylight.
3. A tissue that can store food is [pith / cambium].
4. Water moves upwards through [phloem / xylem] tissue.

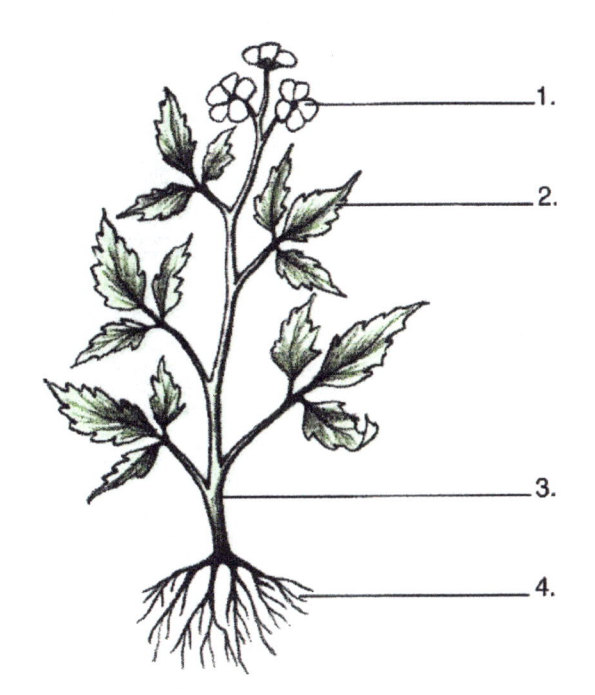

B. On the drawing of the body, write labels for each numbered part. Choose from these terms:

heart   small intestine
thyroid gland   cerebrum

The number of the statements below is the same as the number of the parts in the illustration. Circle the word (or words) between the brackets that makes each statement true.

1. The brain's largest part contains the control center for [thought / breathing].
2. Too little of the hormone thyroxin may cause [tiredness / diabetes].
3. Blood in the right atrium is [high / how] in oxygen.
4. Enzymes digest proteins by breaking them into [fatty / amino] acids.

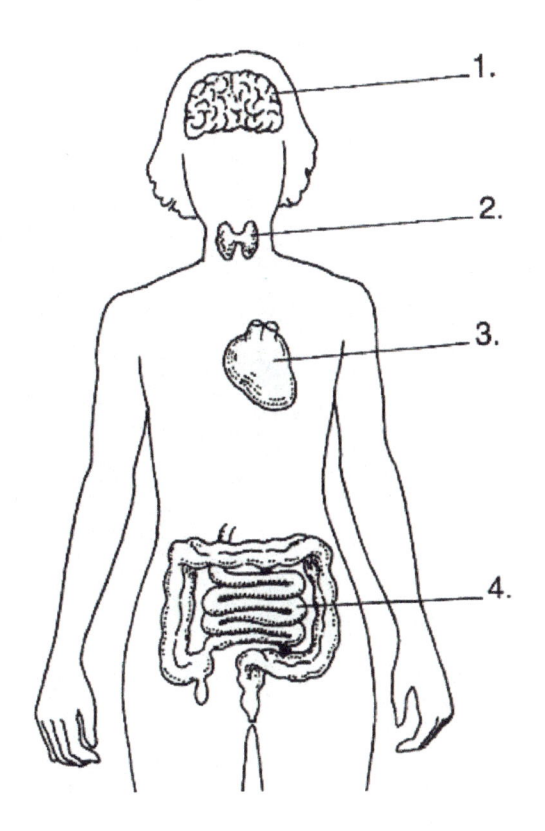

191

UNIT EIGHT
REPRODUCTION IN HIGHER ANIMALS

How Do Animals Reproduce Sexually? U-8 L-1

Exploring Science / Historical Steps

Dr. Just and the Sea Urchins. Look at the photo at the right. Surprisingly, the "pin cushion" in the picture is an animal. It's a sea urchin, a spiny-skinned relative of the starfish.

Much of what scientists know about how an egg changes when it is fertilized by sperm has been learned from sea urchins. Early in the 1900s, a pioneer researcher in this area was a young black man - **Ernest Just**. He had just completed his PhD. Sad to say, because of his skin color, American colleges would not hire him!

The determined Just moved to Europe. He became an expert on fertilization, writing many articles and two books. Scientists knew that the nuclei of a sperm and an egg join at fertilization. But Just discovered that the nuclei needed something else when they joined; they needed cytoplasm. (Recall that cytoplasm is the part of the cell around the nucleus).

The sea urchin looks like an underwater pin cushion. Its mouth is on its underside.

There are still many mysteries about how animals reproduce. Today's scientists are still learning from the lowly sea urchin.

➤ Do sea urchins reproduce sexually or asexually? Explain.

Sexual Reproduction in Animals

Like the sea urchin, most animals reproduce sexually. There are two parents, one male and one female. In animals, the male parent has organs called **testes**, which produce sperm cells. The female parent has organs called **ovaries**, which produce egg cells.

Egg cells of most animals are much larger than sperm cells. Some eggs have a stored food supply, called **yolk** (YOHK). A sperm cell has a head and a whiplike tail. The head is mainly the cell's nucleus. Sperm are released into water or they are produced in a liquid. They use their tails to swim towards eggs. Thousands of sperm may reach one egg, but only one sperm is able to enter it.

When a sperm **fertilizes** an egg, its head enters the egg. The tail is left outside. The sperm nucleus and the egg nucleus join to make one nucleus. This cell, the fertilized egg, is called a **zygote** (ZY-gote).

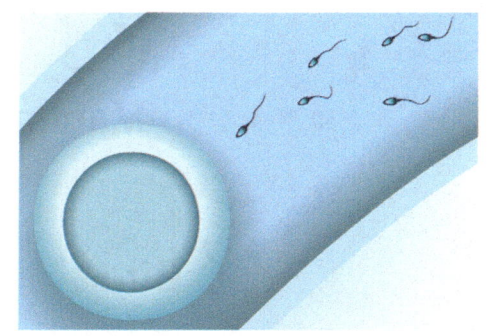

Many sperm cells race to an egg.

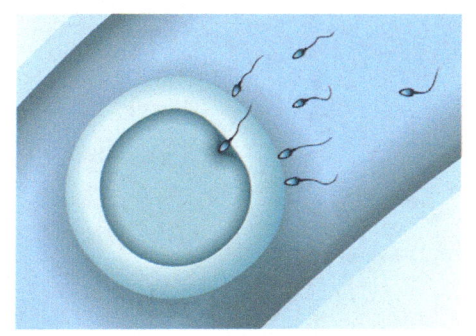

Only a single sperm's head enters the egg.

193

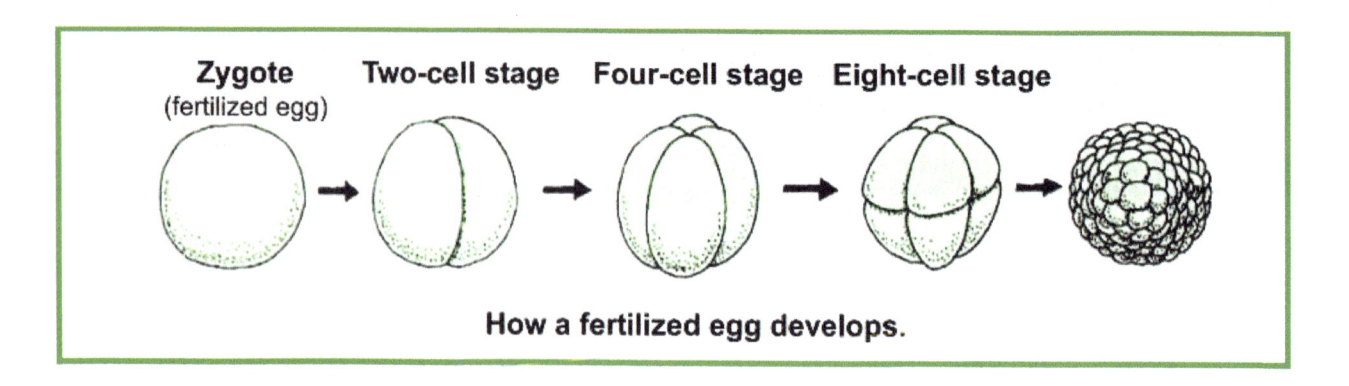

Zygote **Two-cell stage** **Four-cell stage** **Eight-cell stage**
(fertilized egg)

How a fertilized egg develops.

As shown above, the zygote divides into two cells, which do not separate. This two-celled young animal, or **embryo**, divides again, forming 4 cells. These cells then divide forming 8 cells, and so on. Soon, the embryo is a ball of many cells.

Observe the diagram below. The early stages of development of various animals (including humans) look very similar. After the ball of cells is formed, folding and curling occurs. Gradually, the embryos turn into animals that look like their parents.

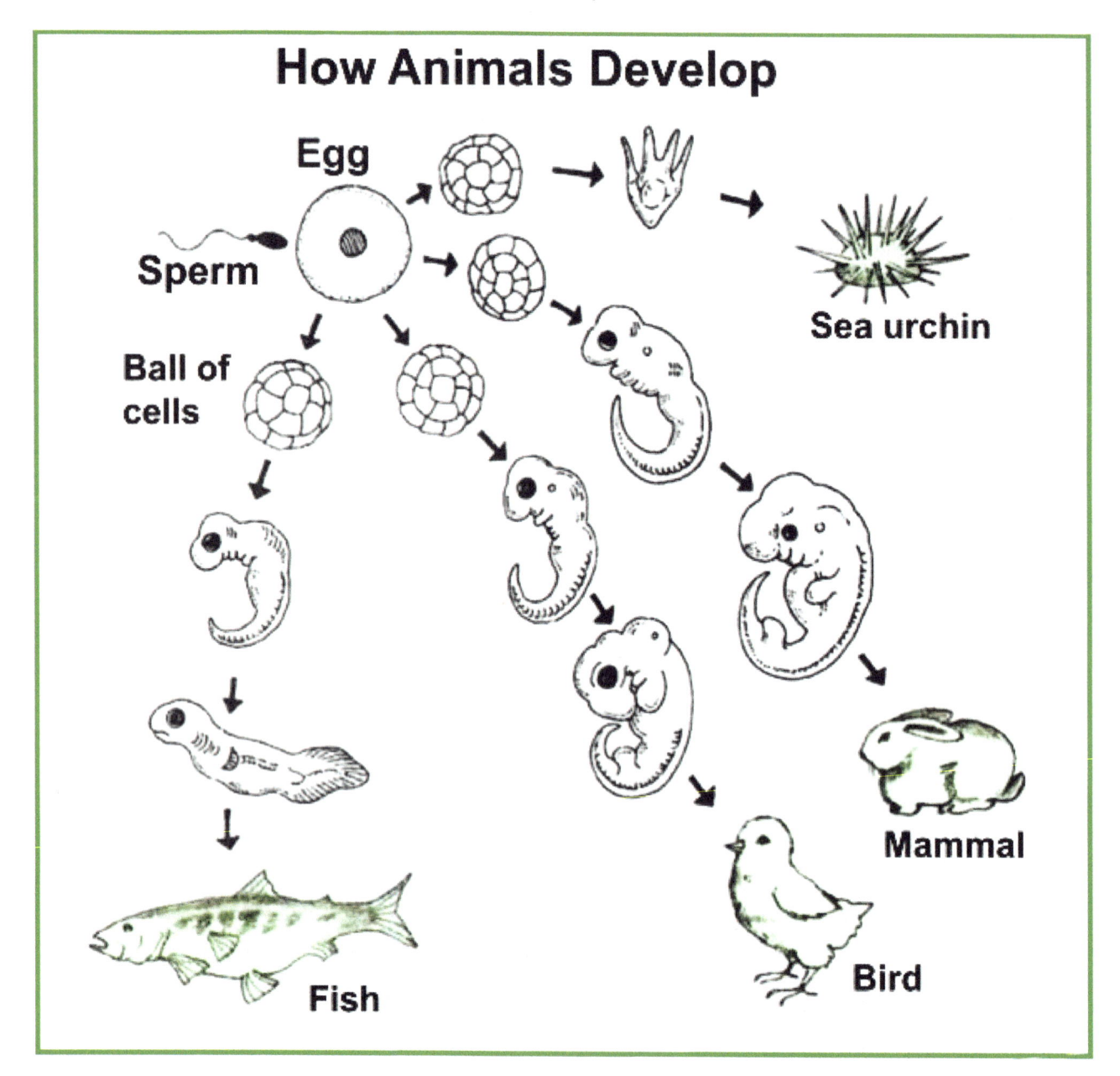

How Animals Develop

Egg → Sea urchin
Sperm
Ball of cells → Fish
→ Mammal
→ Bird

194

> **To Do Yourself** 　　How Do Snails Reproduce?

You will need:

Small aquarium with sand and a few plants, several snails, hand lens, lettuce

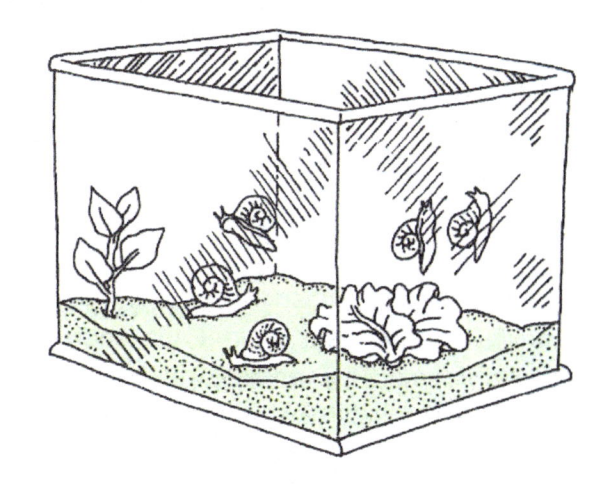

1. Place some snails in an aquarium.
2. Feed your snails some small pieces of lettuce. Remove uneaten portions that begin to decay.
3. Observe your snails to find out when they begin to lay eggs. Carefully observe the aquarium glass with your hand lens.
4. Use the hand lens to observe the eggs' development.

Questions

1. Where did your snails lay their eggs? _____
2. What did the eggs look like? _____
3. How long did it take for the eggs to develop and hatch? _____

---------------------------------- **REVIEW** ---------------------------------- U-8 L-1

I. In each blank, write the word that fits best. Choose from the words below

fertilizes	**divides**	**yolk**	**ovaries**
cytoplasm	**nucleus**	**zygote**	**testes**

The male organs of an animal are the _____. The female organs are the _____. The _____ is a stored food supply. A fertilized egg, or _____, is a single cell that _____ into two cells. The _____ of a sperm cell joins with the nucleus of an egg cell during fertilization.

II. Circle the word (between the brackets) that makes each statement correct.

　　A. The nucleus of a sperm cell is in its [head / tail].

　　B. In fertilization, a [testis / sperm] joins with an egg.

　　C. A ball of cells is a stage of development of the [embryo / ovary].

III. In the drawing on the previous page, which two embryos look the most alike in their later stages? Why do you think this is so?

_____

How Do Fish Reproduce?

U-8 L-2

Exploring Science

Fish That Change Sex. It is easy to see how clownfish got their name. They have brightly colored "costumes" and seem to have sad mouths.

One species of clownfish live in family groups. The family home is near a giant sea anemone (uh-NEM-uh-nee). This animal, related to the jellyfish, looks much like a plant. Its many "stems" are actually "arms" (**tentacles**). The tentacles have "stingers."

A clownfish and an anemone help each other get food. Clownfish do not try to eat the tentacles, and are not harmed by their stingers. In fact, a clownfish will swim in and out of the tentacles and attract other fish into them. Now and then, the tentacles sting and then catch a victim. The anemone then eats most of it. Later, the clownfish eats the parts of the dead fish that float free.

When different types of organisms live together and help each other, this is called a **symbiotic** (sim-be-OT-ik) **relationship**. "Sym" means together, and, of course, "bio" means life. Working together during **symbiosis**, both organisms benefit.

The largest member of the clownfish family is an adult female. She is about five centimeters long. The family also includes an adult male and several "children" - younger and smaller fish.

The sex life of these fish is unusual. If the female is taken away or dies, the male grows larger and turns into a female! Then the largest "child" grows up and becomes an adult male. The family always has just one adult of each gender (sex).

Other kinds of fish sometimes change sex, too. A stoplight fish, as it grows larger and older, changes from female to male. The female is red. What color do you think the male is? That's right - the male is green!

Why do some fish change sex? In each case, and in different ways, the change helps these fish to have more young. You will soon see why fish produce so many young.

➤ You have learned that eggs are larger than sperm. In a fish couple, why might it be better for the female to be larger than the male?

Clownfish are not harmed by the poisonous sting of the sea anemone.

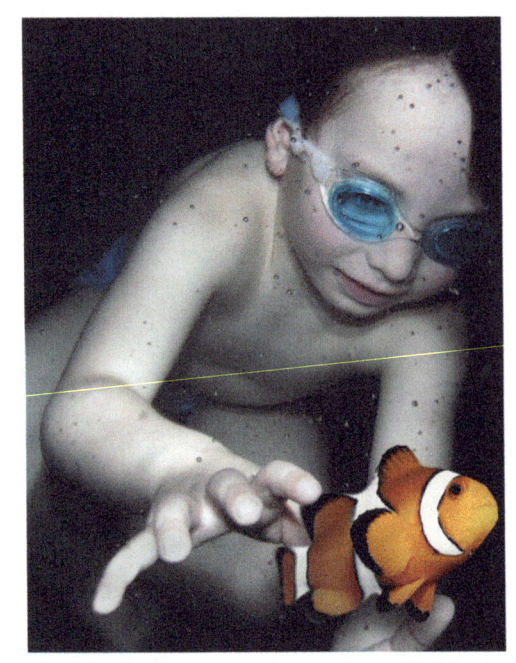

This photo illustrates the fish's size (though it is best not to touch a fish).

The Life Cycle of a Fish

The fish are a group of vertebrates - animals with backbones. They spend their lives in the water. A fish breathes through its **gills**. Water that goes into a fish's mouth goes out through its gills. In the gills, oxygen from the water passes into the fish's blood. Carbon dioxide passes from the blood into the water.

Most species of fish, such as tuna or trout, have the same gender all of their life. In a male fish, the testes produce sperm. From each testis, a **sperm tube** carries the sperm out of the body. In a female, the ovaries produce eggs. From each ovary, an **egg tube** carries the eggs out of the body.

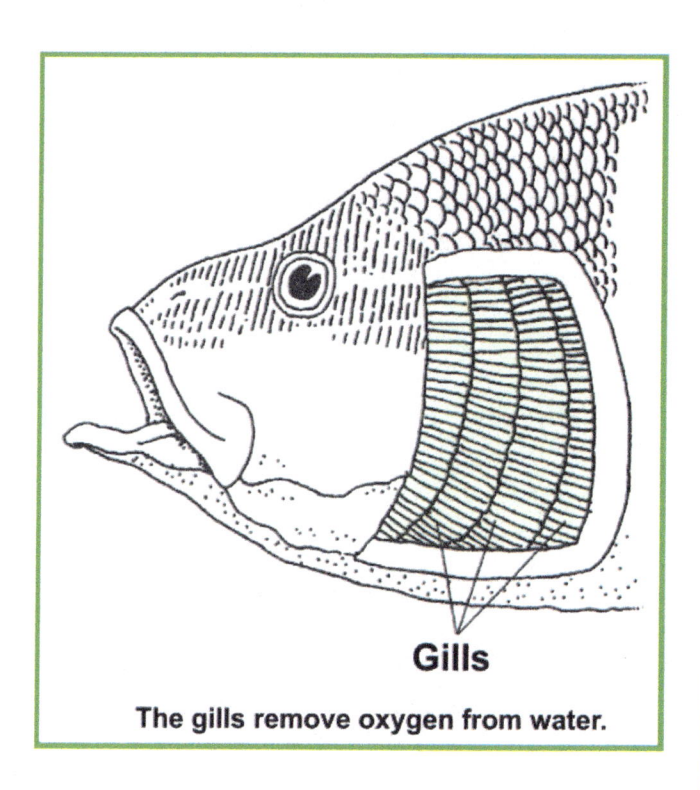

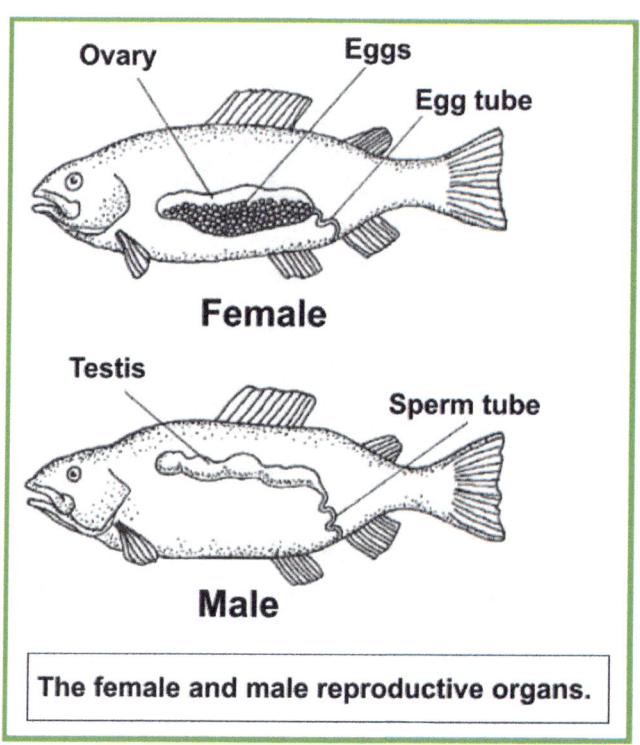

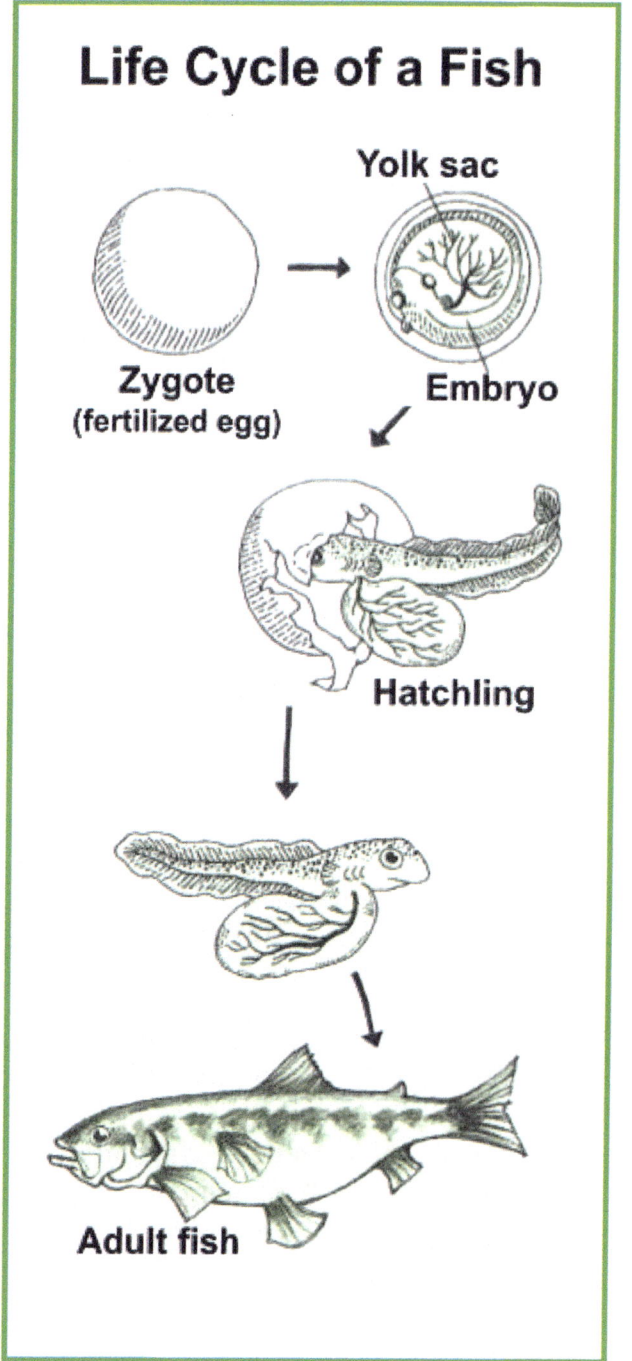

Gills

The gills remove oxygen from water.

Female
Ovary, Eggs, Egg tube

Male
Testis, Sperm tube

The female and male reproductive organs.

Life Cycle of a Fish

Zygote (fertilized egg) → Embryo (Yolk sac) → Hatchling → Adult fish

When most fish mate, the female lays a mass of jelly-coated eggs in the water. The male releases sperm over the eggs. Then fertilization takes place. This is called **external** (ex-STUR-nul) **fertilization**, because the eggs are fertilized outside of the female's body.

The fertilized eggs, or **zygotes**, develop into **embryos**. **Yolk**, the embryo's food supply, is formed inside of a structure called the **yolk sac**. Immediately after the egg hatches, the embryo lives off the yolk. Soon the embryo uses up the yolk and grows big enough to obtain its own food.

In a few kinds of fish, such as guppies, the male deposits sperm into the female's body. This is called **internal** (in-TUR-nul) **fertilization**, because the eggs are fertilized inside the female. The embryos develop inside the mother. When the young are born, they are able to obtain food for themselves.

Female fish lay thousands or even millions of eggs at time. Fish species that have live babies produce only hundreds. Why the difference? Most parent fish give no care to their eggs or their young. Not all eggs will be fertilized in the water, and most of those that are fertilized will be eaten by other fish. So will most of the embryos and the little fish. To make sure that a few will grow to be adults, very large numbers of eggs must be laid.

The few embryos that do grow to be adults complete their life cycle when they produce their own eggs and sperm. A **life cycle** is the pattern of an organism's life - from birth, through growth and reproduction, to death.

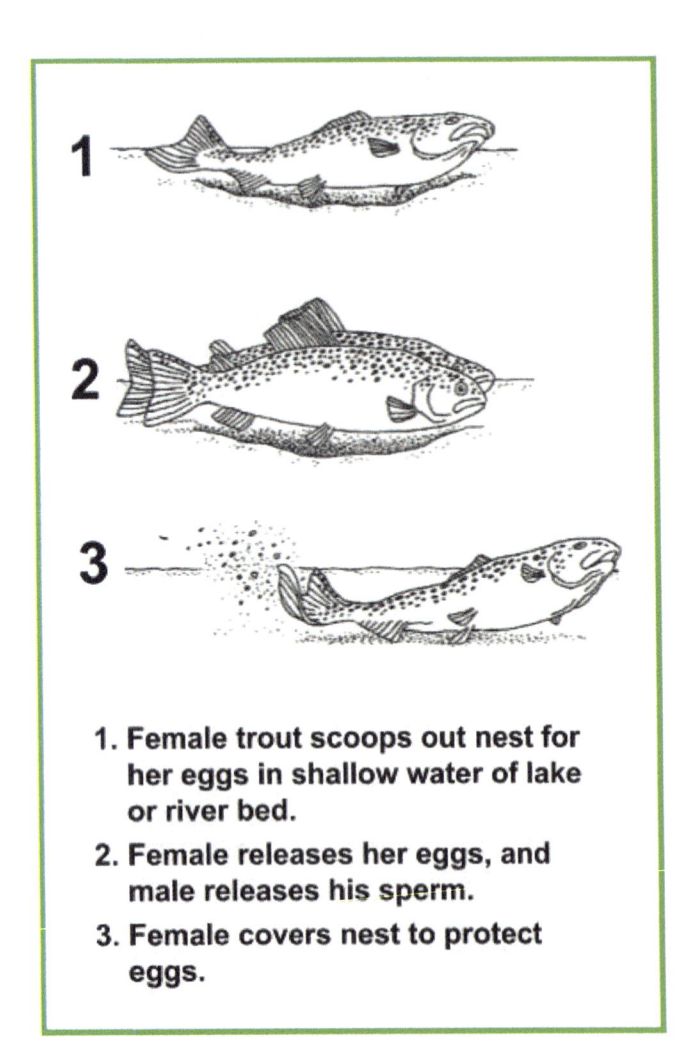

1. Female trout scoops out nest for her eggs in shallow water of lake or river bed.
2. Female releases her eggs, and male releases his sperm.
3. Female covers nest to protect eggs.

Top: A stickleback fish nest
Bottom: The nest of a sunfish

▶ To Do Yourself — How Do Some Types of Fish Reproduce?

You will need:

Small tank or fishbowl, sand, aquarium, plants, several guppies, hand lens, dechlorinating solution, water, fish food

1. Set up an aquarium with sand, plants, dechlorinated water, and guppies.
2. Feed the guppies, as instructed on the food container.
3. Observe the guppies. Males are smaller and more colorful than females. Two weeks after mating and fertilization of the eggs, the female gives birth to 50 to 200 young.
4. Observe and record the reproduction of your guppies over a period of time.

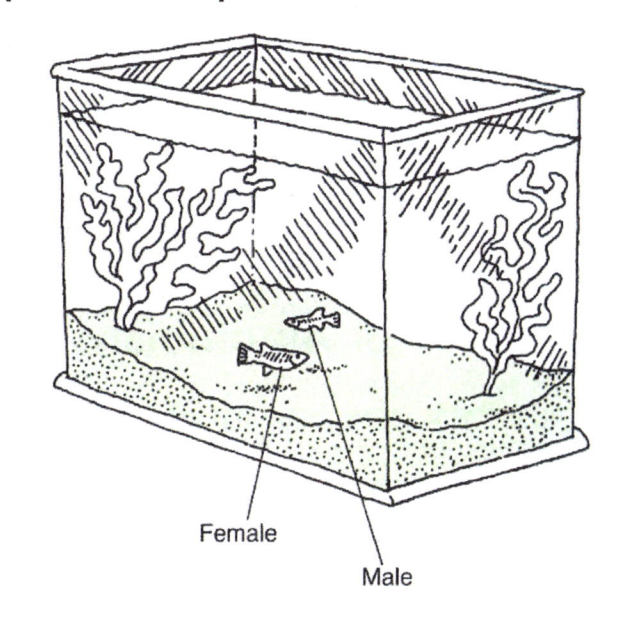

Questions

1. How often did your guppies produce offspring? _____
2. How could you tell that the female was about to give birth? _____
3. How many young were born? Why do you think that there were so many? _____

REVIEW — U-8 L-2

I. In each blank, write the word that fits best. Choose from the words below

| internal | blood | external | zygote |
| egg tube | sperm tube | yolk sac | gills |

Oxygen moves into a fish's blood through its _____. A testis connects to a _____ and an ovary to an _____. Fertilization outside of the female is _____ fertilization. Fertilization inside of the female is _____ fertilization. The fish embryo's food supply is in the _____.

II. Some fish build nests and lay their eggs in them. Would you expect these fish to lay more or fewer eggs than those that lay eggs in open water? Explain.

_____

_____

199

How Do Amphibians & Reptiles Reproduce? U-8 L-3

Exploring Science

Herps Up Close and Personal. Did you ever try to figure out what frogs are saying when they croak? Have you ever seen what some lizards do when they are threatened? Maybe you know how a snake "tastes" the air with its tongue. These are some of the things that thousands of visitors learn about first-hand at the Reptile Discovery Center at the National Zoo in Washington, DC.

The Reptile Discovery Center is literally crawling with snakes, lizards, turtles, frogs and toads. These animals are known as **herps**. The scientists who study them are **herpetologists** (hur-puh-TAHL-uh-jists).

Herpetologists and zoo volunteers provide more than the basics for the animals. They encourage the animals to stay active and to continue their normal behaviors. The enclosures provide choices. Feel like swimming or **basking** (sitting in a warm spot to raise your body temperature)? Food is even hidden or placed in challenging spots to make the animals hunt and problem solve.

The staff at the Reptile Discovery Center help people learn the difference between the facts and the myths about herps. They must be doing a good job. The Center is one of the most popular attractions at the National Zoo. Admission is free.

➤ Almost all herps are harmless. Why do you think that many people fear herps?

When threatened, the frilled lizard opens up an "umbrella" of skin around its neck.

Life Cycles of Amphibians and Reptiles

The **amphibians** (am-FIB-ee-uns) are a group of vertebrates. They include frogs, toads, and salamanders. Most amphibians live part of their life in water - as **tadpoles**. Like fish, tadpoles breathe with gills. As adults on the land, amphibians breathe with lungs. They reproduce sexually.

The **reptiles** (REP-tiles) are another group of vertebrates. They include snakes, lizards, turtles, crocodiles, and alligators. Most reptiles live on land, and all reptiles breathe with lungs. They also reproduced sexually.

Let's follow the life cycle of animals from each of these groups. Observe the illustrations on the following pages as you read.

LIFE CYCLE OF A FROG Frogs mate in the water. A female frog lays hundreds - or even thousands - of eggs in the water. Fertilization is external (outside of the female's body). As the eggs are laid, the male deposits sperm on them. The sperm fertilize the eggs. Each egg contains a good deal of yolk, the food supply for the embryo. A covering of jelly gives a small amount of protection for the eggs while they develop.

A frog's egg hatches into a tadpole. It looks like a fish, with a tail and gills. Slowly, it grows hind legs and then front legs. It loses its gills and its tail. Eventually, it develops lungs, and then hops onto land. Finally, it grows larger - into an adult frog. It has gone through a complete change in body form. This change is called **metamorphosis** (met-uh-MOR-fuh-sis); *meta* means "change" and *morph* means "form."

The Life Cycle of a Frog

Adult frog
(Full grown with no tail)

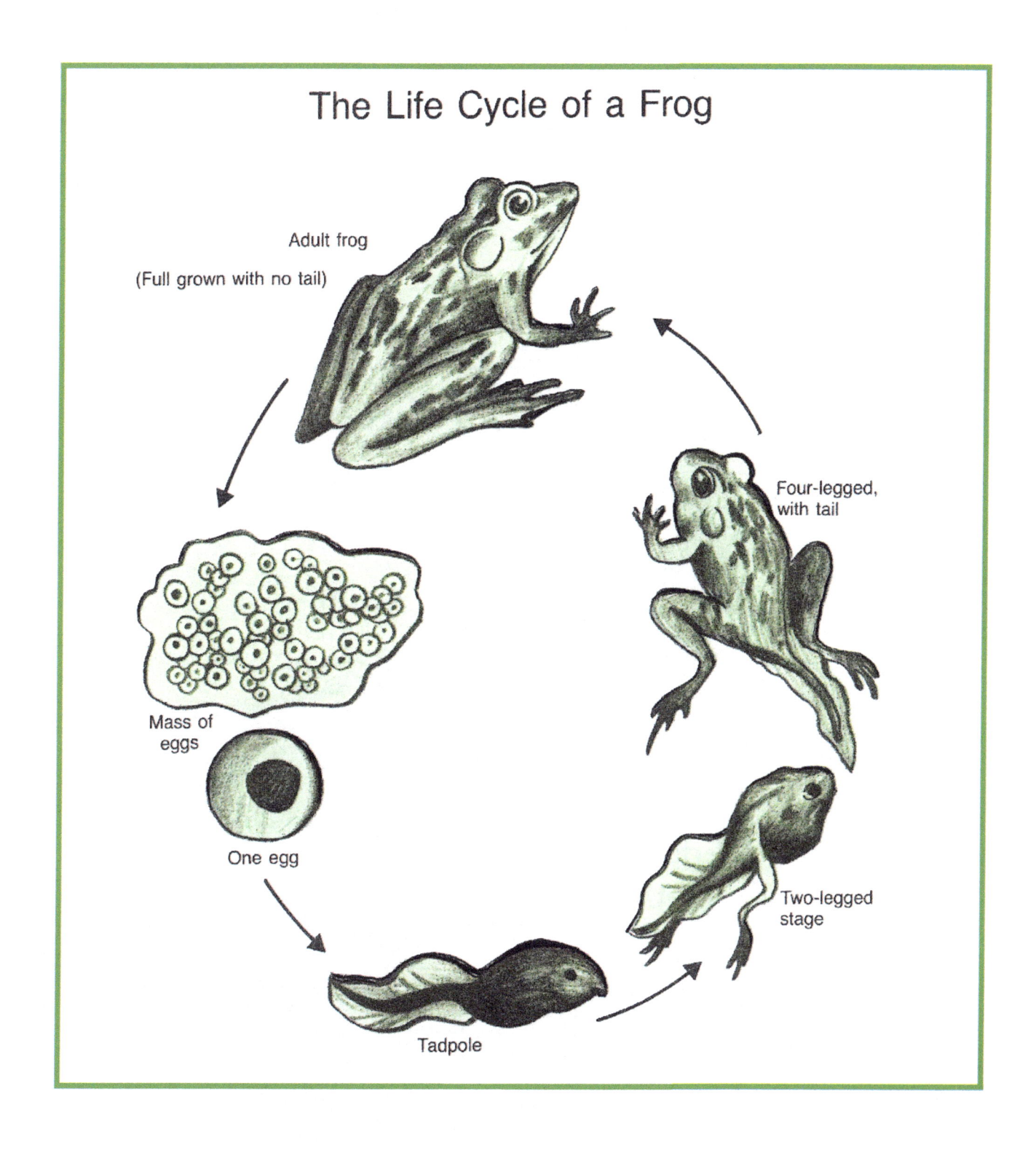

Four-legged, with tail

Mass of eggs

One egg

Tadpole

Two-legged stage

LIFE CYCLE OF A REPTILE At this point, observe the illustration on the next page.

Almost all reptiles mate on land. In reptiles, fertilization is internal - inside of the female's body. After the eggs are fertilized, they become coated with a large amount of yolk. Reptiles' eggs have more yolk than amphibians' eggs. Most reptiles' eggs are laid on dry land; a tough leathery shell keeps the eggs from drying out.

Inside the shell, embryos use the abundant yolk to develop into little animals. They can get their own food as soon as they hatch.

Some snakes and lizards have eggs that hatch while still inside of the mother's body, so these young are born alive. Garter snakes are one example of reptiles that leave the mother's body as small living animals.

201

Reptiles lay fewer eggs than amphibians. Most turtles lay 6 to 12 eggs at a time. An alligator usually lays between 15 and 80 eggs. Snakes have between 12 and 99 young at a time, either born alive or as eggs. Lizards produce between 2 and 30 eggs (or live young) at a time.

Like newly hatched reptiles, those born alive are able to be on their own at birth. The eggs of reptiles are often laid in protected places. After the eggs hatch, alligators and crocodiles protect their young for months. Very few other reptile parents give any protection to their eggs or their young.

The Life Cycle of an Alligator

- Adult
- Egg (leathery)
- Hatchling
- Young (looks like adult)

➤ To Do Yourself — What is the Life Cycle of a Frog?

You will need:

Small fishbowl, hand lens, frog eggs, dechlorinated water

1. Collect about a dozen frog eggs from a pond in the early spring. Place them in a clear glass bowl of dechlorinated water. Keep the bowl in a cool place.
2. Use a hand lens to observe the development of the eggs. Record the changes that you observe, and record the time each change takes place.
3. Release the frogs at the same location once they become small adults.

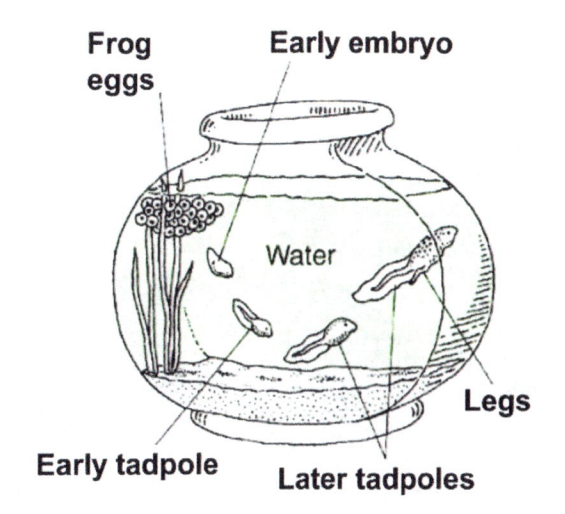

Questions

1. How long does it take for the eggs to hatch? _____
2. What are the main stages in a frog's metamorphosis? _____

REVIEW — U-8 L-3

I. In each blank, write the word that fits best. Choose from the words below

herpetologists   tadpoles   jelly   shells
metamorphosis   amphibians   reptiles

The fishlike young of a frog are _____. Frogs' eggs are covered by _____. The eggs of reptiles are covered by leathery _____. The eggs of _____ are laid on dry land. _____ is the change of body form in a frog.

II. Circle the word (between the brackets) that makes each statement correct.

A. Fertilization in reptiles is [internal / external].

B. The young animal that hatches from a [frog's / turtle's] egg looks like the adult.

C. Animals that breathe with lungs are [tadpoles / snakes].

D. An alligator's egg has more [yolk / jelly] than a frog's egg.

III. Reptiles lay many fewer eggs than fish or amphibians do. Why are the young of reptiles more likely to survive than are the young of fish and amphibians?

_____

How Do Birds Reproduce? U-8 L-4

Exploring Science / Historical Steps

<u>Stay-at-Home Dad</u>. Mom has laid her egg and is going out to sea to hunt for food. Who will stay home and mind the egg? Why, Dad will, of course. He holds the egg on top of his feet. A fold of his "belly" skin covers the egg and keeps it warm. After two months, out pops the chick. Then both parents will take turns holding and feeding the chick.

These are not your average parents. They are emperor penguins, perhaps the most unusual of all birds. For these birds, Antarctica is home. Temperatures above freezing are uncommon. Emperor penguins are adapted to survive - and reproduce - in this extremely cold habitat.

There are well over 200,000 emperor penguins. Their family life remained a mystery until <u>1981</u> when scientists began breeding the tall birds at SeaWorld of California. They watched as a female laid a single egg, and then passed it to the male.

The father does not eat while the chick develops inside of the egg. After hatching, the mother takes over, part time. (Note the two mothers and their chicks in the photo). The very hungry father then goes out to sea for food. At the Penguin Encounter at SeaWorld, the father is fed while he is egg-sitting; no one wants to risk losing him.

In August, <u>2021</u> the U.S. Fish and Wildlife Department recommended that emperor penguins be listed as an endangered species. For emperor penguins to catch fish and to reproduce safely, this species needs a large amount of frozen sea around their continent. A warmer climate means that more sea ice is melting. If we halt global warming, we halt this melting, and we protect emperor penguins!

➤ Why do you think that an emperor penguin, unlike a fish, frog, or snake, produces only one egg at a time?

The Life Cycle of a Chicken

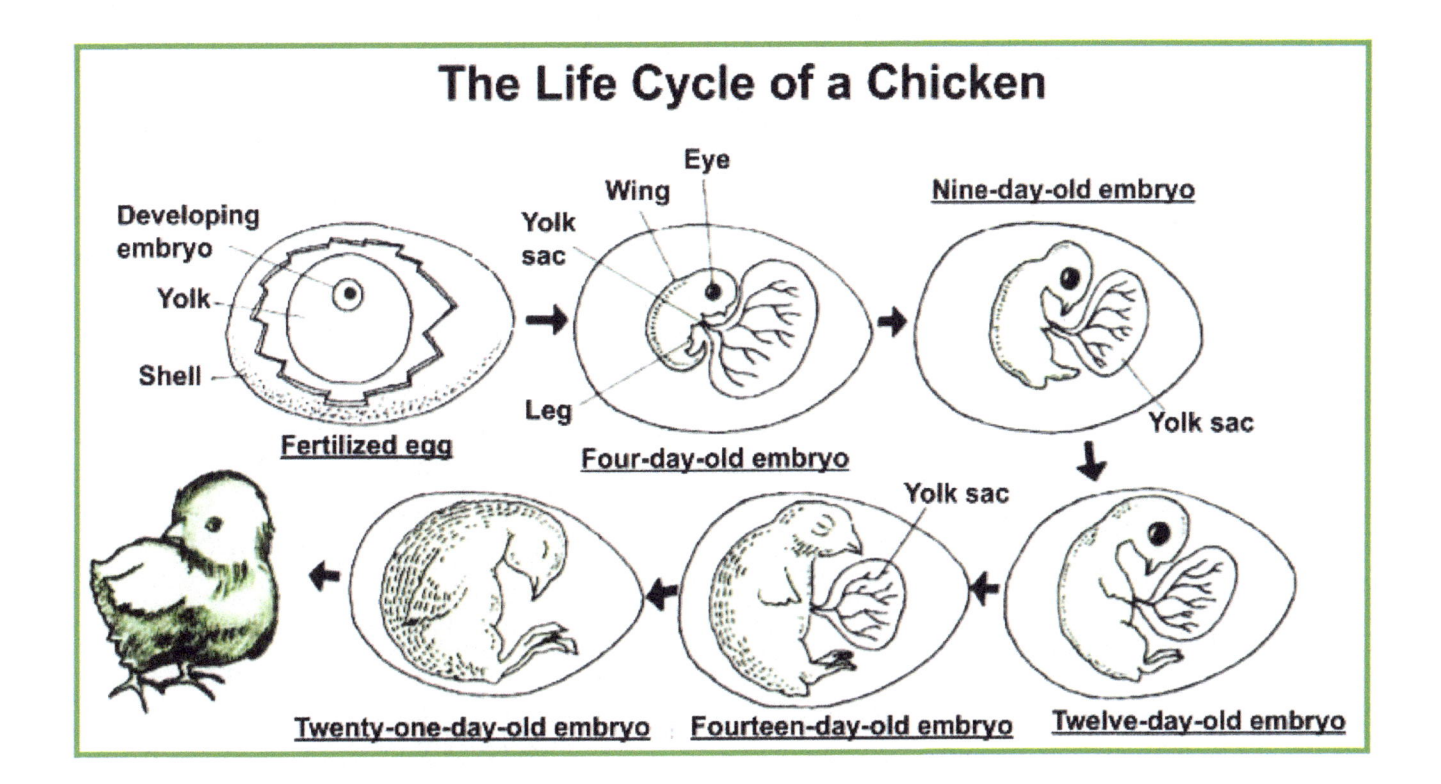

Fertilized egg — Developing embryo, Yolk, Shell

Four-day-old embryo — Yolk sac, Eye, Wing, Leg

Nine-day-old embryo — Yolk sac

Twelve-day-old embryo

Fourteen-day-old embryo — Yolk sac

Twenty-one-day-old embryo

The Life Cycle of a Bird

Penguins belong to the group of vertebrates called **birds**. Most birds have wings and can fly. Birds are covered with feathers and breathe with lungs. They are also warm-blooded. The body temperature of a **warm-blooded** animal stays at a high level, and more or less the same. It does not change if the outside temperature changes. Mammals, including you, are another group of warm-blooded animals.

Fish, amphibians, and reptiles are all **cold-blooded**. Their body temperature changes when the temperature around them changes.

Birds mate through internal fertilization. The ovary of a female bird, such as a chicken, produces sex cells called eggs. The tiny egg cells pass into the top of the egg tube.

Note: In birds, the word "egg" is used both for the reproductive cells made by the female, and for the shelled structure that the female later lays.

The male's testes produce sperm. During mating, the male places sperm inside of the lower end of the female's egg tube. The sperm cells swim up the tube to the eggs, and fertilize them.

Each fertilized egg, or zygote, becomes an embryo as it moves down the egg tube. Along the way, the egg yolk and egg white are added around the embryo. Finally, a shell is added and the egg is laid.

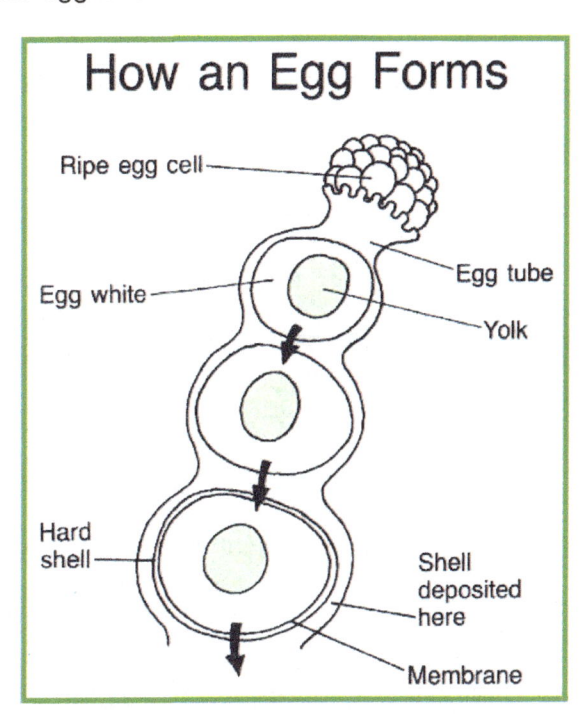

How an Egg Forms

Ripe egg cell, Egg white, Egg tube, Yolk, Hard shell, Shell deposited here, Membrane

As the ripe egg cell moves downward through the egg tube, stored food, a membrane, and a shell are added.

205

All birds, such as penguins, chickens, and robins, keep their eggs warm, or **incubate** (IN-kyuh-bayt) them. A chicken egg is incubated for 21 days, and then it hatches. During incubation, the embryo grows and develops, using both the yolk and the white of the egg for food. For some time after hatching, all bird parents continue to care for their young.

➤ To Do Yourself — What Are the Parts of a Bird Egg?

You will need:

Chicken egg, dish, hand lens

1. Carefully crack open a fresh egg into a dish. Locate the shell membrane, the white, and the yolk. Find the rope-like part of the white that holds the yolk in place. Draw what you observe.
2. Use a hand lens to observe the pores in the egg shell.
3. Examine the yolk. Locate a white speck the size of a pinhead. This is where the embryo starts to develop.

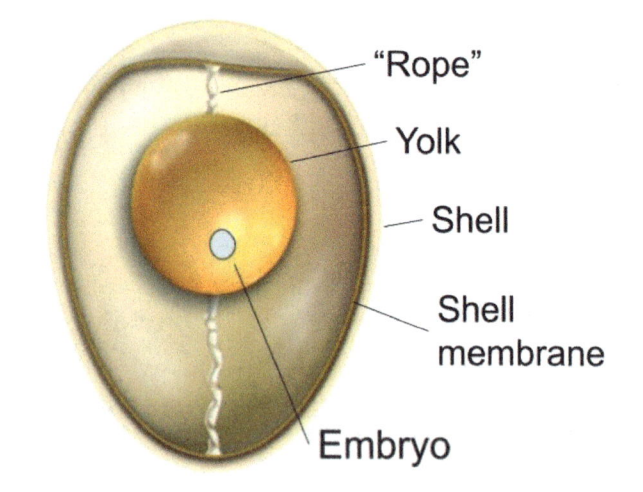

Questions

1. What is the purpose of the egg yolk? _____

2. What is the purpose of the egg white? _____

3. What is the purpose of the pores in the egg shell? _____

4. Why was there no embryo in the egg that you observed? _____

REVIEW — U-8 L-4

I. In each blank, write the term that fits best. Choose from the terms below.

| ovary | egg tube | cold-blooded | warm-blooded | testis | incubate |

A _____ animal's body temperature stays about the same.

A bird's egg cell is produced in the _____. Egg yolk and white form around the embryo while it is inside of the _____. Parent birds _____ their eggs, keeping the developing young warm.

II. An ostrich has feathers, lays eggs, and is warm-blooded. But this animal has no wings and therefore, cannot fly. A bat has hair, gives birth to live young, has wings, and can fly. Which animal is a bird? Explain your answer.

_____

How Do Mammals Reproduce?   U-8 L-5

Exploring Science / Historical Steps

In Vitro or In Vivo? Were you an "in vivo" or "in vitro" baby? These Latin phrases are often used in science. *In vivo* means "in a living thing." *In vitro* means "in glass."

The birth of **Louise Brown** in 1978 made history. She was the world's first "test tube baby." No, Louise was not born from a test tube. But the sperm from her father and the egg from her mother were *joined* in a dish ("in vitro"). In other words, fertilization did not occur inside of her mother ("in vivo"). News reporters thought that it would sound dramatic to use the phrase "test tube baby."

Since this type of fertilization happens in a dish, scientists call it **in vitro fertilization** (or simply **IVF**). For Louise, a few days after fertilization, the embryo was placed in her mother to grow (for 9 months).

In most cases, a human egg is fertilized inside of the egg tube of the mother. But in one out of seven couples, fertilization or embryo growth won't happen on its own. Before 1978, these couples were not able to produce a child. Over a million babies have now been born using in vitro fertilization.

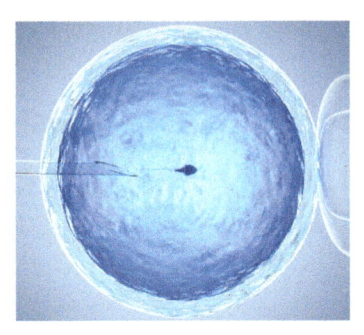

A tiny syringe (at left) injects one sperm into an egg.

➤ Do you think that a human embryo will ever develop entirely outside of a woman's body.

How Mammals Reproduce

Humans are members of the class of vertebrates called **mammals**. The class also includes dogs, cats, monkeys, and whales. All mammals have hair or fur on their bodies and are warm-blooded. Mammals also have **mammary** (MAM-uh-ree) **glands**, which produce milk for feeding their young. Most mammals give birth to one or more live young. However, there are two mammals that lay eggs - the duckbill platypus and the spiny anteater.

For the remainder of this lesson, let's look at the reproduction of a typical mammal - us!

The male's main reproductive organs are the two **testes** (TEST-tees). These produce the sperm cells. The testes are covered with skin and hang outside of the body in a sac called the **scrotum** (SKROH-tum).

A sperm tube from each **testis** (TEST-tis) joins to form a single tube. This tube, the **urethra** (you-REE-thruh), is in the center of the **penis**. For mating to occur, the penis must temporarily become firm (called an **erection**).

Sperm pass through the urethra as they leave the body. [Note: The same tube, the urethra, carries urine out of the body when males go to the bathroom.]

In the female, the organs that produce eggs are the **ovaries** (OH-vuh-rees). About every 28 days, one of the two ovaries releases a single egg. The egg moves into an **egg tube** (also called the **Fallopian** (fuh-LOW-pea-un) **tube**). While the egg is inside of this tube, it can be fertilized.

In the female, young are grown in an organ called the **uterus** (YOO-tuh-rus); it is also called the **womb**. In a young woman it is quite similar in shape to (and about ¾ the size of) an upside-down pear. Obviously, this organ can stretch a great deal in order to hold, and grow, a baby.

The flexible tube from the uterus to the outside of the body is called the **vagina** (va-JIYE-nuh); it is also called the **birth canal**.

Human Reproductive Systems

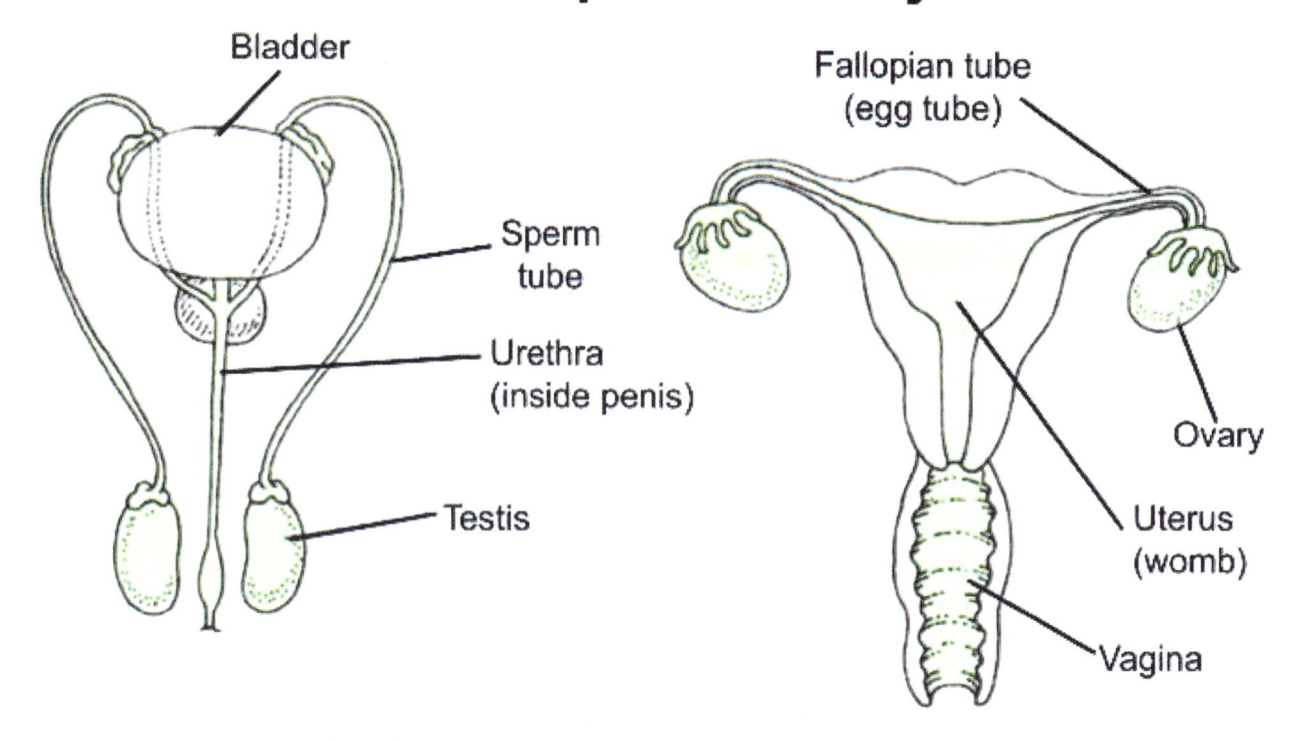

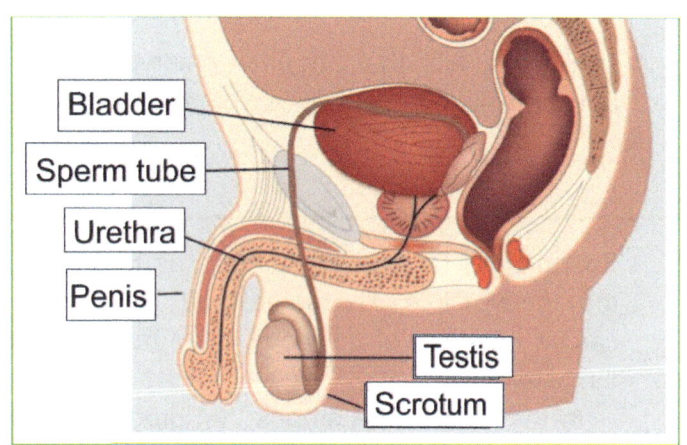

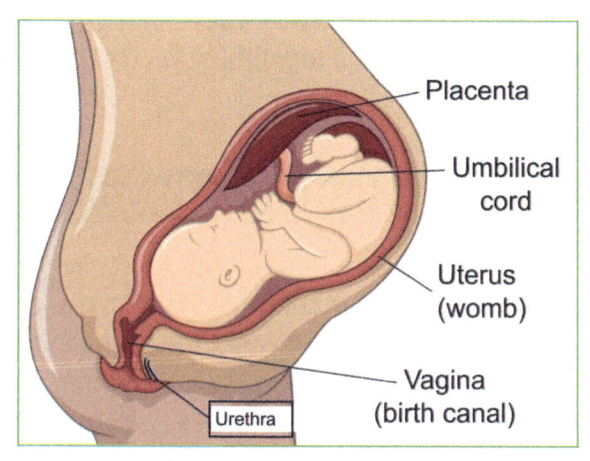

MALE

- Bladder
- Sperm tube
- Urethra (inside penis)
- Testis
- Bladder
- Sperm tube
- Urethra
- Penis
- Testis
- Scrotum

FEMALE

- Fallopian tube (egg tube)
- Ovary
- Uterus (womb)
- Vagina
- Placenta
- Umbilical cord
- Uterus (womb)
- Vagina (birth canal)
- Urethra

[Note: Observe the illustration of the FEMALE above. The vagina is an entirely separate tube from the urethra. In females, the urethra carries only urine. Its opening lies just in front of the opening for the vagina. The urethra drains the urinary bladder. If the bladder were shown in the above diagram, it would be under the uterus, below the baby's head. Do you see why pregnant women frequently need to urinate?]

Once females are old enough to have children, the lining of their uterus goes through a cycle each month. This lining thickens with delicate blood vessels for about 3 weeks. These tiny blood vessels and about 3 tablespoons of blood are then slowly passed out through the vagina - for about 5 days. This "shedding" phase is called **menstruation** (men-STRUAY-shun), or having a **period**. If fertilization occurs, the lining is not shed; that is, there is no menstruation.

During mating, called **intercourse** (when the penis and vagina "join"), sperm cells enter the female's body. The sperm swim through the uterus and into the egg tubes. Millions of sperm may go into the tubes, but only one sperm cell can fertilize one egg.

The fertilized egg, or zygote, moves down the egg tube as it starts to develop. While still a tiny embryo, it floats into the uterus. The embryo attaches itself to the wall of the uterus.

As the human embryo continues to develop, a flat structure called the **placenta** (pluh-SENT-uh) forms on the wall of the uterus. Blood vessels in the placenta bring food and oxygen from the mother to the embryo. These blood vessels also carry wastes away from the embryo. The embryo is joined to the placenta by the ropelike **umbilical cord**.

After about two months, a human embryo is called a **fetus** (FEE-tus). The fetus develops and grows until it is a baby, ready to be born.

When the time of birth comes, the muscles in the walls of the uterus contract, as **labor** begins. The contractions of labor push the baby through the birth canal and out into the world.

During the period of time between fertilization and birth, the mother is **pregnant** (PREG-nunt). The time of pregnancy for humans is about 9 months.

For some other mammals, pregnancy may be shorter or longer. For a rat, pregnancy lasts only 3 weeks. For an elephant, it takes 22 months.

Mammals provide more care for their young than do any other animals.

Chimpanzee mother nursing her baby

[Note: With over 98% of their DNA identical to ours, chimpanzees are our closest relatives.]

REVIEW — U-8 L-5

I. In each blank, write the word that fits best. Choose from the words below.

| umbilical cord | placenta | egg tube | urethra | sperm tube |
| birth canal | uterus | fetus | mammary glands | |

Milk for a mammal's young are produced in the _____.

Sperm cells pass from the sperm tubes into the _____, the tube inside of the penis. An egg can be fertilized while it is in an _____.

The _____ connects the embryo to the placenta. After about two months, the embryo is called a _____. During labor, the _____ contracts and pushes the baby through the _____ (also called the vagina).

II. Circle the word (between the brackets) that makes each statement correct.

  A. Mammals are [warm-blooded / cold-blooded].
  B. The fertilized egg is the [ovary / zygote].
  C. Sperm cells are produced in the [testes / placenta].
  D. A human egg is produced about once every [9 months / 28 days].

III. If a pregnant mother drinks alcohol, the baby can be seriously harmed. In a few sentences, explain how the alcohol gets into the baby. Use these terms in your answer: <u>uterus</u>, <u>placenta</u>, <u>umbilical cord</u>.

How Do Insects Reproduce? U-8 L-6

Exploring Science

Why Jumping Beans Jump. Have you ever seen "Mexican jumping beans?" They aren't really beans. They are seed pods of a Mexican arrow plant. When the pods are warmed, they jump and hop about. Some of the "beans" open and throw out seeds. But other "beans" just fall from the plant to the ground. These are sold to tourists as "jumping beans," or, more correctly, hopping pods. What makes them hop?

The pods that hop don't contain seeds at all. If you open one during the summer, you will find a tiny caterpillar inside. A layer of silk lines the inside of the pod. When the desert sun heats the pod, the caterpillar digs its feet into the silk and jerks its body. This jerking causes the pods to hop. When the pod lands in the cool shade, it stops hopping.

By late Fall, if you open a jumping bean, you will find that the caterpillar has changed. It is wrapped in a covering and seems to be at rest. If you keep an unopened pod long enough, you may see what happens at the end of this insect's "rest." In the Spring, through a tiny hole in the pod, out comes a moth! Why a moth? As you read on, you will find out.

➤ Why are no seeds found inside the pods that hop?

The larva of a moth inside of a pod

Life Cycles of Insects

Moths, ants, flies, and other insects are invertebrates; they have no backbones. But insects do have skeletons on the outside of their bodies. An insect's tough "skin" is its outside skeleton (also called an **exoskeleton**).

Insects reproduce sexually. The male and female mate, and the eggs are fertilized internally (inside of the female). You may have witnessed two butterflies or two dragonflies "connected" as they fly. The two are mating.

In the fertilized egg, the embryo lives on the yolk. The eggs are laid near a food supply that the newly hatched insects eat. After hatching, at some point in their lives, most insects go through a change of body form. This change is called **metamorphosis**. It may be complete or incomplete - as explained next.

COMPLETE METAMORPHOSIS. There are four stages in the complete metamorphosis of an insect: egg, larva, pupa, and adult. Like all moths and butterflies, the "jumping bean" moth goes through complete metamorphosis. This moth lays its eggs in the flowers of the Mexican arrow plant. After the flower forms seeds, the eggs hatch into caterpillars, which are the wormlike **larvae** (or larvas) of the moth. Often insect larvae look somewhat like worms, but with stubby legs. Those that we call caterpillars are the larvae of moths or butterflies. The larva of a fly is called a maggot, and the larva of a beetle is called a grub.

Commonly, larvae focus on eating, and they grow rapidly. The "jumping bean" larvae eat the seeds inside of the arrow plant's pods. Luckily for the plant, not all of its flowers are visited by the moth. Some of the seeds survive and keep the species from dying out.

Complete Metamorphosis

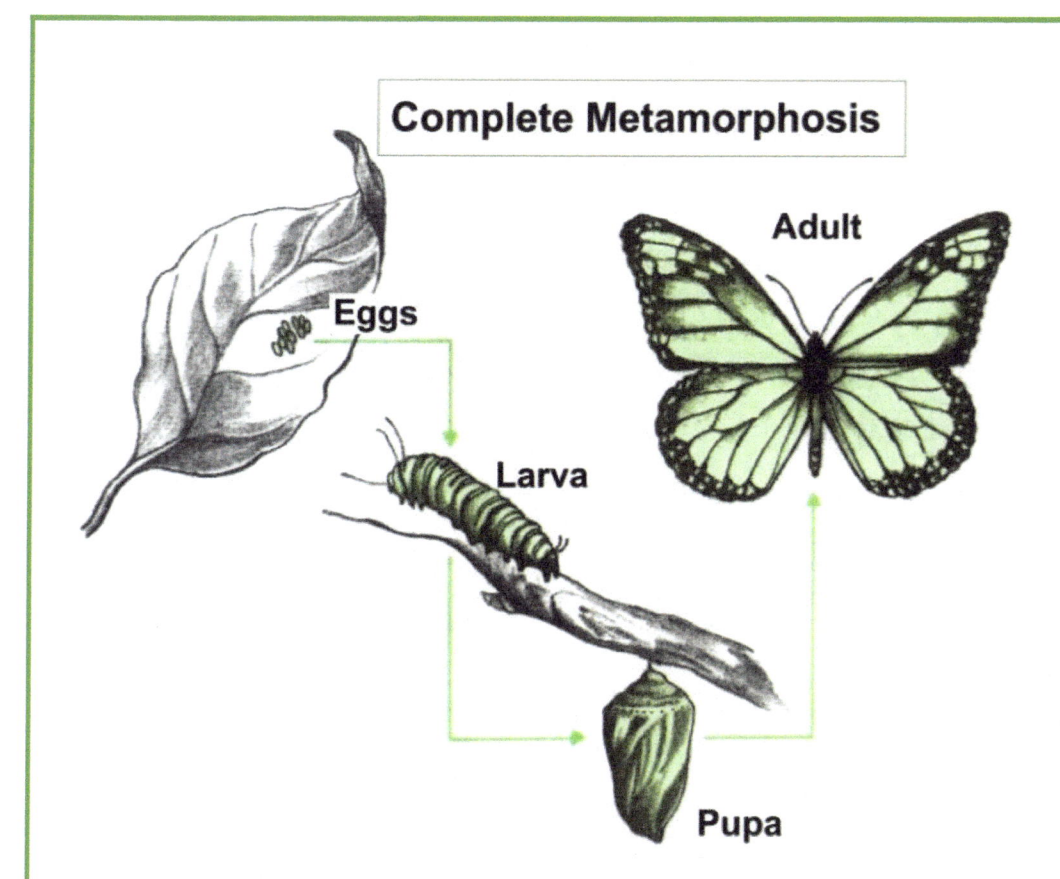

Eggs

Larva

Pupa

Adult

This is the life cycle of the famous **monarch butterfly**. The adults migrate across the U.S. Many thousands gather to mate in one location in Mexico.

Unfortunately, the larvae of other insects can do great damage to crops and trees. For example, in the U.S., the larvae of an invasive, metallic green beetle, the emerald ash borer, has destroyed millions of ash trees.

As a larva grows, it sheds its exoskeleton and replaces it with a slightly larger one. This is called **molting**. Insects molt several times as they grow to become adults.

When a larva stops growing, it stops eating. Then it becomes a **pupa** (PYOO-puh) inside of a hard "skin," or case. In moths, the larva (caterpillar) spins itself a silky shell, or **cocoon** (kuh-KOON). It then becomes a **pupa** inside of the cocoon. We get silk from the cocoons of silkworm moths.

During the pupa stage, the insect seems to be at rest. In fact, many body changes occur during the pupa stage. In time, the pupa's "skin" breaks open, and out comes the adult.

The adult has wings and other body parts completely unlike the larva. The main goal of adult insects is to mate and lay eggs.

A few species of adult insects focus almost entirely on mating; these species eat little or nothing! Many adult moths and butterflies live only long enough to mate and lay eggs. Then the life cycle begins again.

INCOMPLETE METAMORPHOSIS. Some insects go through incomplete metamorphosis. In their life cycles, there are just three stages: egg, nymph, and adult. The egg of a cockroach, a grasshopper, or a dragonfly hatches into a nymph. The **nymph** looks like a small adult, but with undeveloped wings called "wing pads."

As the nymph eats and grows, it molts several times. With each molt, the wing pads grow larger. After the last molt, the wings are fully grown. The adult now produces eggs or sperm, and searches for a mate. The adults mate, eggs are laid, and their life cycle begins again.

Incomplete Metamorphosis

Adult — Wings full grown

Eggs

Nymphs

Wings begin to grow

No wings

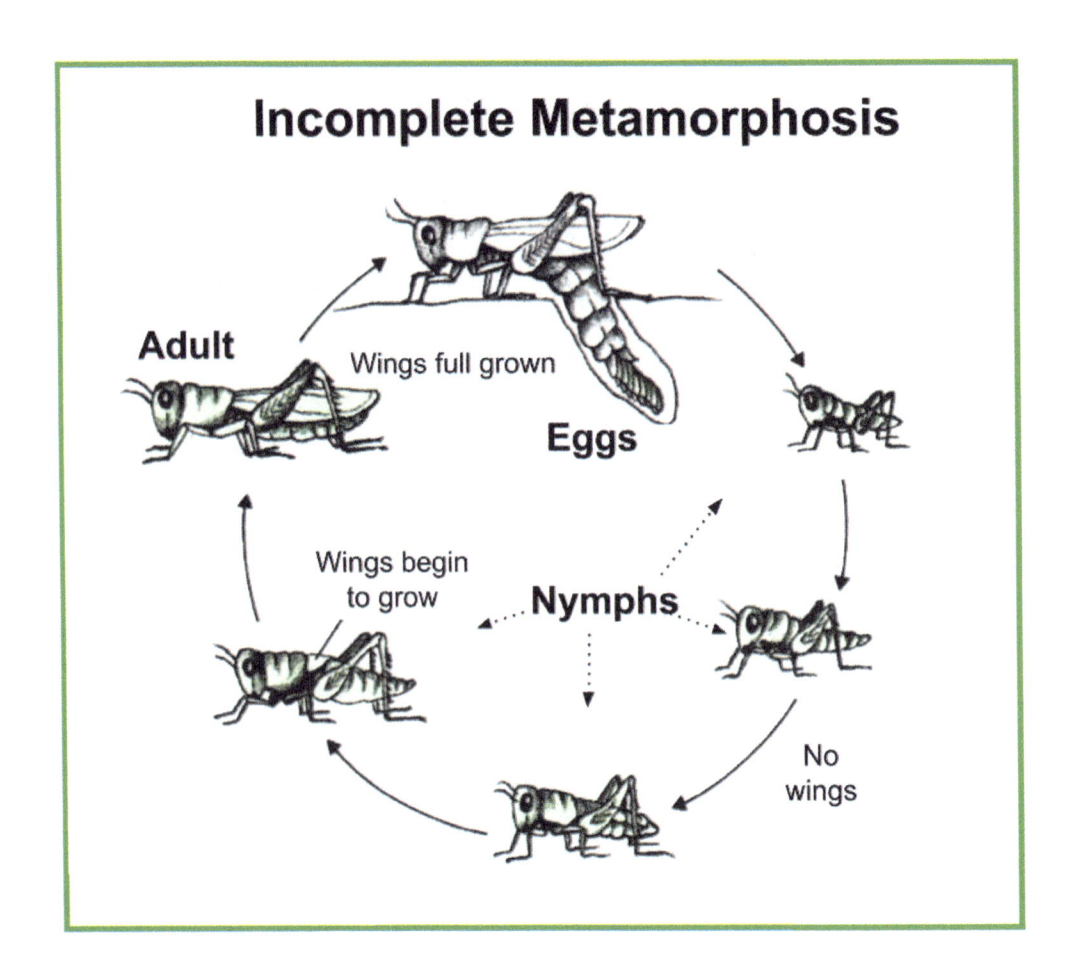

➤ To Do Yourself — How Do Some Insects Reproduce?

You will need:

Caterpillar, jar, lid with air holes, soil, twig, leaves, hand lens

1. Put 2 centimeters of moist soil in a jar. Place a twig in the soil.
2. Place a caterpillar in the jar. Add some leaves from the site where the caterpillar was found.
3. Observe the caterpillar. Record your observations. Draw the changes that you see; record the time each takes.
4. If your caterpillar becomes a butterfly or moth, release it outside.

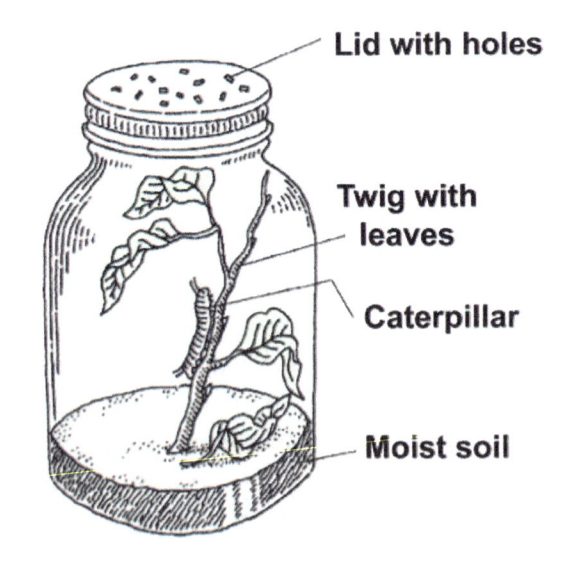

Jar labels: Lid with holes, Twig with leaves, Caterpillar, Moist soil

Questions

1. How many stages did you observe in the development from egg to adult? _____
2. What stage(s) were you not able to observe? _____
3. At what stage would humans say this insect is most "destructive?" _____
4. Is this a complete or an incomplete metamorphosis? _____

REVIEW — U-8 L-6

I. In each blank, write the word (or words) that fits best. Choose from the words below.

| metamorphosis | adult | pupa | invertebrates | larva |
| nymph | molts | embryo | egg | |

Insects are _____ because they have no backbones. The wormlike stage of a moth is the _____. A just-hatched grasshopper that looks like a little adult is a _____. During growth, a young insect sheds its skin, or _____. A change of body form is called _____. The "resting" stage of a moth is the _____. The stage of an insect that lays eggs is the _____.

II. If the statement is true, write **T**. If it is false, write **F**, but <u>ALSO</u> correct the underlined word.

A. _____ An insect's egg is fertilized <u>internally</u>.

B. _____ A grasshopper goes through <u>complete</u> metamorphosis.

C. _____ The stages of a butterfly's life cycle are egg, <u>larva</u>, pupa, and adult.

D. _____ The pupa stage of a moth is covered by a silk case called a <u>nymph</u>.

E. _____ The stages of a cockroach's life cycle are egg, <u>larva</u>, and adult.

III. A centipede looks like a caterpillar in some ways. Both are wormlike and seem to have many legs. A centipede lays eggs that hatch into little centipedes. Do you think that the centipede, like a caterpillar, is the larva of a moth or butterfly? Explain your answer.

_____

_____

_____

Caterpillar

Centipede

Unit 8 -- Review What You Know ---

A. Hidden in the puzzle below are the names of seven stages in the life cycles of different animals. Use the clues to help you find the names. Circle each name you find in the puzzle. Then write each name on the line next to its clue.

A	L	A	R	V	A	D	F
F	G	I	O	Q	D	M	N
E	S	U	Z	X	U	A	Y
T	A	D	P	O	L	E	M
U	V	I	E	F	T	T	P
S	B	R	E	L	Y	D	H
K	N	B	G	W	C	P	C
E	Z	W	G	P	U	P	A

Clues:

1. What a frog's egg hatches into. _____
2. Two-month old human embryo. _____
3. May contain much yolk. _____
4. Wingless grasshopper or roach. _____
5. Produces eggs or sperm. _____
6. A maggot or grub. _____
7. May have a cocoon. _____

B. <u>Write</u> the word (or words) that best completes each statement.

1. Testes are organs that produce
    a. yolk **b.** eggs **c.** sperm   1._____

2. The vertebrates that produce the most eggs are
    a. fishes **b.** birds **c.** mammals   2._____

3. A toad, a crocodile, and a snake are all
    a. amphibians **b.** reptiles **c.** herps   3._____

4. After a zygote has divided once, it is an embryo with
    a. one cell **b.** two cells **c.** a ball of cells 4._____

5. Warm-blooded animals include
    a. rats **b.** salamanders **c.** lizards   5._____

6. An animal that goes through metamorphosis is the
    a. alligator **b.** frog **c.** penguin   6._____

7. A frog's eggs are protected by
    a. egg white **b.** shells **c.** jelly   7._____

214

8. Mammary glands are found in a
   a. platypus   b. turtle   c. grasshopper   8. _____

9. The vertebrates that provide the most parental care are
   a. mammals   b. amphibians   c. reptiles   9. _____

10. When a caterpillar molts it sheds its
   a. legs   b. skeleton   c. wings   10. _____

C. Apply What You Know

1. Study the drawings of the embryos of a trout, a robin, and a pig. Label the numbered parts. Choose from these labels. A label may be used more than once.

   embryo   **umbilical cord**   **shell**   **uterus**
   yolk   **yolk sac**   **placenta**

   Trout
   1. _____
   2. _____

   Robin
   3. _____
   4. _____
   5. _____

   Pig
   6. _____
   7. _____
   8. _____
   9. _____

2. Each statement below refers to one or more of the animals whose embryos are shown. On the line after each statement, write one or more of the following: trout, robin, pig.

   a. The egg from which it developed was fertilized internally. ---- _____

   b. Its food supply comes from the mother's blood. ---------------- _____

   c. It will be incubated while it develops. ---------------------- _____

   d. Its food supply was stored in the egg. ----------------------- _____

   e. It is a cold-blooded animal. --------------------------------- _____

215

3. Study the drawings below of the life cycle of a milkweed bug.

Life Cycle of the Milkweed Bug

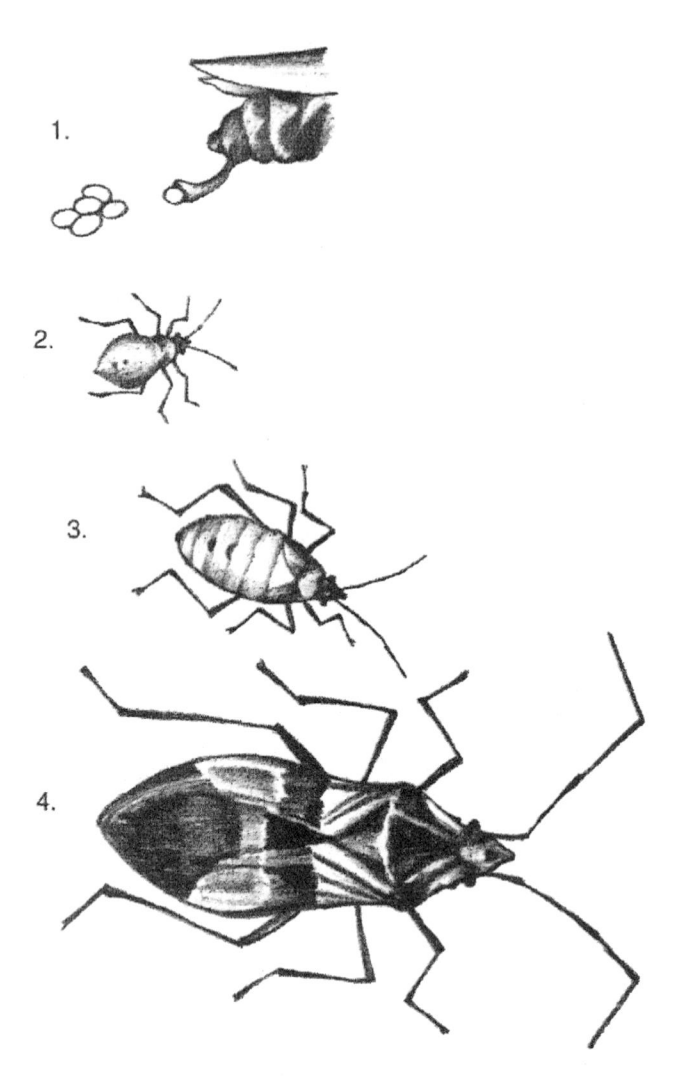

1.

2.

3.

4.

For each statement about the bug that is true, write **T** in the blank.
For each false statement, write **F,** then <u>correct</u> the underlined word.

a. ___ The bug belongs to a group of <u>vertebrates</u>.

b. ___ Between stages 2 and 3, the bug <u>molts</u>.

c. ___ At stage 4, the bug is a <u>nymph</u>.

d. ___ The bug's metamorphosis is <u>complete</u>.

D. Find out more.

1. Read about reproduction in social insects. Some examples are ants, bees, and termites. How are they like and unlike the insects that you have studied?

2. Obtain some snails at a pet shop. Each snail contains both sexes. You can observe their mating and egg-laying. The eggs are large enough to see. Two weeks after laying, the eggs hatch. Make drawings of all of your observations.

Careers in Life Science

Zoos to the Rescue. Why do we have zoos? Viewing wild animals educates and entertains us. In a zoo, we can get near animals and appreciate their beauty.

Humans have a great impact on animals in the wild. It is vital that the habitats of animals be protected. When people actually see wild animals (rather than simply watch videos of them) they are more likely to care about their welfare - and they are more likely to support programs that protect animals.

Zoos also have another important purpose. They help save endangered species. For many animals, zoos may be their last hope for survival.

Do you care deeply about animals? Are you able to handle them more easily than most people? Do you have a feeling of kinship with wild creatures? If so, a career in a zoo might be for you.

Zoo Keepers. Zoo keepers feed, care for, and observe the behavior of animals. Some keepers are experts on the breeding of the animals in their charge. When animal mothers cannot or will not raise their own babies, zoo keepers become foster parents.

Most zoo keepers have two or more years of study in zoology or animal husbandry before they land a job in a zoo. Some keepers begin as zoo volunteers while still in middle or high school. All learn most of their job as apprentices to experienced keepers.

Veterinarians. A veterinarian, or "vet," is an animal doctor. If you have a pet, you may have taken it to a vet. Zoo vets plan diets for the animals, treat their illnesses, and perform surgery. Many zoo vets and other scientists do research on the reproduction of endangered species. Left in the wild, some of these species would likely become extinct. Instead, some are now increasing their numbers inside of zoos.

Zoo vets start their preparation the way that all vets do, with four years of college. At least three more years of study in a school of veterinary medicine are then required.

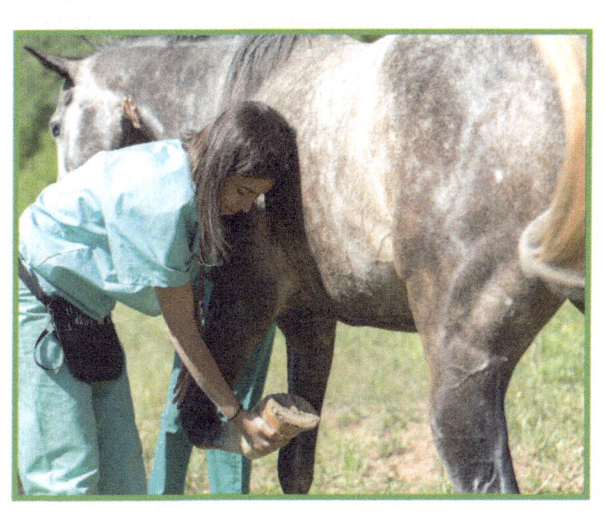

A vet specializing in reptiles

A vet specializing in horses

217

UNIT NINE
HEREDITY AND EVOLUTION

How Do Parents Pass Traits to Offspring? U-9 L-1

Exploring Science

The Jim Twins. "I looked into his eyes and saw a reflection of myself. I wanted to scream or cry, but all I could do was laugh." That's what Jim Springer said when he first met Jim Lewis, his twin brother.

Both Jims knew that they had been born twins. But Springer thought his twin brother had died as a baby. Lewis knew that he and his twin were adopted at birth by different families. At age 39, Lewis, who lived in Lima, Ohio began to search for Springer. Lewis found Springer in Dayton, Ohio, a nearby city.

The Jim twins are identical. This means that they look almost exactly alike. As they talked, the Jims found that they were also alike in other ways. At school, both loved math and hated spelling. Both studied carpentry and mechanical drawing. Both had a dog named Toy. Both had a first wife named Linda and a second wife named Betty. One of the Jim twins named his first son James Alan. The other Jim named his first son James Allan!

Scientists have studied many sets of identical twins who were raised apart. Most of those twins have some amazingly similar life experiences.

But they also have differences. Both Jim twins worked part time jobs as law officers. But they had different main jobs. Springer was a records clerk, while Lewis was a security guard. Studying twins like the Jims helps scientists learn about traits that parents pass to offspring.

Another set of identical twins. See any differences?

➤ Some sets of twins are not identical. They look no more alike than ordinary brothers and sisters. Suppose that a pair of unlike twins are raised apart. Do you think that they have as many of the same experiences as identical twins? Give a reason for your answer.

➤ Want more? Twins are dealt with in detail in lesson 6.

From Parents to Offspring

What do you look like? Is your hair brown, red, black, or blond? Is it curly, kinky, wavy or straight? What color are your eyes and your skin? These are some of your characteristics or **traits**. So are your height and the shape of your nose. For the Jim twins, and all identical twins, traits like these are the same, or nearly so. For most people, their traits identify them as individuals.

Where do you get your traits? You receive many traits from your parents. Traits passed from parents to offspring are **inherited**.

How can you inherit your "father's eyes" or your "mother's nose?" The answer is in the center, or nucleus, of your cells. In the nucleus of every cell there are important threadlike parts called **chromosomes** (KROH-muh-sohms). Each chromosome is made up of thousands of tiny **genes** (JEENS). It is the genes that determine the traits that you inherit. The study of genes is called **genetics** (Juh-NET-iks).

Chromosomes and genes can be thought of as instructions. They are the instructions for building all of the parts of an organism. Often, genes are compared to a recipe. A recipe is the instructions for making a certain food, such as an apple pie.

Almost all cells, such as bone, skin, and muscle, are called **body cells**. **Sex cells** are the only cells that are not called body cells. In each body cell, the chromosomes are paired.

One chromosome in each pair comes from the father. The other chromosome in the pair comes from the mother.

Every kind of organism has a certain number of chromosomes in its body cells. For humans, the number is 23 pairs, or a total of 46 chromosomes. For fruit flies, the number is 4 pairs, or a total of 8.

In the sex cells, chromosomes are not paired. An egg cell or a sperm cell has just half the number of chromosomes as a body cell. For humans, there are 46 chromosomes in a body cell. Therefore, there are 23 chromosomes in an egg or a sperm. For fruit flies, there are 8 chromosomes in body cells. How many chromosomes are there in a fruit fly's egg or sperm? For corn, the number of chromosomes in a body cell is 20. How many are there in an egg or a sperm of the corn plant?

Each sex cell has a complete set of instructions for building the organism. Since two sex cells join, each body cell actually has two complete sets of instructions.

When were the two sets of instructions for building *you* put together? The answer: when one set from a sperm joined with one set from an egg! 23 chromosomes in the nucleus of a sperm cell (from your father) joined with 23 chromosomes in the nucleus of an egg (from your mother). So your first body cell (the zygote) had 46 chromosomes, or 23 pairs.

This is how body cells get chromosomes in pairs - one in each pair comes from each parent.

Of course, the genes that make up chromosomes come in pairs, too. Since you get half of your chromosomes from each parent, you get half of your genes from each parent. This also means that you get half of your heredity from each parent.

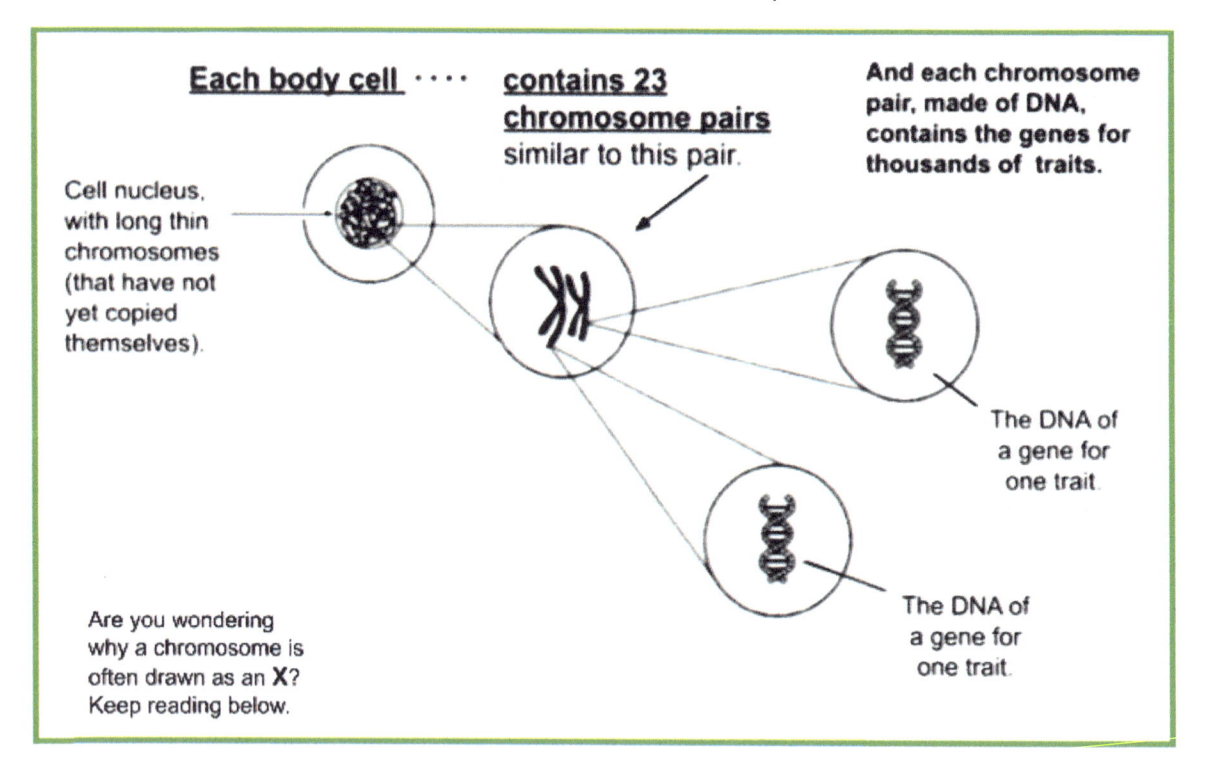

Each body cell · · · · **contains 23 chromosome pairs** similar to this pair.

And each chromosome pair, made of DNA, contains the genes for thousands of traits.

Cell nucleus, with long thin chromosomes (that have not yet copied themselves).

The DNA of a gene for one trait.

The DNA of a gene for one trait.

Are you wondering why a chromosome is often drawn as an **X**? Keep reading below.

Why are chromosomes so often shown in the shape of an **X**? Actually, chromosomes are usually quite long and thin. They are so thin that they are not able to be seen under the high power of a compound microscope.

But something happens when a cell is about to divide (to become two cells). The long, thin chromosomes copy themselves. (How this happens is explained in the next lesson).

Immediately after they have copied themselves, the original chromosome coils up (like a slinky toy); the copy also coils up (like another slinky). Now, in this coiled shape, they are both thick enough to be visible under a compound microscope. At this time, these two stay attached to each other at one spot (near their middle). Hence, they now have the appearance of an "**X**".

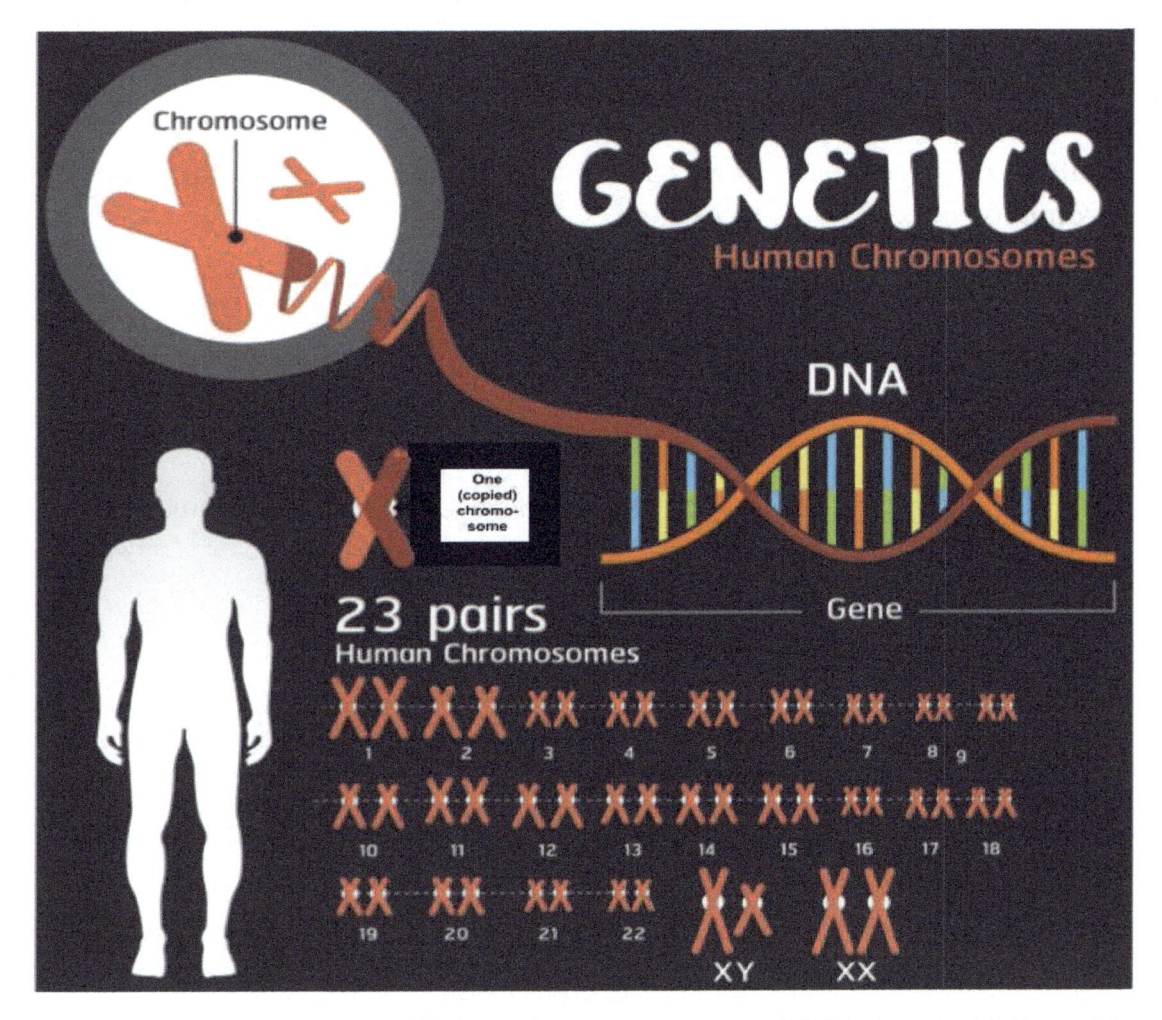

Genetics is the study of genes. Note the "X" shape of most chromosomes. [DNA is described in the next lesson.]

➤ To Do Yourself   What Is An Inherited Trait?

You will need:

A set of hair strands (1 from two parents, and 1 from their offspring), scissors, microscope, slide and coverslips, forceps (tweezers), dropper, water

1. Obtain a set of hair strands from your family, or good friends, or teacher.
2. Cut the strands so that you can place them side by side on the microscope slide. Place a drop of water and a cover slip on the slide.
3. Examine each hair strand under the microscope.
4. Compare the parent's hair with that of the offspring. Write your comparisons in the chart below.

Slide	Color	Texture	Shine	Amount of curl
Parent 1	_____	_____	_____	_____
Parent 2	_____	_____	_____	_____
Offspring	_____	_____	_____	_____

Questions

1. Are you able to determine which hair characteristics were inherited from parent 1 and parent 2?

_____

221

------------------------------ **REVIEW** ------------------------------ U-9 L-1

I. In each blank, write the word that fits best. Choose from the words below.

identical　　　　　**body cells**　　　**genes**　　　　　**traits**
chromosomes　　　**sex cells**　　　　**inherited**

Brown hair and blue eyes are examples of _____. A trait passed

on from parent to offspring is _____. There are 46

_____ in your body cells. There are thousands of

_____ in each chromosome. Cells in which chromosomes are not

in pairs are the _____.

II. Circle the word (between the brackets) that makes each statement true.

　A. The number of chromosomes in a human egg and a human sperm are
　　　[the same / different].

　B. A body cell of a pea plant has 26 chromosomes. Therefore, a sex cell of the pea plant
　　　has [26 / 13] chromosomes.

　C. A fertilized egg has [half / twice] the number of chromosomes as a sex cell.

　D. The [body cells / sex cells] of a cat have chromosomes in pairs.

III. Some behaviors are due to inherited traits. That is, we are born being able to do them. Other behaviors must be learned. Give some examples of things that people do that are not due to inherited traits.

_____

_____

_____

_____

IV. Looking ahead. For now, simply give these two questions some thought - and circle your response. At the end of this entire unit, see if your feelings have changed or if they remain the same. These are not right or wrong type questions. As science learns to control more about life, people will need to decide what is done with this new knowledge.

- Do you think that humans should be able to "change" our genes? **YES / NO / SOMETIMES**

- How about changing the genes of <u>other</u> organisms? Should humans be allowed to change these? For example, in lesson 3 of this unit you will learn that researchers are now able to "trick" a bacteria's genes; the bacteria then produce a *human* protein that controls how tall a person will become. Should this be allowed? **YES / NO / SOMETIMES**

What Happens When Cells Divide? U-9 L-2

Exploring Science / Historical Steps

Thinking Like a Scientist
- To Solve The Secret of Life!

Taylor was excited during her visit to Breonna's home; such a big backyard! Taylor was a "big-city" girl and had never seen a real garden. Breonna pointed out the green layer of small weeds growing between the tomato plants. Those had to go!

Breonna asked Taylor to grab the hand-held cultivator from the shed at the back of the yard. Taylor was baffled. She wondered what a cultivator looked like? As she entered the shed, she tried to picture the _shape_ of a tool whose _job_ it was to remove weeds. [Pause. Mentally picture a cultivator. Now, turn ahead 3 pages.]

Taylor wasn't aware of it, but at that moment she was thinking like a scientist. She was pondering the shape of something based on its job. Instead of the word "'shape," scientists say structure. Instead of the word "job," scientists say function.

Let's look at one of the best examples of this type of thinking in the history of biology. For many decades, scientists sought the structure of a vital molecule - DNA. They knew some of DNA's main functions. They were determined to learn its actual structure.

Why was this so important? Knowing the structure of DNA would reveal more about how cells actually copy themselves. It is the ability to copy (to reproduce) that makes cells - and entire organisms - alive!

The final pieces of the DNA puzzle were completed by a woman named **Rosalind Franklin** and three men - **Maurice Wilkins**, **James Watson** and **Francis Crick**. In the early 1950s, Franklin and Wilkins did most of the key lab work, but the puzzle was actually solved by Watson and Crick. These two based their idea on experiments that were completed by Franklin, Wilkins, and many scientists before them.

On the morning of February 28th, 1953, Watson and Crick built a model. It was their guess of DNA's structure. Soon, all four scientists agreed that it was a perfect fit for DNA's functions. That day, during lunch at the Eagle (a pub near their lab), Crick excitedly announced, "We've Discovered the Secret of Life!"

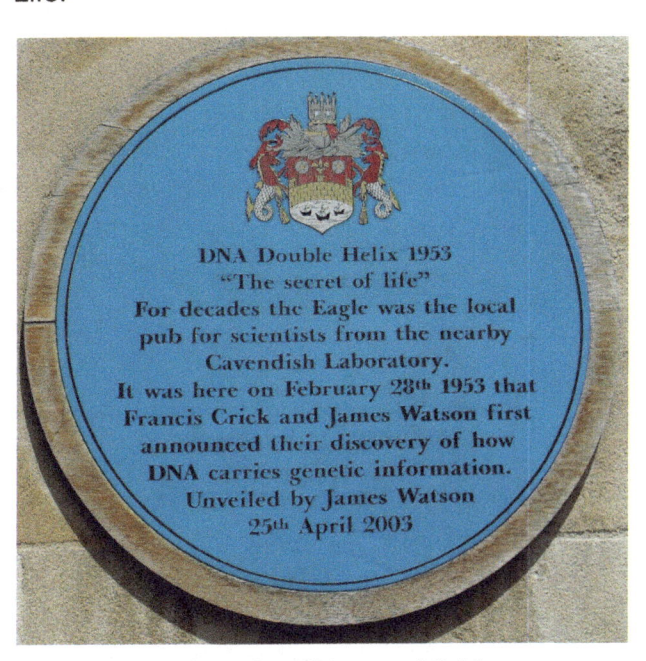

Plaque honoring Watson and Crick outside the Eagle, a pub in Cambridge

People everywhere now recognize the structure of DNA - a twisted ladder called a **double helix**.

➤ In 1958, at age 37, Rosalind Franklin died of cancer. In 1962, the three men mentioned were awarded the Nobel Prize. The Nobel Prize is only awarded to living scientists. Do you think that sounds fair?

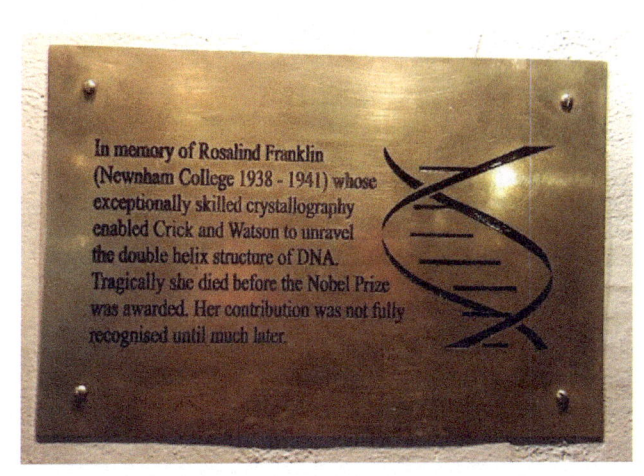

Plaque honoring Rosalind Franklin inside the Eagle, a pub in Cambridge, England

223

When Cells Divide

GENES, CHROMOSOMES, AND DNA.

Each chromosome is made of thousands of genes. And the genes of all living things are made of the chemical **DNA**.

If you could look at DNA under a very powerful microscope, what would it look like? It would look like a long, twisted ladder.

Look at step 1 of the diagram below. You can see that each step (or rung) of the DNA ladder has two halves. You can also see that there are only four different kinds of these "half rungs." (In the diagram, the difference between each of the four kinds of "half rungs" is illustrated by the dark and light colors, as well as the pointed and curved ends).

Different genes result from different arrangements of the rungs. Billions of arrangements are possible.

For cells to copy themselves, the chromosomes - made of genes - made of DNA - must copy themselves.

The following four steps, illustrated in the diagram below, show how DNA copies itself.

1. The DNA ladder untwists.
2. The rungs of the ladder split in two.
3. Each side of the original ladder replaces the missing side.
4. The new ladders return to the twisted shape. Now there are two (identical) copies of the original ladder.

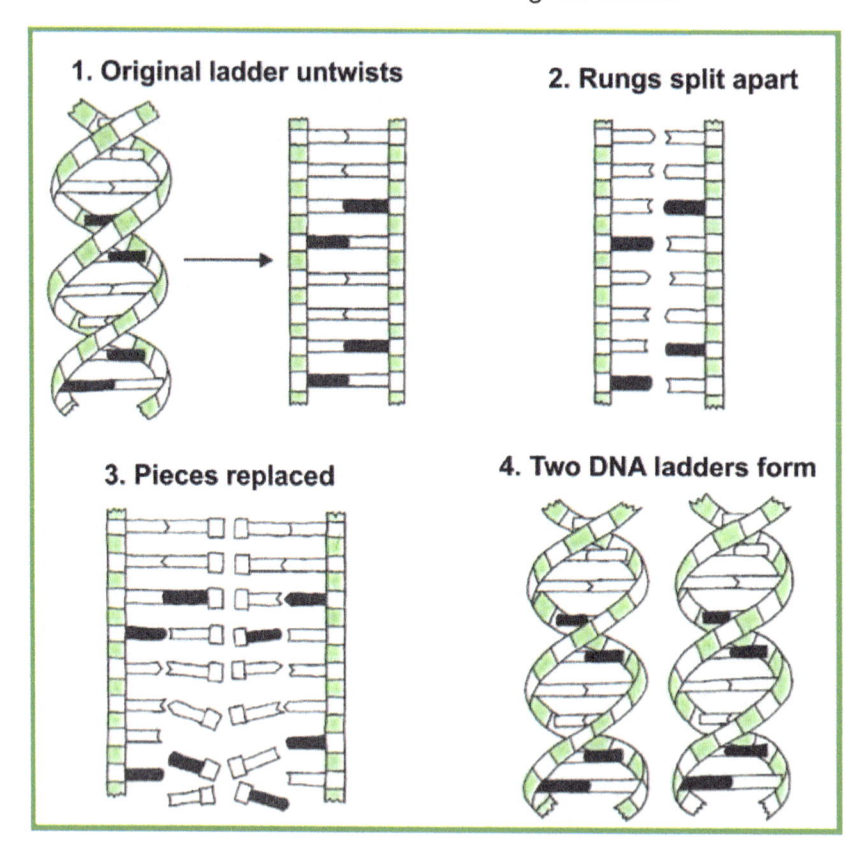

1. Original ladder untwists

2. Rungs split apart

3. Pieces replaced

4. Two DNA ladders form

MITOSIS (See the illustration on the next page).

Remember that all cells except sex cells are called *body* cells. The process in which a body cell's nucleus divides to become two nuclei is called **mitosis** (my-TOH-sis).

Pretend that an organism has only 3 pairs of chromosomes. At the start of mitosis, each chromosome copies itself. As the nucleus divides into two, one copy of each chromosome goes into each new nucleus. Each new body cell's nucleus is now exactly like the original cell's nucleus. Each has all of the same kinds of genes in it.

As you grow, your body cells divide to make new cells. Mitosis takes place during these cell divisions. Also, cells in your body may wear out or become damaged. Mitosis helps make new body cells to replace the old cells.

The Four Steps of Mitosis

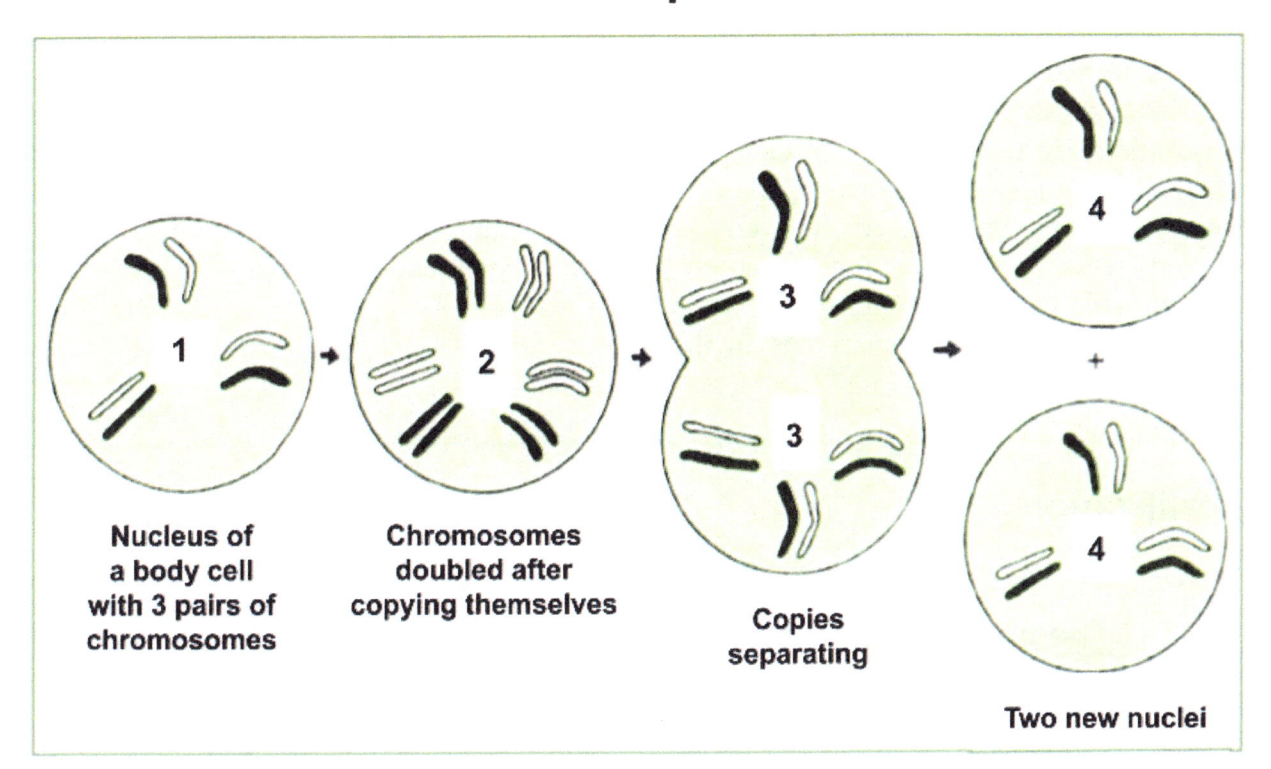

1 Nucleus of a body cell with 3 pairs of chromosomes

2 Chromosomes doubled after copying themselves

3 Copies separating

4 Two new nuclei

MEIOSIS (See the illustration below).

When *sex* cells form, the nucleus divides in a different way. Remember that *body* cells have chromosomes in pairs (one that came from the father, and one from the mother). When it's time to make a *sex* cell, a body cell in an ovary (or in a testis) divides to make two sex cells.

The pairs of chromosomes in the nuclei of these body cells *separate* from each other. One chromosome from each pair goes into each new sex cell's nucleus. Each new sex cell now has *half* as many chromosomes as the body cell. This kind of division is called **meiosis** (my-OH-sis).

In our imaginary organism, there are 3 *pairs* (6 total) chromosomes in a body cell. So, after meiosis there are only 3 *individual* chromosomes in each sex cell that is formed.

In human *body* cells, there are 46 chromosomes. After meiosis, there are 23 chromosomes in each *sex* cell (each egg or each sperm) that is formed.

Meiosis

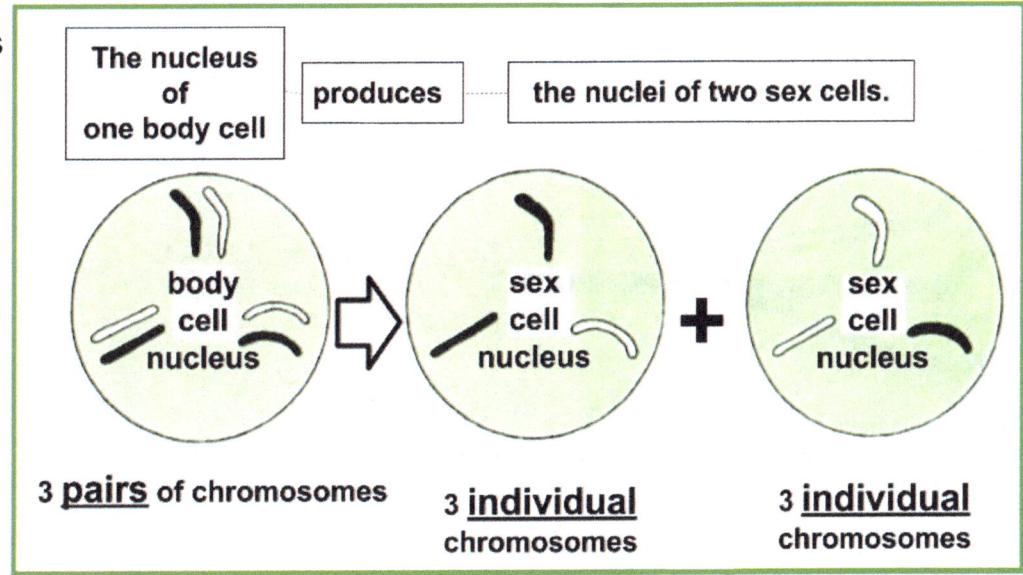

The nucleus of one body cell — produces — the nuclei of two sex cells.

body cell nucleus → sex cell nucleus + sex cell nucleus

3 **pairs** of chromosomes   3 **individual** chromosomes   3 **individual** chromosomes

FERTILIZATION (See the illustration below). What happens to chromosomes at fertilization - when an egg and a sperm join? Our imaginary organism's egg cell has 3 chromosomes. Its sperm cell has 3 chromosomes. When the sperm cell fertilizes the egg cell, the result is a zygote (fertilized egg) with 6 chromosomes.

The same thing happens when a human sperm fertilizes a human egg. The human egg and sperm each have 23 chromosomes. When they join, there are 46 chromosomes in the zygote.

AND BACK TO MITOSIS

Once the zygote is formed, normal mitosis returns. The zygote is a *body* cell. It has pairs of chromosomes. In a human zygote, there are 23 pairs, or 46 total chromosomes. When this zygote divides (by mitosis), each cell in the embryo that develops is also a body cell. All of the divisions of the zygote into other body cells are accomplished by mitosis.

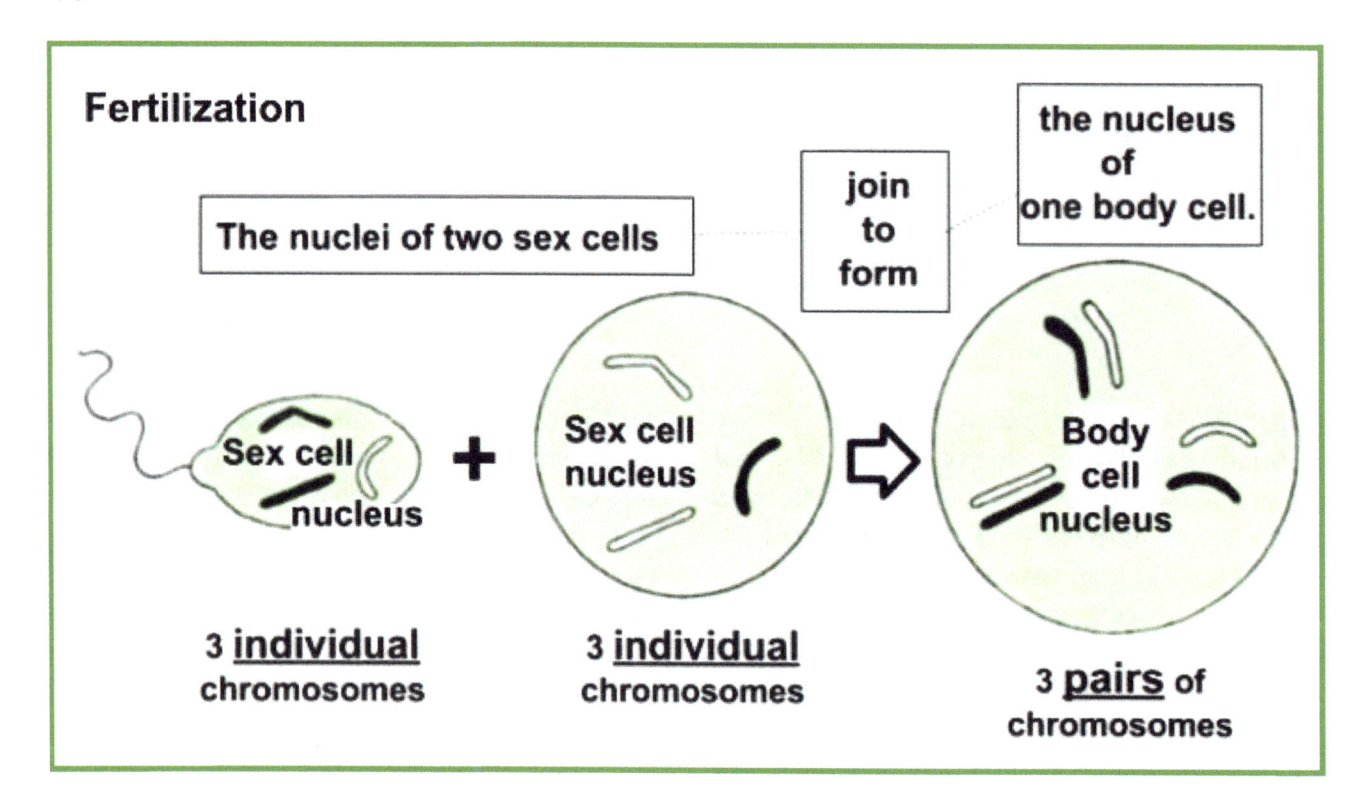

Fertilization

The nuclei of two sex cells join to form the nucleus of one body cell.

Sex cell nucleus + Sex cell nucleus ⇒ Body cell nucleus

3 **individual** chromosomes | 3 **individual** chromosomes | 3 **pairs** of chromosomes

---

*A cultivator - as mentioned in the opening story of "Exploring Science / Historical Steps".

REVIEW — U-9 L-2

I. In each blank, write the word that fits best. Choose from the words below.

DNA   two   chromosomes   zygote   mitosis   sex cells   four   fertilization

The chemical that makes up genes is _____. The nucleus of a body cell divides by _____ to form another body cell. Because DNA can copy itself, _____ can also copy themselves. Sex cells join during _____. In a DNA molecule, each step (rung) is made up of two halves, and there are a total of _____ possible types of these "half rungs."

II. Circle the word (between the brackets) that makes each statement true.
   A. Sperm cells form as a result of [mitosis / meiosis].
   B. Skin cells form as a result of [mitosis / meiosis].
   C. After [fertilization / mitosis], the number of chromosomes in a cell has not changed.
   D. An egg cell, with just half as many chromosomes as a body cell, forms during [meiosis / fertilization].
   E. A process in which the number of chromosomes is doubled is [meiosis / fertilization].

III. Each body cell of a corn plant has 20 chromosomes. How many chromosomes are there in a sperm cell of corn? In an egg cell? How many are there in a zygote of corn? Explain.

_____
_____
_____
____ ____ ____ ____ ____ ____ ____ ____ ____ ____

Another Big Step Regarding DNA! [More details for those interested in the "high school version."]

In 2003 scientists completed a major project called the **Human Genome Project**. We now know the order of all of the rungs (actually called **base pairs**) on all of the 46 DNA "ladders" (that is, on all 46 human chromosomes). There are 3 billion rungs in all!

Knowing the order in which these rungs occur is opening doors to new knowledge about how our bodies work, how things sometimes go wrong, and how we might fix many of these problems.

We also know that humans have about 20,000 genes. How many rungs does it take to make one gene? Most genes are made of about 27,000 rungs - but some have many more!

How Does DNA Make Proteins? U-9 L-3

Exploring Science / Historical Steps

Growing up Short... Or Not So Short.

Height is a human trait. Human traits are the result of human proteins. You might say that proteins are "what we're made of."

Twelve-year-old students may have very different heights. This is normal. But some students are unusually short because they do not have enough of a protein called human growth hormone (abbreviated **HGH**). When these girls and boys grow up, they will be much shorter than average.

To help some short children grow taller, doctors give them shots of the HGH protein.

From the 1950s through the 1970s, there was only one way for doctors to get HGH proteins for these shots. The HGH protein was taken from the pituitary glands of people who had died. (Remember that growth hormone is made in the pituitary gland). It took 50 to 100 human pituitaries to help one girl or boy grow only a little taller. The supply of the glands was not nearly enough to meet the demand. By the 1980s, scientists had learned a great deal about genes. They began *using* genes to make HGH proteins.

A certain human gene "tells" the pituitary to make HGH protein. Scientists learned how to attach this particular *human* gene to the gene material *in bacteria*! The bacteria followed the human gene's instructions - and made HGH proteins! And, as the bacteria copied themselves, they copied this HGH gene. Soon there were millions of bacteria making HGH proteins!

Using bacteria to produce HGH proteins (and other needed proteins such as insulin) is one way that scientists are solving many problems caused by faulty genes. This type of work is often called **genetic engineering**.

➤ Can you think of other gene-based problems that might be helped by genetic engineering?

A wide range of heights are normal in young people. For children with low growth hormone levels, HGH is an option.

DNA makes RNA makes Protein

To make HGH protein, scientists first had to learn how our cells make proteins. They discovered that it is a rather complicated process.

The next paragraph is the middle school version of the key steps involved.

Each gene's DNA contains a message; that message is how to build ONE protein. The DNA shares this message with a molecule called **RNA**.

RNA looks much like one side of a DNA ladder. The new RNA, now called **messenger RNA** (**mRNA**), takes this message to a ribosome in the cytoplasm. The ribosome "reads" the message and then builds that one protein.

In summary:

> **DNA** →
>   makes **mRNA** →
>   makes **protein**

If you are interested in the high school version of DNA's role in making proteins - keep reading.

You know that:

1) our body cells have 46 chromosomes, and that 23 come from each parent;

2) each chromosome is made up of thousands of genes;

3) each gene is made of DNA - the long molecule that is shaped like a twisted ladder;

4) DNA is made of rungs (steps) - that are each made of two halves. From this point on, we will use the word **base** when referring to a "half-rung". When two bases are linked together (to make a complete step on the ladder), they are called a **base pair**.

DNA contains the *messages* for making traits. What exactly do we mean by this? When you think of messages, you think of words. When you see certain letters of the alphabet - in a certain order - your brain has received the *message* that you are reading a specific word. For example, if you see a D, then an O, then a G - in that order - your brain thinks DOG. You know that you are reading about a furry animal called a dog!

Where are DNA's "words"? The words of DNA are the bases. More specifically, the order of the bases.

When three bases on the DNA molecule are in a specific order, the DNA molecule is "telling" the cell to select one specific type of amino acid. And for each of the 20 different amino acids, there is a different set of three bases.

Since it takes thousands of amino acids to make a single protein - DNA is a very long molecule. In fact, each portion of DNA that we call a *gene* tells the cell how to put together all of the amino acids to make just one specific protein.

Since human cells consist of thousands of different proteins, it takes thousands of genes to build a human. In fact, humans have about 20,000 genes!

Let's look at some examples of genes:
- One gene contains the message to make one type of *muscle* protein.
- Perhaps you remember that enzymes are also proteins. One individual gene contains the message to make each type of enzyme. For example, the enzyme amylase breaks starch molecules into sugar molecules. One gene carries the message how to make amylase.
- One gene contains the message to make one type of *antibody* (a germ-fighting protein) - in order to fight one specific germ.

To better understand how the bases of a DNA ladder guide a cell to make a protein, we must first look closer at the base pairs of the DNA ladder. Recall this image from the last lesson.

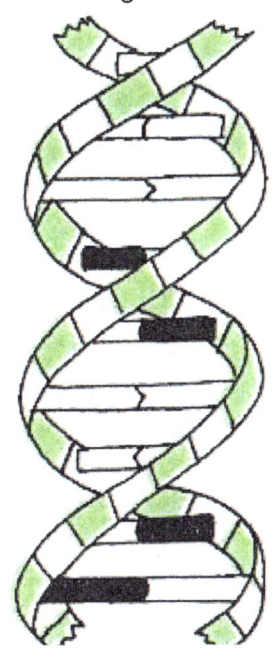

You know that each *base pair* is made up of two bases. But there's more. Notice that there are only four different shapes of bases. The illustration shows that one base ends in a *point*, one base ends in a pointed *notch*, one base ends in a *curve*, and one base ends in an *inward curve*.

Scientists have given each of these four bases their own name. The names are shown on the next illustration. To make it simpler, each base is usually referred to by only the first letter of its name. These four types of bases are simply called **A**, **T**, **C**, and **G**.

Notice how each base will only pair up with one other base - to make a *base pair*. In order for an **A** to form a base pair, it *must* join with a **T** (and only with a **T**). Likewise, the other two bases, **C** and **G,** are able to pair up only with each other. So a **C** joins only with a **G** to form a base pair.

In summary, to make a base pair (a complete rung on the DNA ladder), an **A** will only join with a **T** - and a **C** will only join with a **G**.

So how can only four bases carry the instructions to make thousands of different proteins? Within a DNA molecule, each base may occur any number of times in a row. For example, one side of a section of DNA may have an **A**, followed by another **A**, followed by a **T**, followed by three **C**s in a row, followed by five **A**s, followed by a **G**! Do you see how the possible arrangements of these four bases are limitless - just like the possible arrangements of the letters in the alphabet (to make words) are limitless?

A summary of the key parts of DNA

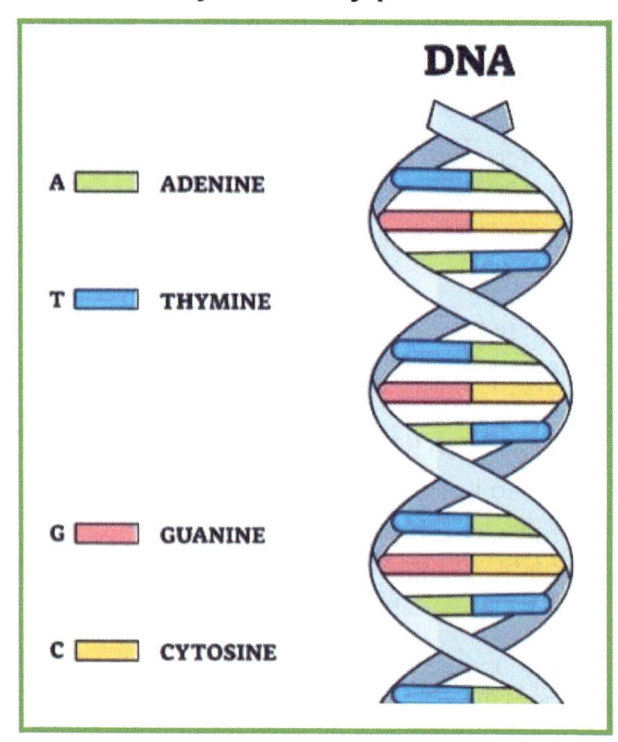

A — ADENINE
T — THYMINE
G — GUANINE
C — CYTOSINE

===
===

At long last, let's now see how the cell builds a protein. Refer often to the diagram on the next page - step by step - as you read further.

1) First, a section of DNA splits open. How much DNA must split? The amount that is equal to one gene - which is thousands of base pairs!

2) For the second step, one side of the newly opened DNA "pulls in" new "partner" bases to build one side of a "ladder" (quite similar to one side of a DNA). Instead of the word "partner," scientists say the bases are "complementary."

The new "one-sided" molecule is called **RNA**. Do you see how the bases that build this RNA are determined by the bases on the DNA?

Unlike DNA, RNA will always have this one-side-of-a-ladder shape. That is, RNA never becomes a full ladder (a "double helix"). Instead of saying "one side of a ladder," scientists say that RNA is "*single stranded*."

Once the RNA is built, it floats away from the DNA. Then the two sides of the opened DNA simply close back up.

The newly made RNA now has the needed message.

3) For the third step, the RNA floats out of the nucleus and into the cytoplasm. It takes its message to a **ribosome** in the cytoplasm. Since the RNA is now carrying a *message* from the DNA, this RNA is commonly called **messenger RNA**. Messenger RNA is abbreviated **mRNA**.

4) Now for the final step (four). Look ahead at the additional large diagram on the following page. Study it from left to right. The key structures to note are the mRNA (at the left), and the newly formed protein (at the top).

Ribosomes are able to "understand" the order of mRNA's bases. Each set of three bases on the mRNA "tells" the ribosome to bring in a particular amino acid - and to link it to the previous amino acid. These "sets-of-three" bases on a mRNA molecule are what scientists call the **genetic code**. When all of the amino acids have been linked together - a specific protein has been built!

A Summary of the Steps of Protein Production

1) In the nucleus, the bases of DNA that are needed in order to make one gene - split open.

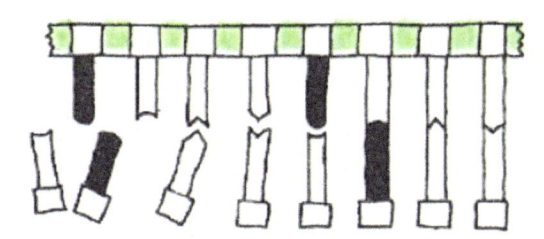

2) One side of the DNA (at the top) guides the production of an RNA molecule (at the bottom).

3) The new (single stranded) RNA is complete. It leaves the nucleus. It is now called <u>messenger RNA</u> (mRNA). This mRNA moves to a ribosome in the cytoplasm.

[For details of step 4, see the illustration the next page
 - while you reread the last paragraph of the previous page.]

4) Guided by the mRNA, proteins are built by the ribosomes.
Slowly study the illustration on the next page - from left to right.

The key structures to note are …

the <u>**mRNA**</u> (entering at the left),

and the <u>**protein forming from amino acids**</u> (at the top).

231

Ribosome making a protein

Ribosomes are in the cytoplasm

CELL

Amino acid

Protein forming from amino acids

Ribosome top half

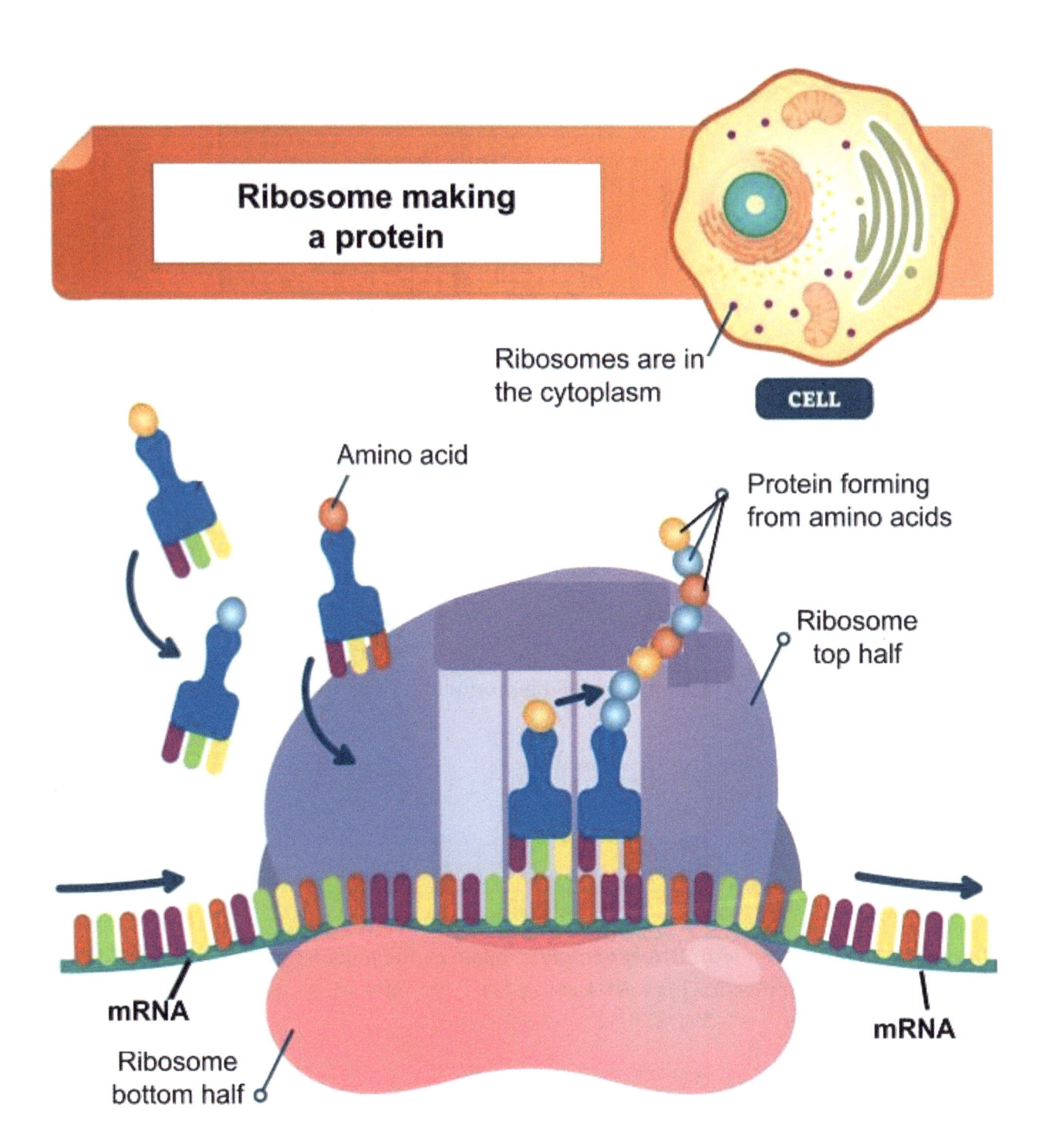

mRNA

mRNA

Ribosome bottom half

> **To Do Yourself**   How Might You Build a Model of DNA? of RNA?

You will need: **Creative thinking!**

What materials might you use to build a model of the double-helix structure of DNA, or a single-stranded RNA? Straws? Pipe cleaners? Styrofoam balls? Toothpicks? Clay? Glue? Tape?

Some background:
   Model building gave Watson and Crick an advantage when it came to determining the structure of DNA. They decided to try to build a model after learning that another famous scientist, **Linus Pauling**, used model building to discover some of the basic parts of proteins. Had Rosalind Franklin or Maurice Wilkins tried **model building**, the odds are good that they would have been the first to determine the structure of DNA.

   If you decide to build an **RNA** model, there is a bit of additional information that you need to know. When cells build RNA, they don't use thymine (the **T**). In its place, cells use a very similar molecule called **uracil** (which is abbreviated as **U**).

1. Decide which model you will build.
2. Brainstorm what materials would work best. Gather your materials.
3. Plan how to put your materials together.
4. Add labels to some of the key parts, such as the bases. You might use **A** or the full word **adenine**, **T** or **thymine**, **C** or **cytosine**, **G** or **guanine**. [Remember: If you choose to build RNA, use **U** (uracil) instead of **T** (thymine).]
5. If you would like to label the molecules that build the spiraling outside parts of DNA and RNA, research the following term: **sugar-phosphate backbone**.

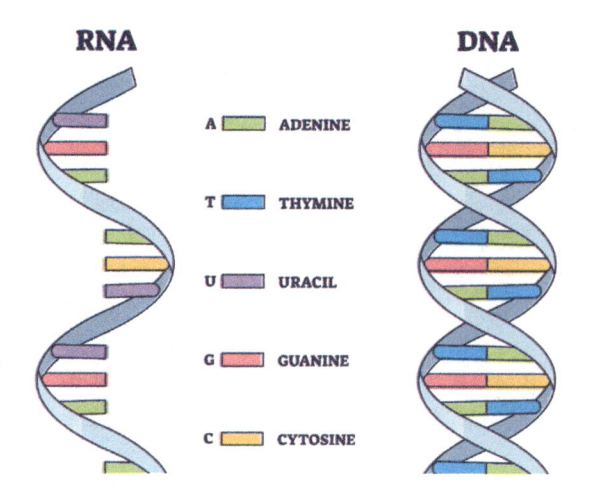

RNA   **DNA**

A — ADENINE
T — THYMINE
U — URACIL
G — GUANINE
C — CYTOSINE

REVIEW   U-9 L-3

I. In each blank, write the word that fits best. Choose from the words below.

   protein   HGH   mRNA   DNA
   fat   gene   ribosomes   RNA

In the nucleus of a cell, _____ molecules store the message for making proteins.

Each DNA molecule has many genes, one for each _____ that will be made. DNA shares the message (to make a protein) by first building a molecule that is

abbreviated as _____. Once this molecule has the *message* to make a protein, its

abbreviated name becomes _____. In the cytoplasm, proteins are actually put

together in the parts of cells called_____. Some people's

bodies make too little of a protein called _____. As a result, these people stay rather short.

II. Circle the word (between the brackets) that makes each statement true.

   A. To put a human gene into a bacterium, scientists use [grafting / genetic engineering].

   B. A message-*storing* molecule inside of the nucleus is called [DNA / mRNA].

   C. A message-*carrying* molecule that travels to a ribosome is called [DNA / mRNA].

What Are Dominant and Recessive Genes? U-9 L-4

Exploring Science / Historical Steps

Speckled Corn and Jumping Genes. Have you seen Indian corn? Its grains can be purple, yellow, or speckled. The same ear can have grains of different colors. Some people use Indian corn to decorate their homes.

Scientist **Barbara McClintock** had another use for Indian corn. She studied the way its colors are inherited. She began her work with corn in the 1920s.

One of McClintock's questions had to do with the speckled color. After years of work growing and analyzing Indian corn, she discovered that some of its genes can jump from one chromosome to another! When a gene does this, specks may appear on the corn grain.

At first, other scientists paid no attention to McClintock's discovery of jumping genes. Her work was "ahead of its time" as the saying goes.

We now know that jumping genes are common in many organisms, including humans! A great deal of important research is being done based on McClintock's proof that some genes jump. This work is helping scientists better understand some diseases. It is also helping scientists learn how living things change over time.

So it turned out that McClintock's discovery was *very* important. One October day in 1983, at age 81, McClintock heard on the radio that she had won the Nobel Prize! This prize is the highest honor that a scientist can receive.

➤ Want more? Search the term transposon. Trans means 'across.' Pos means 'position.' Genes that move to other positions - jumping genes - are called transposons.

Barbara McClintock's study of Indian corn led to her discovery of "jumping genes."

234

Dominant and Recessive Genes

The laws of heredity were first stated in the 1800s by an Austrian monk and scientist, **Gregor Mendel**. Like McClintock, Mendel made his discoveries with plants.

Mendel worked with garden peas. Some pea plants grow tall. Others grow short. Height is an inherited trait in peas. Each pea plant has two genes for height, one from each parent. A plant that has two genes for tallness is "**pure**" for that trait. It is called **pure tall**. A plant that has two genes for shortness is **pure short**.

Mendel observed that when both pea plant parents are pure tall, all of their offspring are tall. When both parents are pure short, all of the offspring are short. He used the capital letters **TT** to stand for the two tallness genes in a pure tall. The small letters **tt** stand for the two shortness genes in a pure short. To repeat, all of the offspring of two **TT** parents are also **TT**, or pure tall. And all of the offspring of two **tt** parents are **tt**, or pure short.

What would happen if one parent was pure tall (**TT**) and the other parent was pure short (**tt**)? Mendel took pollen (which contains the plant's sperm) from some pure talls. He placed this pollen on the tip (the stigma) of pistils of pure short plants. Recall that the pistil's ovary contains the plant's eggs.

He also placed pollen from pure short plants onto the stigmas of pure tall plants.

We say that Mendel **crossed** the pure talls with the pure shorts. Then he waited for the offspring to grow.

Keep in mind that each parent plant in Mendel's crosses was *pure* for height. The tall plants had two genes for tallness (**TT**). The short plants had two genes for shortness (**tt**).

When a pea plant has one gene for tallness and one gene for shortness, it is called a **hybrid** (HY-brid). We write **Tt** to stand for a hybrid. The offspring of Mendel's crosses of the pure plants were all hybrids. Each offspring had received one gene for tallness from one parent, and one gene for shortness from the other parent.

What do you think the hybrids looked like? Were they tall, or short, or in-between? Mendel was surprised. They were *all* tall! Only the genes for tallness showed up. The genes for shortness seemed to disappear.

Here is how Mendel explained what happened. Tallness in pea plants is a **dominant** (DOM-uh-nunt) **trait**. A dominant trait is one that shows up in a hybrid. The trait that does not show up in a hybrid is **recessive** (rih-SES-iv). For pea plants, shortness is a recessive trait.

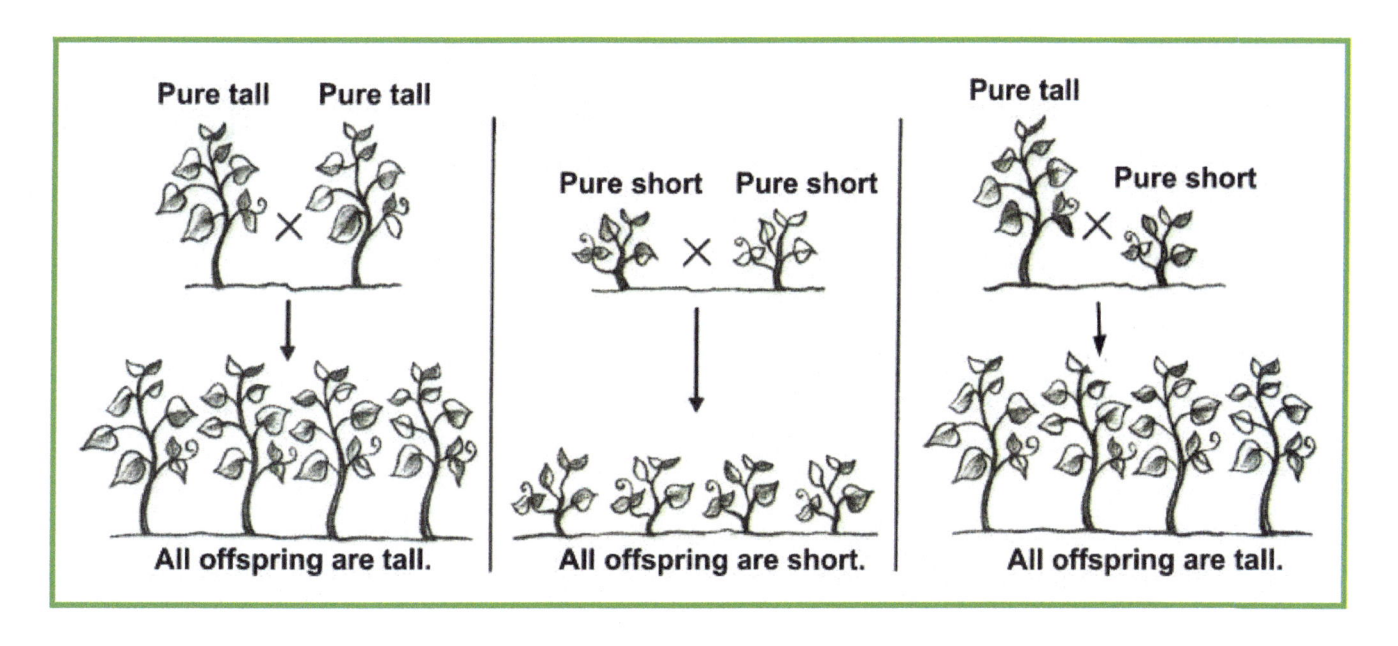

Pure tall × Pure tall	Pure short × Pure short	Pure tall × Pure short
All offspring are tall.	All offspring are short.	All offspring are tall.

235

Next, Mendel crossed two hybrid tall plants (**Tt**). What would the offspring look like this time? Mendel was surprised again. Most of the offspring were tall. But some were short. On average, one offspring out of four was short.

Mendel reasoned that shortness was hidden by the tallness in the hybrids. But when two hybrids (**Tt**) are crossed, some of the offspring are pure short (**tt**), with no dominant (**T**) gene. In a hybrid (**Tt**), a dominant gene hides the recessive gene. But the recessive gene is still there, and it can show up in a hybrid's offspring if the offspring get only the recessive (**t**) genes.

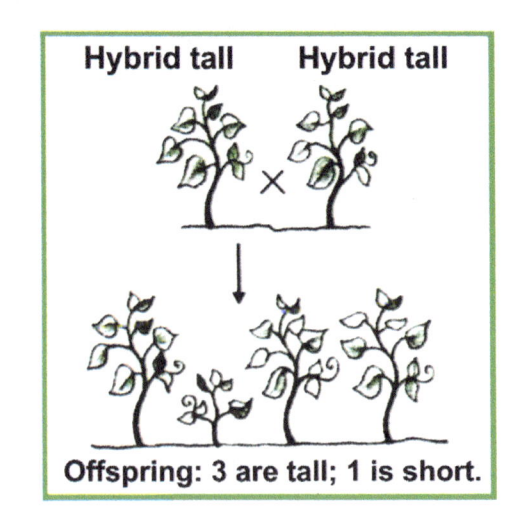

Hybrid tall   Hybrid tall

Offspring: 3 are tall; 1 is short.

➤ To Do Yourself   How Are Genes From Parents Combined in Offspring?

You will need:

20 red beans, 20 white beans, 2 jars.

You will make a model of the way dominant and recessive genes are inherited.

1. Place 10 red beans and 10 white beans into each of 2 large jars.
2. Shake each jar thoroughly. Close your eyes and pick one bean from each jar. Record your selection in a table.
3. Repeat step 2 until all of the beans have been picked.

 -Each red bean will stand for a dominant gene.

 -Each white bean will stand for a recessive gene.

- Red combines with red to produce red offspring (pure).
- Red combines with white to produce red offspring (hybrid).
- White combines with red to produce red offspring (hybrid).
- White combines with white to produce white offspring (pure).

Jar 1   Jar 2

(One parent)   (One parent)

Questions

1. Why was one bean selected from jar 1 and one from jar 2? _____

_____

2. Why were the two beans "combined?" _____

3. What does this activity show about genes? _____

_____

--- **REVIEW** --- U-9 L-4

I. In each blank, write the term that fits best. Choose from the terms below.

pure tall   **pure short**   **Mendel**   **hybrid**
dominant   **McClintock**   **recessive**

The first scientist to state the laws of heredity was _____.

In peas, a _____ plant has two genes for shortness. A plant with one gene for tallness and one gene for shortness is _____. In peas, tallness is a _____ trait and shortness is a _____ trait.

II. Mendel crossed peas pure for green seeds with peas pure for yellow seeds. All of the offspring had yellow seeds. Which seed color (in peas) is dominant? Explain.

_____

_____

_____

III. In humans, curly hair is dominant over straight hair. Is it possible for two curly-haired parents to have a straight-haired child? Explain.

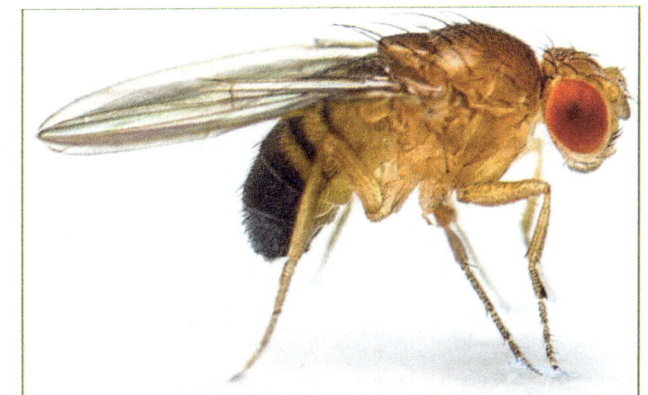

_____

Four fruit flies on a slice of banana.

Close up of a fruit fly.

Studies using fruit flies (Drosophila melanogaster) have contributed much to genetics. Why? They are so tiny that thousands may be kept in a small amount of space, and their food is inexpensive. Fruit flies complete their life cycle (egg, to larva, to pupa, to adult) in only 2 weeks, so huge numbers of results from "crosses" of flies with various traits may be obtained quickly.
• **Want more?** Research the geneticist Thomas Hunt Morgan - the most famous person to use fruit flies.

Can Heredity Be Predicted?

U-9 L-5

Exploring Science / Historical Steps

Arch, Loop, or Whorl: What's Your Type?
Look at your thumb with a magnifying glass. Compare what you see with the illustrations below. One type of fingerprint is an arch, one is a loop, and one is a whorl. Which type matches yours? While these patterns are common in humans, no two humans have exactly the same fingerprints - not even identical twins!

For many years detectives have used fingerprints to track down criminals. Prints from babies' feet have also helped to prevent mixing up newborn babies in hospitals.

Then, in the 1980s, two big advances in identifying individuals occurred. In 1984, a researcher named **Alec Jeffreys** discovered what became known as "DNA fingerprinting." This has nothing to do with the fingers. It uses DNA to identify an individual from a tiny sample of their tissue (such as a drop of blood).

A year later, another researcher, **Kary Mullis**, found a way to make millions of copies of a tiny piece of DNA. With this method, only an incredibly small amount of DNA was needed to carry out the complicated steps of DNA fingerprinting.

➤ It is now possible to easily have your DNA "mapped" and then compared to others who have done so - across the globe. Can you think of advantages and disadvantages of this technology?

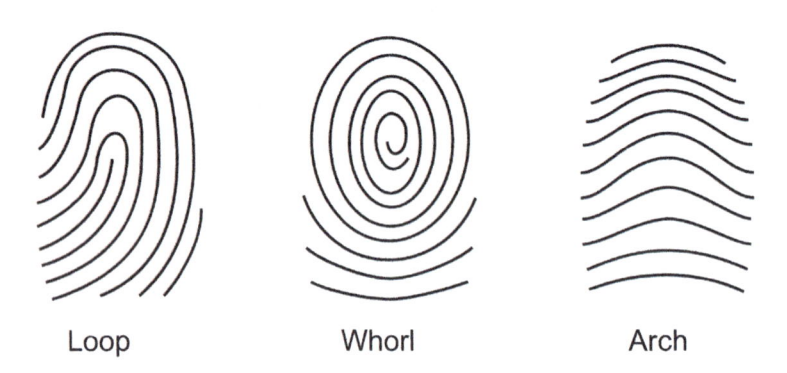

Loop   Whorl   Arch

The loop, the whorl, and the arch are the three most common types of fingerprints.

Predicting Heredity

If a parent has a certain gene, a child may, or may not, inherit that gene. Some human genes are either dominant or recessive. Others are neither. Let's see how these kinds of genes are passed on.

DOMINANT AND RECESSIVE GENES.
Suppose that two people with brown eyes marry. Can they have a child with blue eyes?

The gene for brown eyes is dominant. And the gene for blue eyes is recessive. A capital **B** stands for the brown-eye gene. A small **b** stands for the blue-eye gene. For a person with two brown genes we write **BB**. For a person with two blue genes we write **bb**. For a person with one brown gene and one blue gene we write **Bb**.

The person with **BB** is called pure brown. The person with **bb** is called pure blue. And the person with **Bb** is a hybrid. The sex cells of a pure brown person are all **B**. The sex cells of a pure blue person are all **b**. And the sex cells of a hybrid person have a 50-50 chance of being **B** or **b**.

How can the sex cells of two hybrd brown-eyed parents combine? To show this, we use a chart called a **Punnett** (PUN-et) **square**. A Punnett square shows the different ways that two parents' sex cells can combine. For each number below, look at the same number in the diagrams.

(1) We draw a square with four boxes in it. For two hybrid parents, write **B** and **b** for the father's genes across the top. Write **B** and **b** for the mother's genes down the side.

(2) In the upper left corner box, we write what would happen if the first sperm (**B**) combined with the first egg (**B**). The result would be **BB**, a pure brown. A child with this combination will have brown eyes.

(3) In the upper right corner box, we write how the second sperm (**b**) would combine with the first egg (**B**). The result is **Bb**, a hybrid brown. A child with this combination will have brown eyes.

(4) In the lower left corner box, we write how the first sperm (**B**) would combine with the second egg (**b**). Again, the result is hybrid brown, **Bb**.

(5) Finally, In the lower right corner box, we write how the second sperm (**b**) and the second egg (**b**) would combine. This child has **bb**, or pure blue eyes.

So, to repeat, can two brown-eyed parents have a blue-eyed child? Yes, if both parents are hybrid (**Bb**) for eye color. The *chance* of these parents having a blue-eyed child is 1 in 4, as shown by the Punnett square (as well as the diagram on the following page).

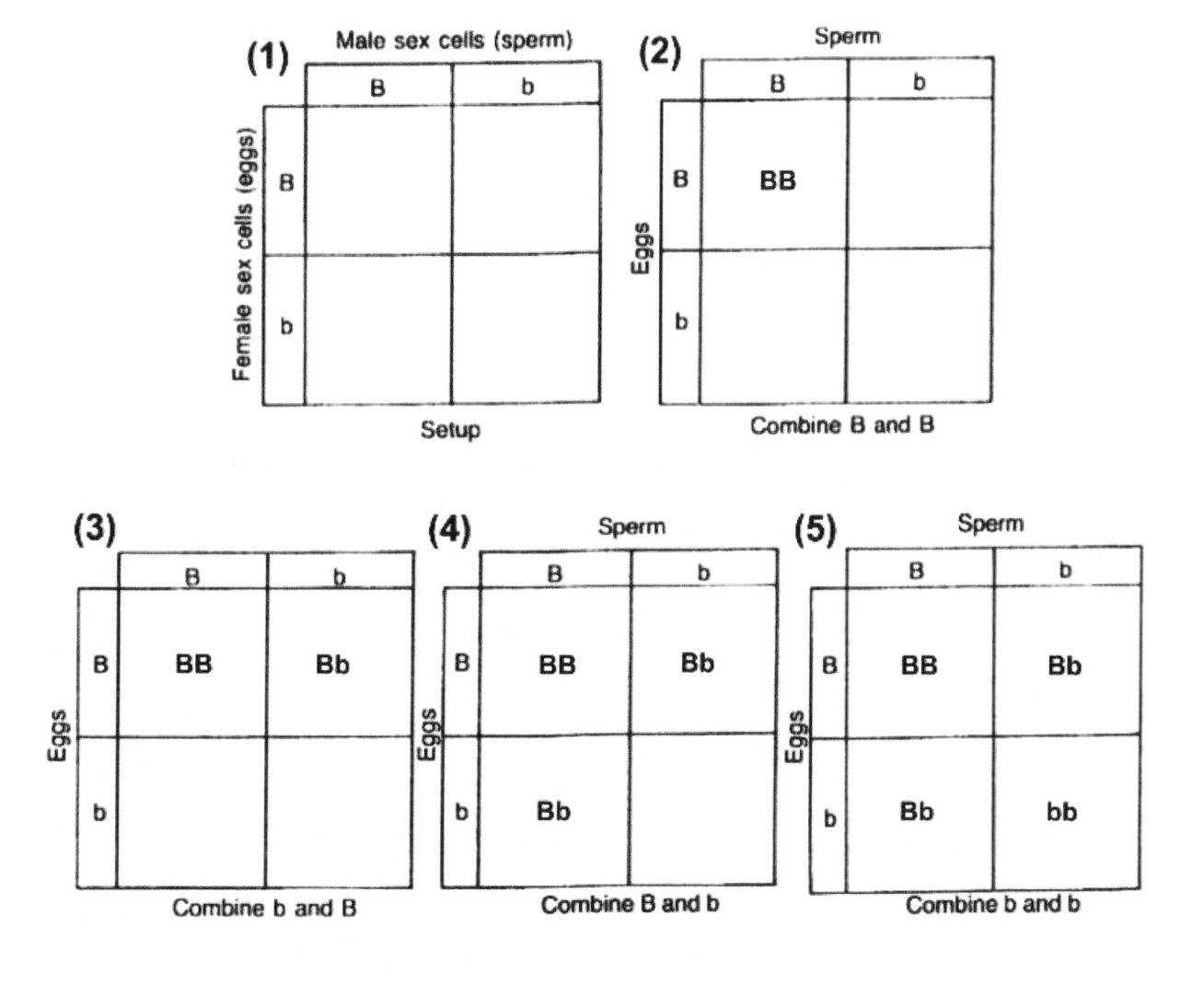

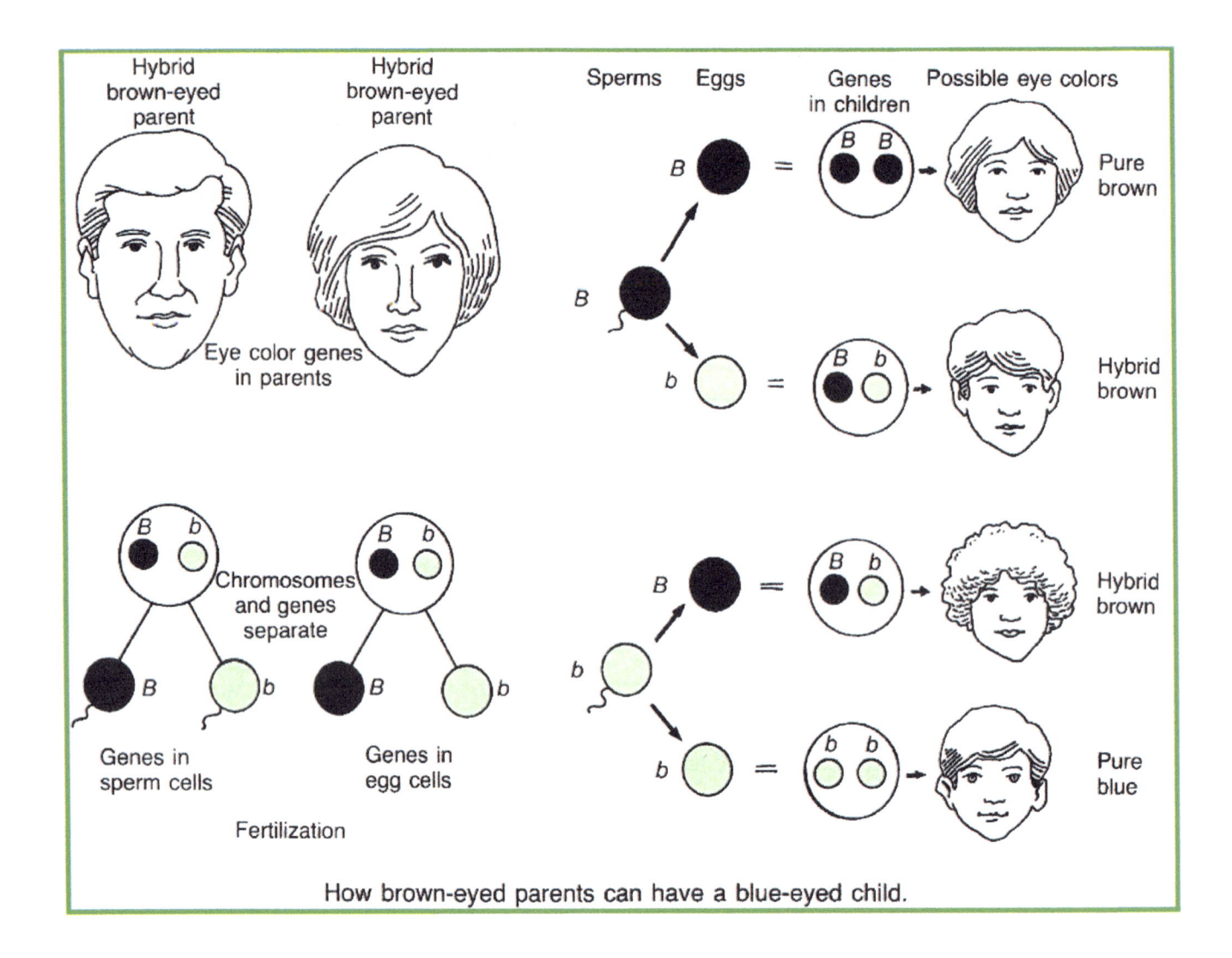

How brown-eyed parents can have a blue-eyed child.

Let's look at another example of how eye color might be inherited. Draw yourself a Punnett square. One parent is hybrid brown (**Bb**). The other parent is pure blue (**bb**). What are the odds of each eye color in the children?

Your Punnett square should show that the chance of these parents having a child with blue eyes is one in two. The chance of them having a child with brown eyes is also one in two.

===

BLENDING TRAITS. [See next page].

The genes for some traits are neither dominant nor recessive. In four-o-clock flowers, the genes for color can be red (**R**) or white (**W**). When a pure red flower (**RR**) is mated with a pure white flower (**WW**), all of the offspring are pink (**RW**). In this kind of gene combination, the traits are mixed - or **blended**.

Blending takes place in much of human heredity. Genes for dark and light skin color are neither dominant nor recessive. In fact, for skin color, there are *several* pairs of genes (not just one pair) that determine this trait. Thus, many different shades of human skin color are possible.

Some people have neither brown nor blue eyes. Hazel, green, and gray eyes may be due to blending, or it may be due to the work of several pairs of genes.

240

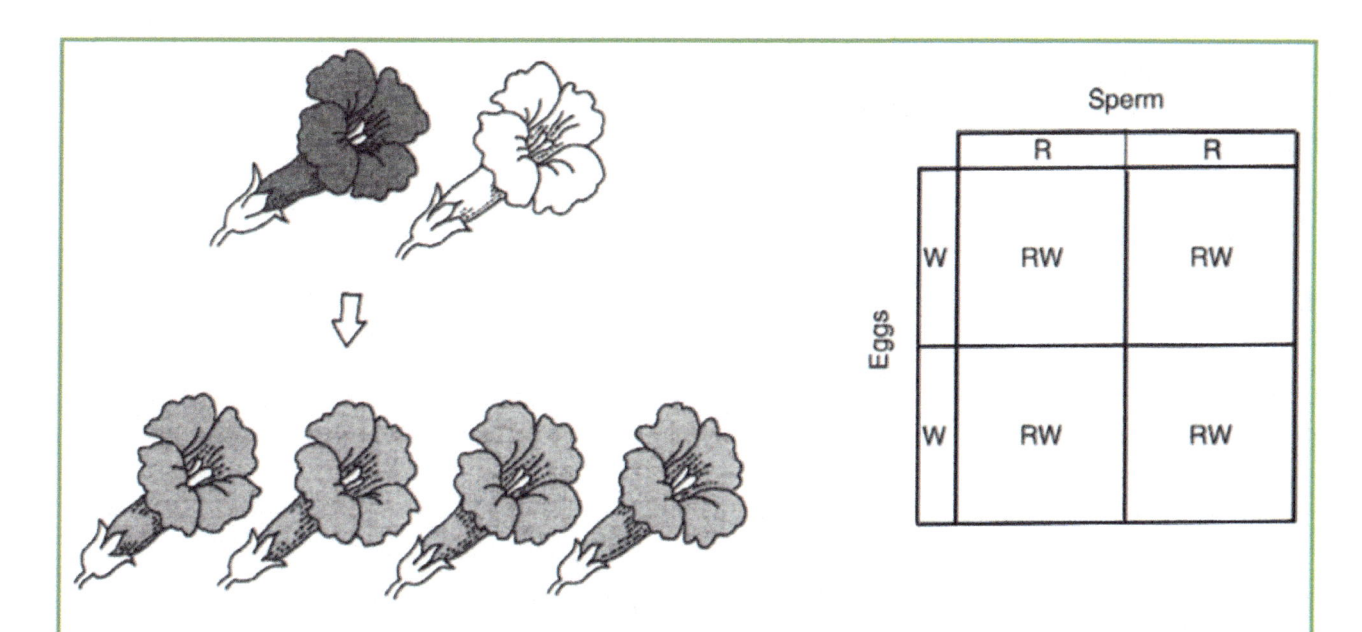

In some flowers, when one parent is red and the other is white, the offspring are all pink.

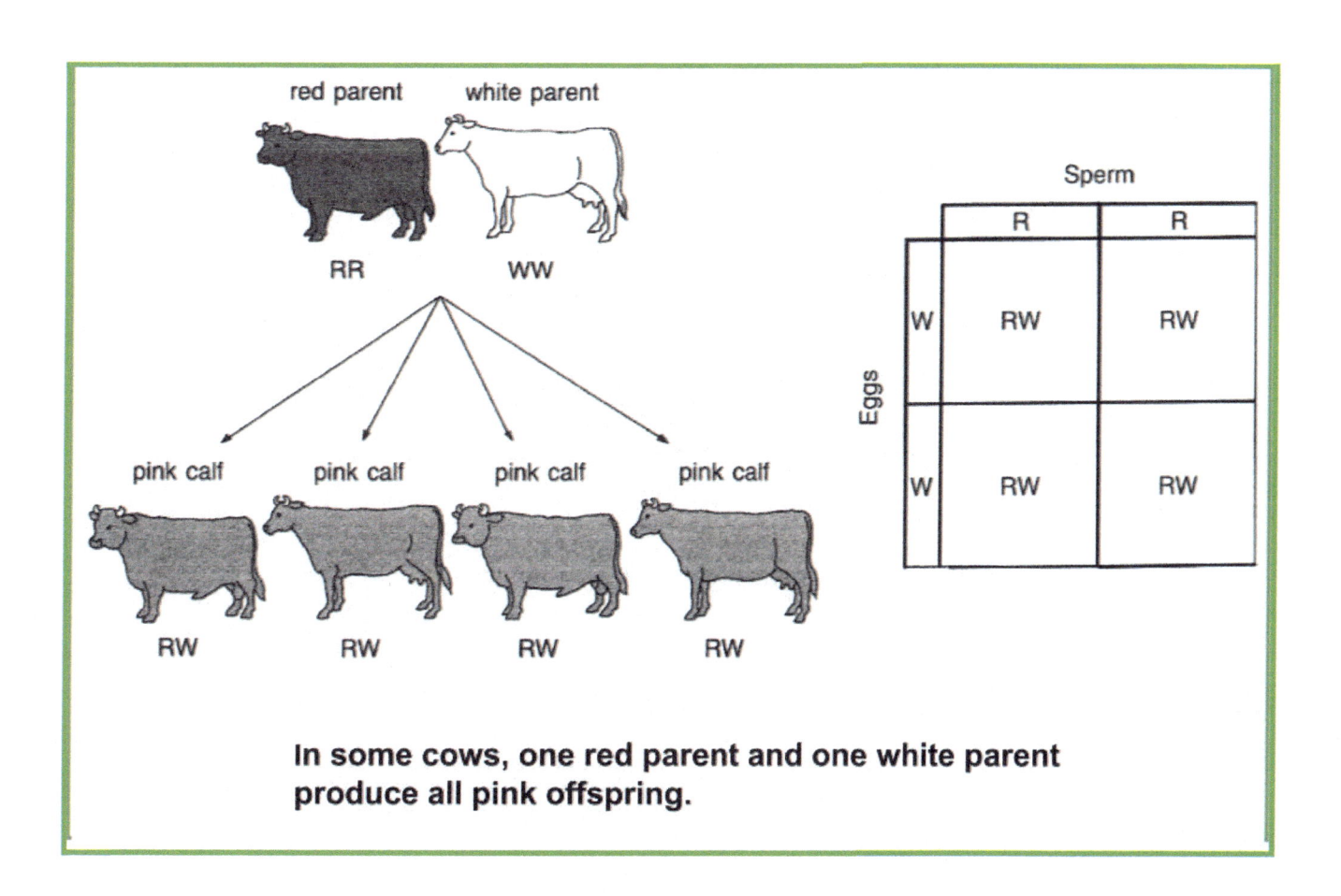

	Sperm	
	R	R
W (Eggs)	RW	RW
W	RW	RW

In some cows, one red parent and one white parent produce all pink offspring.

red parent (RR) × white parent (WW) → pink calf (RW), pink calf (RW), pink calf (RW), pink calf (RW)

	Sperm	
	R	R
W (Eggs)	RW	RW
W	RW	RW

241

> **To Do Yourself**   **How Can Crossbreeding Produce New Kinds of Organisms?**

You will need:
Tangerine, grapefruit, tangelo, plastic knife, paper plates, napkins

1. Make a table to record the following traits in each of the fruits: size, color, seed size, juiciness, taste (sweet or sour), flavor, smell.
2. Observe the grapefruit and the tangerine for each of the traits in step 1. Record the traits in your table.
3. Determine which of the traits are favorable and which are unfavorable.
4. Then, examine the tangelo for the traits in step 1. Repeat steps 2 and 3. The tangelo is a cross between the grapefruit and the tangerine. Record the traits in the table.

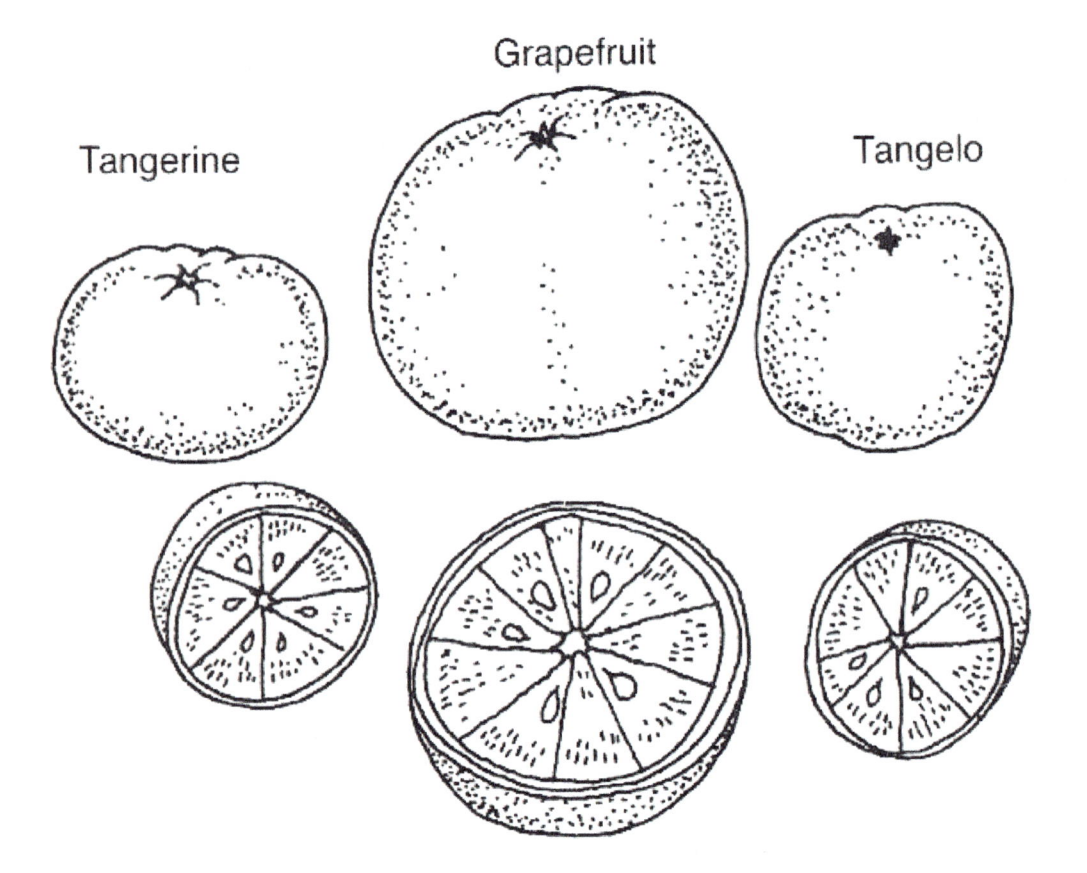

Tangerine   Grapefruit   Tangelo

Questions

1. What were the favorable traits in the grapefruit? In the tangerine? In the tangelo?

___

2. What were the unfavorable traits in the grapefruit? In the tangerine? In the tangelo?

___

3. Why are fruits such as the tangelo developed? ___

REVIEW — U-9 L-5

I. Circle the word (between the brackets) that makes each statement correct.
   A. In a Punnett square for eye color, the letter **b** stands for a gene for [brown / blue].
   B. The letters **BB** in the square stand for a person who has [brown / blue] eyes.
   C. A hybrid offspring in the square would be written [**Bb** / **bb**].
   D. Among the offspring of two hybrid brown-eyed parents, the chance of a blue-eyed child is 1 in [3 / 4].

II. In the Punnett square below, show what offspring may result from a pure blue-eyed father and a hybrid brown-eyed mother. Then answer the questions.

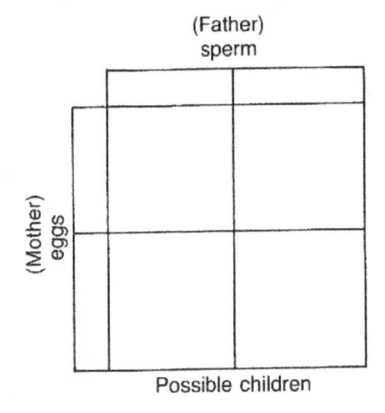

(Father) sperm / (Mother) eggs / Possible children

   A. Can any of the children have brown eyes? If so, what are the chances? _____
   B. Can any of the children have blue eyes? If so, what are the chances? _____
   C. Suppose the parents have just one child. Will the child have brown eyes or blue eyes?
      Explain your answer. _____
   D. Suppose there are four children. Will two have brown eyes and two have blue eyes?
      Explain your answer. _____

III. In the Punnett square below, show what offspring may result from a cross between hybrid (**RW**) pink four-o-clocks. Then answer the questions.

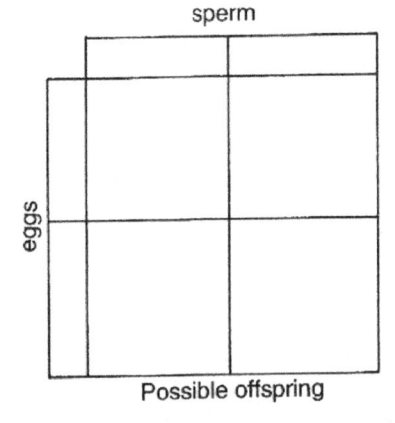

sperm / eggs / Possible offspring

   A. Can any of the offspring be red? If so, what are the chances? _____
   B. Can any of the offspring be pink? If so, what are the chances? _____
   C. Can any of the offspring be white? If so, what are the chances? _____

How Is Sex Inherited?

U-9 L-6

Exploring Science / Historical Steps

You're biased because you're a girl!
You're biased because you're a boy!

These may seem like silly arguments. However, to a scientist, being called biased is very serious. A **biased** person lets their own experiences influence how they view things. Scientists must view things **objectively** - meaning without bias.

It is very difficult to avoid biases. For example, a person raised in a small town might be biased (in a negative way) about life in a big city. Of course, the opposite might also occur.

A good experiment is designed without biases. Scientists must also be unbiased as they judge an experiment's conclusion.

Scientists have tried for years to create unbiased experiments in their study of the brains of girls and boys. Does a person's **gender** (sex) mean that they have differences in their brains?

This became an even more exciting field when new tools that could "see" brain activity were invented. You have probably heard of MRI and f**MRI** machines. The "f" indicates a special type of MRI that can tell which part of the brain is "**f**unctioning" at a specific time. MRI's have a long history.

Experts in physics made discoveries related to MRI's in the 1930s. Eventually these discoveries were applied to medicine. The first MRI scan was used on a cancer patient in 1978. Millions of scans are now completed each year. In fact, even more advanced brain imaging machines - with names like MEG, and PET - have been developed.

So, have brain scans shown differences based on gender? Despite years of research, the verdict is still out. Soon after one study makes a claim, the next study disputes it.

The gender that a person feels "fits" them is called their **gender identity.** A person's gender identity does not always match their body. Many people consider themselves to be "straight" (attracted to the opposite gender); some people consider themselves to be "gay" (attracted to the same gender). And some people refer to themselves as **nonbinary;** they feel that they do not fully fit one gender or the other.

Scientists certainly know that the brain is incredibly complex, and that all types of humans exist. Like scientists, all of us need to strive to be unbiased about each other.

➤ Do you think that better brain imaging machines will someday confirm brain differences based on people's gender identity?

Genes and Gender

"Is it a girl or a boy?" This is the first thing everyone wants to know about a new baby. We don't know for sure how the brains of girls and boys are different. But we do know why some babies are born with female body parts and others are born with male body parts. The reason is found in the chromosomes' genes.

You know that you have 46 chromosomes in your body cells. In a lab, all of the chromosomes in a body cell can be lined up into 23 _pairs_.

Scientists give each chromosome pair a number. Whether you have female or male body parts depends on chromosome pair number 23 - often called the **sex chromosomes**.

Sex chromosomes come in two possible shapes. One shape is an "**X chromosome**." The other shape is called a "**Y chromosome**." Females' body cells have two **X** chromosomes (**XX**). Males' body cells have one **X** chromosome and one Y chromosome (**XY**).

On the next page, look at a diagram of all of the chromosomes in the nucleus of one human body cell. Keep in mind that humans receive 23 chromosomes from **each** parent. Therefore, humans have a total of 46 chromosomes. Count them in the illustration! Pay special attention to chromosome # 23.

244

Below are the chromosomes that this person received **from their mother**.

- Note: From their mother, humans can only receive a chromosome # 23 that looks like an "**X**."

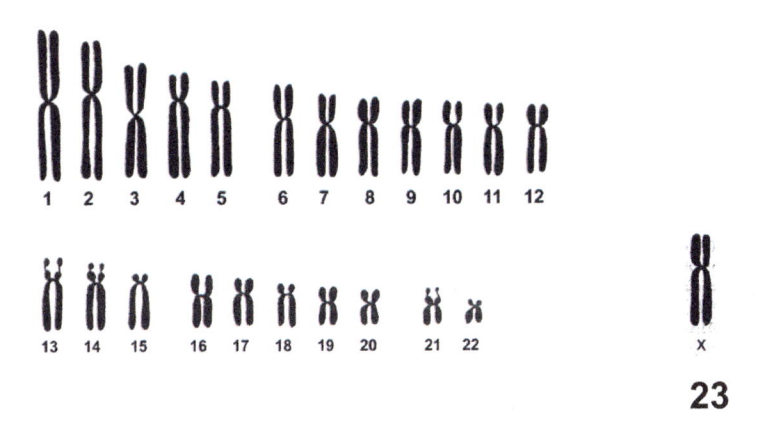

Below are the chromosomes that the same person received **from their father**.

- If this person is a <u>male</u>, from his father he received a chromosome # 23 that looks like a "**Y**"
- If this person is a <u>female</u>, from her father she received a chromosome # 23 that looks like an "**X**."

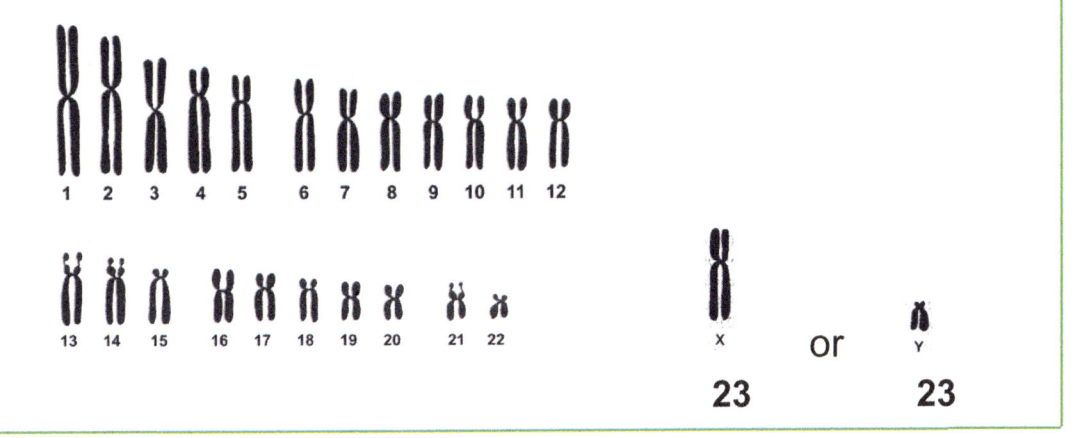

From what you just read, do you see that <u>every</u> egg cell has one **X** chromosome, but only <u>half</u> of the sperm cells have an **X** chromosome? The other half of the sperm cells have a **Y** chromosome.

If a sperm with an **X** chromosome fertilizes an egg, the zygote has two **X** chromosomes (**XX**). This child will be a female.

If a sperm with a **Y** chromosome fertilizes an egg, the zygote has one **X** chromosome and one **Y** chromosome (**XY**). This child will be a male.

For each baby in a family, what are the chances that it will be female or male? The chances are even. To show why, we again use a Punnett square.

We see that half of the possible egg-sperm combinations will result in female offspring (**XX**). The other half will result in male offspring (**XY**).

Father (Sperm cells)

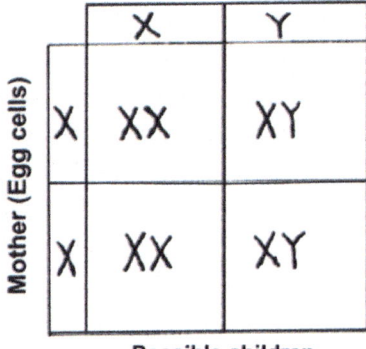

	X	Y
X	XX	XY
X	XX	XY

Mother (Egg cells)

Possible children

Keep in mind that a Punnett square only shows the *chances* of having girls or boys. In reality, in any one family, most, or even all, of the children can be of one gender. But in the whole population of humans on Earth, the number of girls and boys are about equal.*

245

WHAT ABOUT TWINS? We can use what we know about **X** and **Y** chromosomes to answer some other questions. Why are identical twins always the same gender? And why are nonidentical (**fraternal**) (fruh-TUR-nul) twins sometimes of different genders and sometimes of the same gender?

As you know, after a sperm fertilizes an egg, an embryo forms. On rare occasions, an embryo splits in half soon *after* fertilization. Each half becomes a separate embryo and, eventually, a separate individual. Since they came from the same sperm and egg, both embryos have all of the same chromosomes.

Having all of the same chromosomes means that the twins have all of the same genes, too. Because of this, twins that develop from one fertilized egg are **identical**. And because identical twins have the same genes, they have the same traits. Therefore, they are always of the same gender.

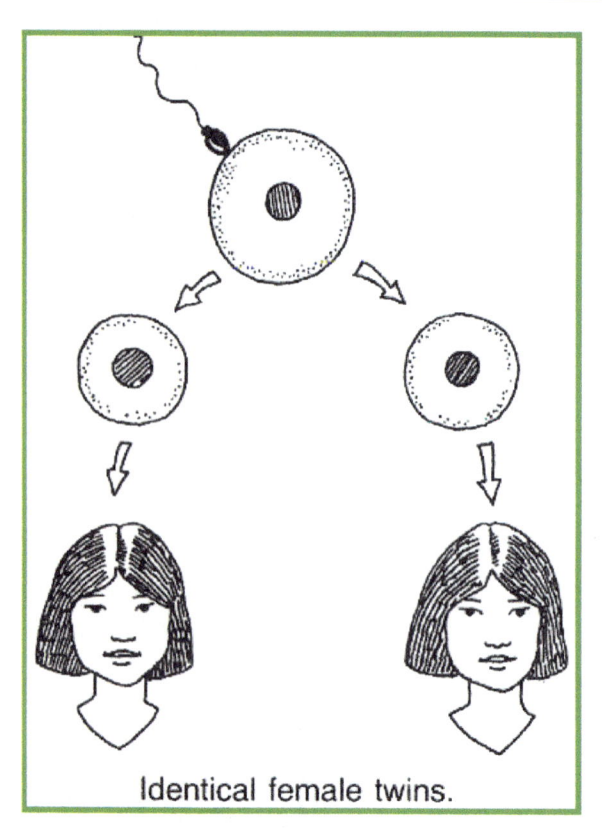

Identical female twins.

Identical twins come from the same egg cell and sperm cell.

What about fraternal twins? Sometimes, a woman's ovaries release *two* eggs at once. These may both get fertilized. Each egg is fertilized by a *different* sperm. In such cases two embryos develop, each from its own egg and sperm. These are **fraternal**.

Suppose that one egg is fertilized by an **X** sperm and the other egg by a **Y** sperm. These twins consist of a girl and a boy. Can you explain why? Of course, both eggs could be fertilized by **X** sperm cells, or both by **Y** sperm cells. What gender would the twins be in each case? Why can they be of the same gender, but still not be identical - that is, not have all of the same traits?

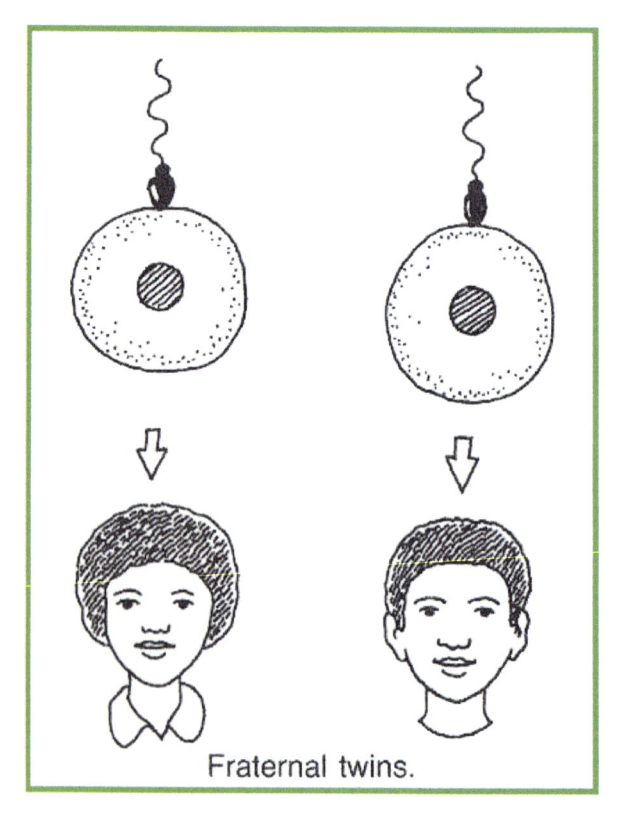

Fraternal twins.

Fraternal twins come from different egg cells and sperm cells.

➤ To Do Yourself — What Are The Chances of Producing a Male or a Female?

You will need:

15 red beans, 5 white beans, 2 jars.

You will make a model of the way **X** and **Y** chromosomes pass from parents to offspring.

- Each <u>red</u> bean stands for an **X** chromosome.
- Each <u>white</u> bean stands for a **Y** chromosome.

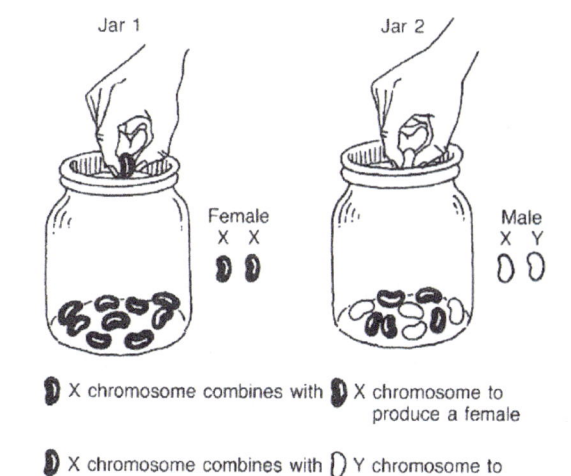

1. Place 10 red beans in jar 1.
2. Place 5 red beans and 5 white beans in jar 2.
3. Close your eyes and select one bean from jar 1 and one bean from jar 2. Make a table to record your selections.
4. Repeat step 3 until no more beans remain. Record your results.

Jar 1 — Female X X
Jar 2 — Male X Y

● X chromosome combines with ● X chromosome to produce a female

● X chromosome combines with ◯ Y chromosome to produce a male

Questions

1. How many females and males were produced? _____

2. If you made the same selections 100 more times, what would be the results?

_____

3. What does this tell you about the chances of producing a female or a male? _____

_____

* From 1952 - 1967, Tancy and Jack (from KY) had six children - all boys.
Their first six grandchildren were all girls!

**Each child born has a 50% chance of being male, and a 50% chance of being female.
In the population as a whole, the genders are nearly balanced.**

(Note: In the mid 1900s, large families were more common than they are today!)

247

REVIEW — U-9 L-6

I. In each blank, write the term that fits best. Choose from the terms below, but use one twice.

identical	male	fraternal	female	Y
sex chromosomes		hormone	sex cells	X

Your body cells contain 23 pairs of chromosomes including one pair of _____

_____. The chromosome pair **XX** is found in _____

body cells, and the pair **XY** is found in _____ body cells. A sperm cell may have

either an _____ chromosome or a _____ chromosome. An egg cell always has

a(n) _____ chromosome. Twins that come from one egg are _____

. Twins that come from two eggs are _____.

II. Jacob and Sarah are brother and sister. They are also twins.

    A. Are they identical or fraternal? Explain. _____

_____

    B. Which twin came from a sperm with a **Y** chromosome? _____

    C. Which twin came from a sperm with an **X** chromosome? _____

III. A family has two boys and one girl. They are expecting another child. What are the chances that it will be a boy or a girl? Explain.

_____

_____

IV. A famous set of quintuplets, the Kienasts, consist of three girls and two boys. Each child has traits that are different from all of the other children. Can you explain how they were formed?

_____

_____

_____

V. In the diagram below, are these twins fraternal or identical? Explain.

_____

_____

_____

_____

What Are Genetic Disorders?

U-9 L-7

Exploring Science / Historical Steps

Dr. Ferguson Led the Way!

Do you remember hemoglobin (from Unit 4, Lesson 5), the oxygen-carrying protein molecule in red blood cells? With normal hemoglobin, a person's red blood cells are round. In some people, there is a defect in the gene that makes hemoglobin.

In the millions of people with this defective gene, many of their red blood cells are shaped like sickles. They have **sickle-cell disease**.

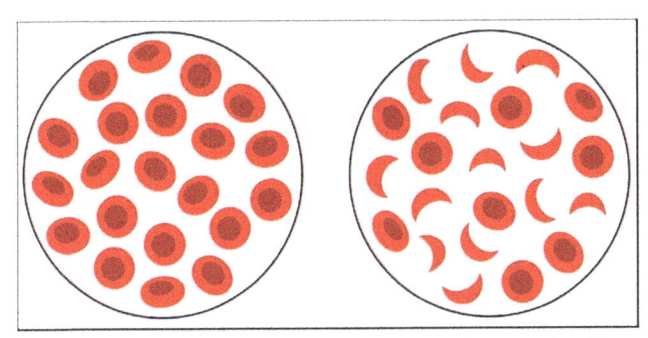

Normal red blood cells   **Sickle and normal cells**

Unlike normal red blood cells, sickle cells easily break apart as they move through capillaries. So, people with sickle-cell disease do not have enough red blood cells. With too few red blood cells, not enough oxygen gets to the body cells. With too little oxygen, people with sickle-cell disease often feel tired. Many victims have painful, swollen joints and general poor health. Tragically, many sufferers' lives are shortened.

In the 1950s, **Dr. Angella Ferguson** developed a method to determine whether a newborn baby has this disease. She also found ways to help people with this disease feel better and live longer. Sometimes, living longer can mean living to see progress toward a cure - at least for one's offspring. It may even mean living to be cured yourself! The next paragraphs provide a good example.

Most body cells, when they divide, simply make more of that same type of cell. Muscle cells divide to make more muscle cells. Skin cells divide to make more skin cells.

Angella Dorothea Ferguson February 15th

Dr. Ferguson (born Feb 15, 1925) has received two certificates of merit from the American Medical Association for her work on sickle cell disease.

However, in the 1960s scientists learned about some very different types of cells - called **stem cells**. Stem cells have the special ability to produce *many* different kinds of body cells. For example, stem cells in the bone marrow don't simply make bone cells, some make the body's red blood cells.

In the 1980s and 1990s, scientists learned a great deal about stem cells. Then, in 2012, an amazing new technique called **CRISPR** gave scientists the ability to "cut and paste" genes; this is known as "**gene editing**."

By 2019, scientists had learned how to use CRISPR to fix the gene that causes sickle cell disease! With CRISPR, some stem cells in the bone marrow of a brave, 34-year-old sickle cell patient named **Victoria Gray**, were "fixed." These stem cells made millions of new healthy red blood cells - with normal hemoglobin! When the special stem cells made enough normal red blood cells, Victoria was cured!

No doubt, many of the children and grandchildren of those who were helped by the work of Dr. Ferguson, will enjoy lives free of sickle-cell disease.

➤ Want more? Research **Jennifer Doudna** and **Emmanuelle Charpentier**, key scientists (and Nobel Prize winners) in regards to CRISPR.

Genetic Disorders

Over 6,000 genetic disorders are known. A genetic disorder is a disease that a person can inherit from his or her parents.

Some genetic disorders mainly affect certain groups of people. Sickle-cell disease occurs more frequently among natives of Mediterranean and African nations, and among African Americans. Some other genetic diseases occur most commonly among white people.

Some genetic diseases are more common among males than females. We can use Punnett squares to show how some of these disorders are inherited.

===

SICKLE-CELL DISEASE. Sickle-cell disease is caused by a recessive gene. A small letter **s** stands for the gene. To have the disease, a person must have a pair of the genes (**ss**). A person who is hybrid for the disease has one recessive, defective gene (**s**). This hybrid person also has one dominant, normal gene (**S**). A hybrid person (**Ss**) is healthy, but they do have one sickle-cell gene; they are called a **carrier**.

Look at the Punnett square. It shows how the genes of two hybrid parents (**Ss**) can be passed to their children. There is one chance in four that their child will have sickle-cell disease. What are the chances that they will have a child who is a carrier (**Ss**), but does not have the disease?

People who think that they may have the sickle-cell gene can be tested. They can learn if they have a defective gene (**s**).

HUNTINGTON'S DISEASE. This is a rare and truly terrible disease. It seriously harms the victim's mind and body. Research is underway to improve CRISPR and eventually beat Huntington's disease. For now, sadly, there is no cure.

Huntington's disease is caused by a dominant gene (**H**). A person who is hybrid also has a normal, recessive gene (**h**). In the case of Huntington's, a hybrid (**Hh**) *has* the disease; the defective gene is dominant, and thus shows up.

The Punnett square at the right shows how Huntington's disease may be passed to children from one hybrid (**Hh**) and one normal (**hh**) parent. Study the diagram. Do you see that there is one chance in two that a child of this cross will have the disease?

Inheritance of Sickle Cell Disease

	Genes in father's sperm	
Genes in mother's eggs	**S**	**s**
S	SS	Ss
s	Ss	ss

There is one chance in four that these parents will have an affected child.

Inheritance of Huntington's Disease

	Genes in father's sperm	
Genes in mother's eggs	**h**	**h**
H	Hh	Hh
h	hh	hh

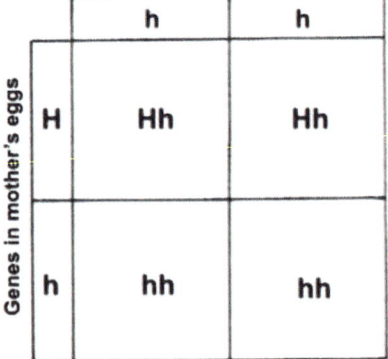

There is an even chance that the parents will have an affected child.

Sadly, the signs of Huntington's disease rarely show up early in life. The symptoms usually begin after age 40. By that time, without knowing it, a person who has the gene may have passed it on to their children.

People with a family history of Huntington's disease can be tested for the presence of this gene. For young adults, the outcome of this test can help them decide whether to have children.

===

HEMOPHILIA. In the disease called **hemophilia** (hee-muh-FIL-ee-uh) the victim's blood does not clot normally. People with hemophilia are sometimes called "bleeders." Their blood does not contain a protein needed for blood clotting.

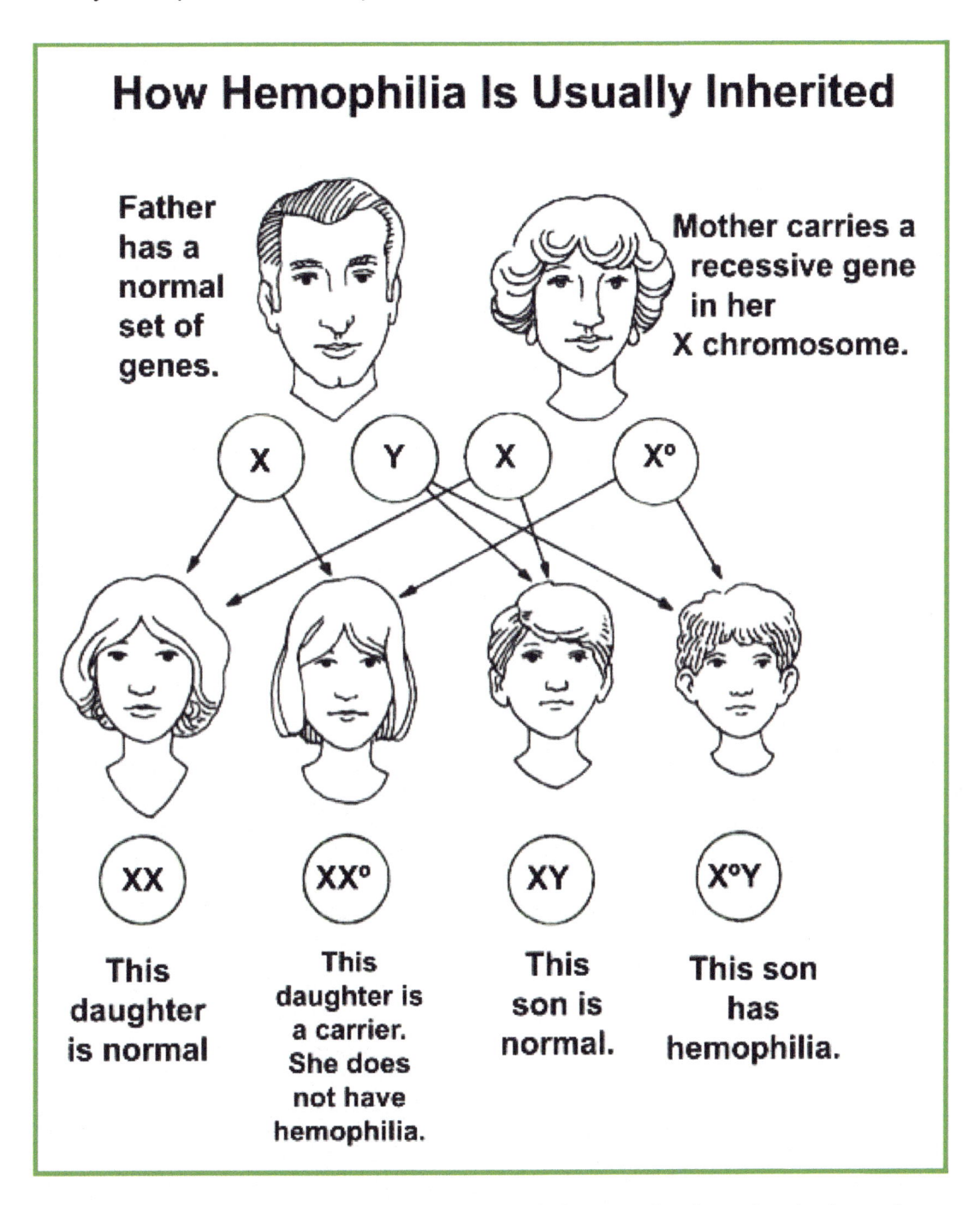

How Hemophilia Is Usually Inherited

Father has a normal set of genes.

Mother carries a recessive gene in her X chromosome.

X   Y   X   X°

XX — This daughter is normal

XX° — This daughter is a carrier. She does not have hemophilia.

XY — This son is normal.

X°Y — This son has hemophilia.

Hemophilia occurs in males as shown above. It is rare for females to have hemophilia.

People with hemophilia take shots containing the blood-clotting protein. Without the shots, even tiny cuts will not stop bleeding.

The gene for hemophilia is attached to an **X** chromosome. It is a recessive gene. A female who has the gene on *one* of her two **X** chromosomes is normal. But having this gene makes her a carrier of the disease. If a female has the defective gene on *both* of her **X** chromosomes, she has the disease.

A male, remember, has only *one* **X** chromosome. If he has the hemophilia gene on his **X** chromosome, he has the disease. There is no second, normal gene to "hide" the hemophilia gene.

To make a Punnett square for hemophilia, we write X^0 for an **X** chromosome that has the hemophilia gene on it. We write **X** for a normal **X** chromosome. And we write **Y** for a **Y** chromosome. A male who has the disease is X^0Y. A normal male is **XY**. A female who is a carrier is XX^0. A normal female who is not a carrier is **XX**. A female hemophilia victim is rare. But when that does happen, she is X^0X^0.

Study the Punnett square. It shows the most common way in which a person inherits hemophilia. Most bleeders are males whose mothers were carriers. Do you see why there are fewer females than males with hemophilia?

Inheritance of Hemophilia

Father's chromosomes
(Normal = No disease)

	X	Y
X^0 Mother's chromosomes (Carrier on one of her two X's)	X^0X	X^0Y
X	XX	XY

Two of the children can be normal, one can be a carrier, and one can be affected.

------ Some Other Genetic Disorders --------

Disease	Some Effects
Tay-Sachs disease.....	nerve damage, early death
PKU disease............	mental deficiencies
Cystic fibrosis............	abnormal breathing, abnormal digestion
Down's syndrome.......	mental deficiencies
Muscular dystrophy.....	wasting away of muscles
Turner's syndrome......	dwarfism, heart defects

--------------------------------- **REVIEW** -------------------------------- U-9 L-7

I. Circle the word (between the brackets) that makes each statement correct.

  A. The gene for sickle-cell disease is [dominant / recessive].

  B. A person who has the sickle-cell *trait* has [one gene / two genes] for the disease.

  C. To have sickle-cell *disease*, a person must have [one gene / two genes] for the disease.

  D. The gene for Huntington's disease is [dominant / recessive].

  E. A person who has Huntington's disease received the gene from [one / both] parent(s).

  F. The gene for hemophilia is found on the [X / Y] chromosome.

  G. Most bleeders are males whose [fathers / mothers] were carriers of hemophilia.

III. **Cystic fibrosis** is a common genetic disease among white people. It affects breathing and digestion and shortens life. This disease is caused by a <u>recessive</u> gene. Can two healthy parents have a child with cystic fibrosis? **Y** or **N**
Draw a Punnett square with **c** for the gene for the disease, and **C** for a normal gene.

II. A man with hemophilia (X^0Y) marries a normal woman (**XX**).
Will the sons have hemophilia? _____
Will the daughters have hemophilia? _____
Will the daughters be carriers? _____

Draw a Punnett square to explain.

What is Evolution by Natural Selection? U-9 L-8

Exploring Science / Historical Steps

Now You See It, Now You Don't. Imagine that you are a moth. You like to rest on tree trunks. How do you protect yourself from hungry birds? You can hold still and keep quiet. Or, you can be the same color as the tree. In other words, you can look like part of the tree. If birds cannot see you, you are safe.

In the mid 1800s, light-colored peppered moths were common in England. The moths were a perfect match for the light-colored trees on which they lived. Once in a while, a change occurred in a moth's genes for color. Then a dark moth was born. When the dark moth rested on a light tree, it was easily seen. A bird quickly ate it. The few dark moths that were born did not live long enough to reproduce. They could not pass the dark genes to offspring. So almost all of the peppered moths were light.

Then many factories were built in England. As soot from the factories filled the air, the tree trunks turned dark. The light moths became the ones easily seen and eaten. They became the ones that did not live long enough to reproduce. The few dark moths were now protected. They survived, reproduced, and passed their dark genes to their offspring. More and more dark moths - and fewer and fewer light ones - were born.

By 1900, most of the peppered moths in England were dark. Few light moths were left.

➤ Hares are relatives of the rabbit. In a very cold, snowy climate, which hare is more likely to live long enough to reproduce - a white hare or a brown hare? Explain.

(Left) Light-colored peppered moths could not be seen easily by birds.

(Right) When soot darkened trees, light-colored moths could be seen easily by birds and were eaten. Dark-colored moths survived.

Natural Selection

The English population of peppered moths changed from one of mostly light moths into one of mostly dark moths. The evidence is overwhelming that the living things on Earth have changed over time. This change is called **evolution**.

A theory to explain how new forms of living things come to be, and how old forms die out, was first proposed by **Charles Darwin**. In 1831, a ship called the *Beagle* set sail from England. Darwin was aboard as the ship's naturalist. For five years, the *Beagle* traveled around the world. It stopped in South America, in Australia, and at many islands.

During the voyage, Darwin studied many forms of life. He collected many fossils and living things. He took these back to England for further study. He was struck by the great variety of species of both the present and the past. In 1859, Darwin put forth the idea that present species evolved from species of the past. He felt that he knew how this happened. He called the method **natural selection**. Darwin's theory has five parts:

(1) Most living things produce more young than can grow to adulthood. A fish may lay millions of eggs. A plant may produce thousands of seeds.

(2) Such large numbers of young must compete for food, space, mates, and hiding places. Only some of the young survive. The rest lose the struggle and die.

(3) No two living things are exactly alike. Traits differ, or vary, even within the same species. Look at the people around you. Each has some different traits. Likewise, peppered moth individuals vary in color.

(4) The traits of some organisms make them better fitted, or **adapted**, to their environment. Well-adapted individuals are better able to survive. This is now often called the "survival of the fittest."

(5) Organisms that survive may live long enough to reproduce. The traits that helped them to survive will pass to their offspring.

Darwin knew that offspring could inherit traits from parents. But no one knew about genes at that time. In the 1900s, genes were found to be the carriers of traits. Sometimes, a gene goes through a change; we call such a change a **mutation** (myoo-TAY-shun). When this organism reproduces, the changed gene passes to its offspring. For example, an original dark moth was the result of a mutation; its offspring inherited this gene for dark color.

Study the pictures that illustrate how giraffes likely evolved. Do you see how natural selection accounts for the long necks of today's giraffes?

Millions of years ago, giraffes inherited necks of different lengths.

Natural selection led to the survival of offspring with long necks.

Today, only long-necked giraffes have survived.

➤ To Do Yourself   What Is the Survival Rate of Some Organisms?

You will need: Seeds (pea, bean, or radish), soil, cut-off juice carton, water, plastic spoon

1. Plant 9 seeds in some soil in the bottom half of a juice carton. Label with the date and the number of seeds planted.
2. Water to moisten the soil, but do not soak. Place the carton in sunlight and keep the soil moist.
3. After one week, count how many of the seeds sprouted. Record your results.
4. Let the seedlings grow for another week and count the plants that survive. Again record your results. Keep the soil moist as the seedlings grow.
5. Compare your results with the class's results.

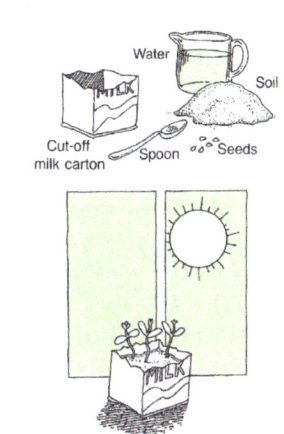

Questions

1. Did the plants compete with each other as Darwin suggested? _____
2. How does producing many seeds help a plant species to survive? _____

REVIEW — U-9 L-8

I. In each blank, write the word that fits best. Choose from the words below.

vary	selection	struggle	mutations
adapted	evolution	reproduce	heredity

Darwin's famous book is <u>On the Origin of Species by Natural Selection</u>. Obviously, his idea of *how* species evolve is called natural _____. A _____ for existence occurs because large numbers of young compete. Among individuals of a species, traits _____ ; that is, they differ.

Some traits make a living thing better fitted, or _____, to their environment. Organisms better fitted to survive will live long enough to _____.

Changes, or _____, in genes can adapt a species to a changed environment.

II. Suppose a thousand tadpoles (young frogs) hatch at the same time in a small pond.

    A. Can all of them survive? Explain. _____

_____

    B. Which ones do you think are more likely to grow to become adult frogs?

_____

Are Fossils Evidence of Evolution? U-9 L-9

Exploring Science / Historical Steps

Are Dinosaurs "For the Birds?" The weather was hot. Everyone was thirsty and covered with dust. But they were having fun! So claimed the many students that went on dinosaur digs with noted dinosaur expert, **Dr. Bob Bakker**. In his long career, Dr. Bakker supported some bold ideas; very few scientists agreed with him on a key one - that birds descended from dinosaurs. He even felt that some dinosaurs likely had feathers. Sure enough, fossils were later discovered that proved this to be true!

Microraptor - a dinosaur with feathers

Having a hypothesis and looking for evidence is part of **paleontology**. Paleontologists study fossils in order to learn about life in the past. To uncover dinosaur remains, Dr. Bakker traveled the world. He spent much time in the western United States, where dinosaurs roamed millions of years ago. He encouraged students to be part of his team. Armed with shovels and picks, they helped him look for dinosaur fossils.

A skeletal display, influenced by Dr. Bakker, indicating an active, fast-moving dinosaur.

In 1986, Dr. Bakker put his ideas in a now famous book, The Dinosaur Heresies. At that time, many experts thought that most dinosaurs were cold-blooded, slow moving reptiles. Bakker argued that dinosaurs (like their descendents, the birds) were likely warm-blooded and could run at a fast clip. He pointed out that dinosaurs had hip bones and a muscle structure similar to some birds of today.

His theories (and those of others) led museum operators to change the public's view of dinosaurs. The skeletons on display were rearranged. Many now show dinosaurs standing more upright and alert. Often their tails are held out, not dragging the ground.

➤ Some scientists now combine their knowledge of DNA with expertise in paleontology. What do you think this field of study is called?

Change Through Time

Dinosaurs are no longer alive. They died out, or became **extinct** (ik-STINGKT). Scientists believe that life has existed on the earth for well over 3 billion years. During that time, large numbers of species have lived for a time and then become extinct.

Traces or remains of living things from long ago are called **fossils** (FAHS-uls). Many fossils do not look like any living animals or plants. So we know that they belong to species that are now extinct.

Some fossils are **prints** of parts - such as feet, shells, and leaves. The prints were made in soft material that later turned to rock. Scientists have made models of ancient tree ferns (very tall, tree-like ferns) from fossils of extinct kinds of plants. The diagram below shows what huge tree ferns looked like in a North American forest 300 million years ago.

When the climate changed, the fern forests died out. In time, layers of dead mosses and ferns became beds of coal. Prints of ancient tree ferns are often found when coal is dug from the ground.

The **hard parts** of dead organisms can also become fossils. Bone, teeth, and wood may remain after soft body parts decay. These hard parts may turn into stone. If they become stone, they may last for millions of years.

In a few cases, a whole animal has been preserved as a fossil. An insect sometimes became stuck in tree sap. The sap hardened into clear **amber** (AM-bur). The insect was preserved inside of the amber. Fossils in amber can also last for millions of years.

Huge extinct mammals related to elephants called woolly mammoths (MAM-uths) have been found in ancient ice. (See Unit 1, Lesson 12 for an image of a woolly mammoth). Some fossils in ice have been preserved for thousands of years.

By studying fossils, scientists can infer how some kinds of animals have changed through time. Sixty million years ago, a horse was only about the size of a small dog. Its feet had four toes. Scientists believe that this was the ancestor of today's horses.

Over millions of years and many generations, the horse became larger. Its middle toes became a hoof. Today's horse stands about five times as tall as the horse of 60 million years ago.

Fossils are strong evidence that living things have changed over time. In other words fossils are evidence of **evolution** (ev-uh-LOO-shun).

This is how a coal-forming forest in North America looked 300 million years ago.

Time		Leg
3 million years ago		
25 million years ago		
60 million years ago		

By studying fossils, scientists have learned much about the evolution of the horse.

> **To Do Yourself**   How Can You Make a Fossil Print?

You will need: Clay, small bowl, plaster of Paris, spoon, water, shell, hand lens

1. Press a shell into clay to make a print, and then remove the shell carefully. You have made a <u>cast</u>.
2. Observe the cast with your hand lens and note the details. Then add a thin layer of petroleum jelly to the cast.
3. Place 4 to 5 spoonfuls of plaster of Paris in a small bowl. Carefully add several spoonfuls of water and mix.
4. When the mixture is smooth and thick, pour some of it in the cast. Let the plaster harden.
5. When the plaster is hard, remove it carefully from the cast. This is a <u>mold</u>. Observe the details with your hand lens. Compare these to the cast.

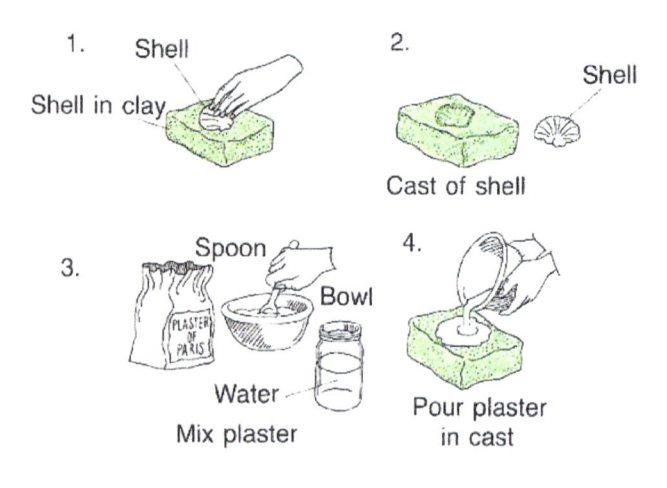

Questions

1. What part of the activity is like a fossil? _____

2. How are fossils used to identify organisms? _____

REVIEW   U-9 L-8

I. In each blank, write the word that fits best. Choose from the words below.

| **horses** | **extinct** | **fossils** | **evolution** | **dinosaurs** | **mammoths** | **insects** |

A species that has died out is _____. The remains or traces of life long ago are called _____. Change through time that results in new types of living things is _____. Today's elephants may be related to extinct woolly _____. Birds are present-day relatives of extinct _____.

II. Circle the word (between the brackets) that makes each statement true.

  A. The earliest ancestor of the horse had [two / four] toes.

  B. Forests that lived long ago and turned to coal were [ferns / redwoods].

  C. Most fossils are preserved [hard / soft] body parts.

  D. Tree sap that turned into [stone / amber] sometimes contains fossils of insects.

III. Fossils of ancient sea shells have been found on mountains in Kentucky. What might this tell us about how the earth has changed through time?

_____

What Other Evidence is There for Evolution? U-9 L-10

Exploring Science / Historical Steps

Where are the Missing Links? Often, those who doubt evolution ask that question. In fact, some important links *have* been found. One, named **archaeopteryx** ("ancient wing"), was discovered in Germany in 1861 (only two years after Darwin published his famous book On The Origin of Species). The incredible find was a 150-million-year-old bird-like fossil. Its beak and feathers were obvious, but there were some strange parts to this "bird." Like a bird it had hollow bones. But its tail was extremely long. And its beak had teeth! And claws stuck out from the end of its wings!

Long tails, teeth and claws are *reptile* traits! Reptiles existed before birds. Archaeopteryx is an obvious "link" between reptiles and birds.

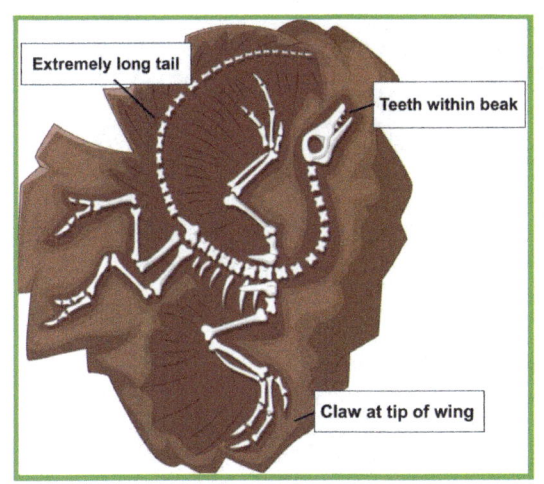

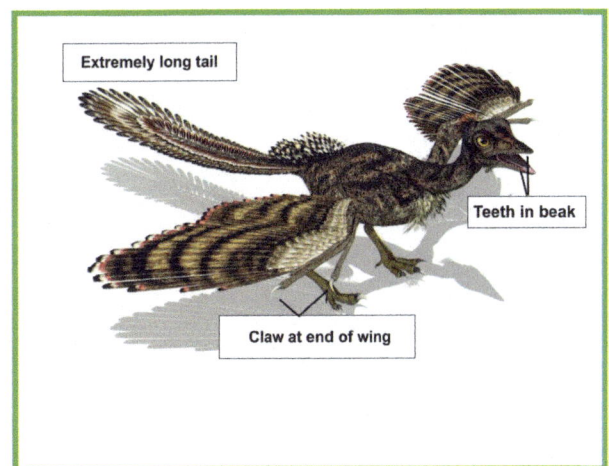

Archaeopteryx fossil, and its likely appearance in life

===

Overlapping traits between types of organisms are evidence that one type led to the other. This type of evidence isn't found only in fossils. First, review Unit 2, Lessons 2 and 3. As you continue reading, look at the photos below.

Scientists are convinced that fish led to amphibians. Here are a few reasons why. Fish breathe with gills; young amphibians (such as tadpoles) breathe with gills. Fish's tails are flattened side to side; so are amphibians' tails. Fish have no neck; amphibians have no neck.

The fin bones of fish match the leg bones of amphibians. Fish make jelly-like eggs (though, in some, these are fertilized inside of the female); amphibians make jelly-like eggs.

Do you see why scientists are convinced that the ancestor of the amphibians was a fish-like organism?

➤ Want more? View the video "Your Inner Fish" by **Neil Shubin**, the discoverer of **Tiktaalik**, a fossil "link" between fish and amphibians.

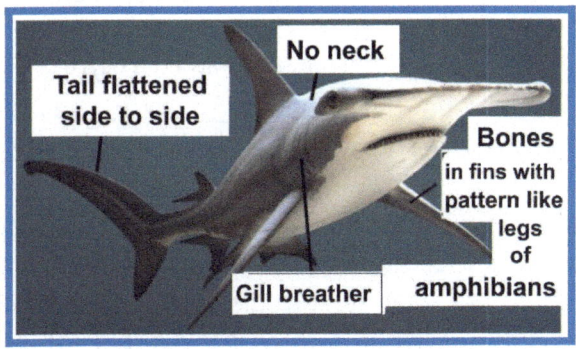

Fish-like traits in amphibians are evidence that amphibians evolved from fish.

Even More on Evolution

Biologists are confident that an original living cell led to <u>all</u> forms of life on earth. Below is a simplified summary of how some major groups of organisms evolved.

If all organisms are descendents of the same original organism, shouldn't all living things still have *some* parts that came from that organism? In fact - they do!

<u>All</u> living things are made of cells. In these cells, all organisms have DNA (or RNA) to store the instructions for building proteins. All living things are built primarily of proteins, which are built from amino acids.

Also, the cells of all organisms carry out many of the same chemical reactions. (For example, the steps of the cell respiration equation occur in nearly all organisms).

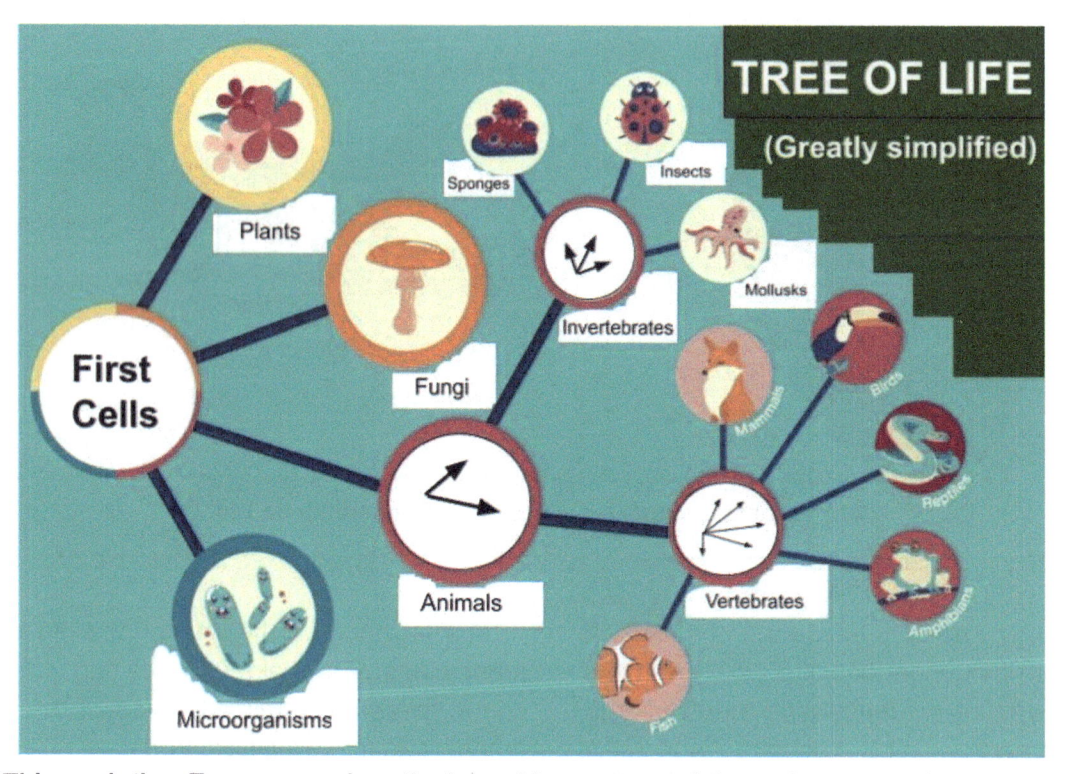

This evolution *Tree* summarizes the branching pattern of the major steps of evolution.

Speaking of cell respiration, there is an amazing piece of evidence for evolution that is related to this process. Observe the illustrations on the next page. Recall that in all cells <u>except</u> those of bacteria and archaea, cell respiration is carried out inside of tiny structures called mitochondria. Guess what scientists found inside of <u>all</u> mitochondria? **<u>DNA</u>**!! Yes, mitochondria have their <u>own</u> DNA - separate from the DNA of the cell that they occupy! *Why?*

Beginning in <u>1967</u>, the work and writings of **Dr. Lynn Margulis** gradually convinced scientists of the answer. Over a billion years ago, one very tiny single-celled bacteria-like organism (with its <u>own</u> DNA) was "gulped" by a larger version (with <u>its</u> own DNA) - but the tiny one was not digested.

Instead, the two cells began working together; the big one provided shelter and "food," while the tiny one shared the energy that it produced (through cell respiration). The small one never left the big one; instead it reproduced inside of the big one.

In other words, the tiny cell (with its <u>own</u> DNA) *became* the **mitochondria** for the big cell (with <u>its</u> own DNA)! The mystery of why mitochondria had their own DNA was finally understood! The chloroplasts in plant cells came about in a similar way. And yes, chloroplasts also have their own DNA!

Recall that the word **symbiosis** describes two organisms that help each other to survive. Dr. Margulis' special type of cell-joining became known as **endosymbiosis**. *Endo* means "inside," *sym* means "with," and *biosis* means "living."

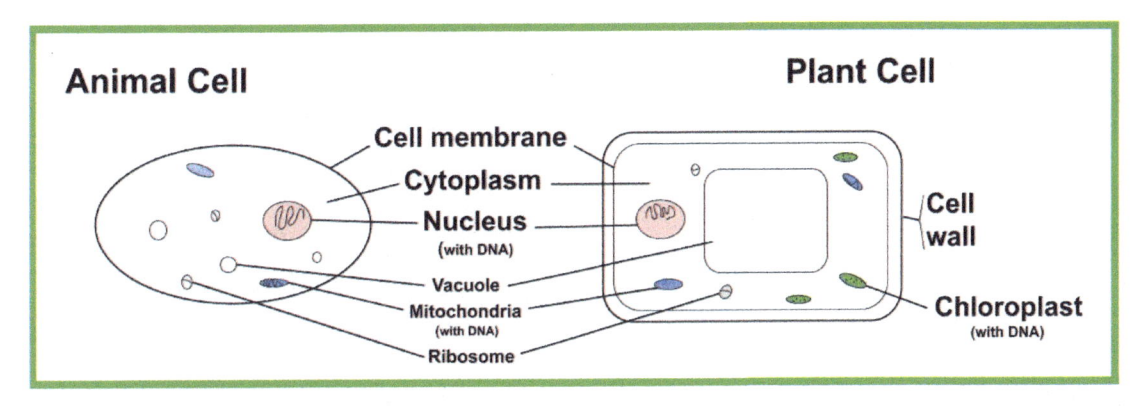

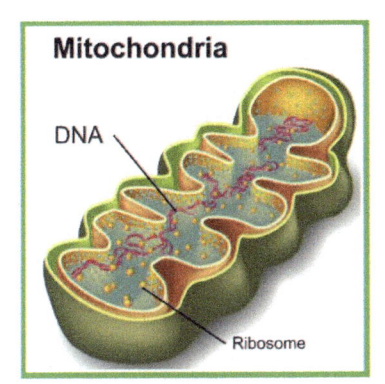

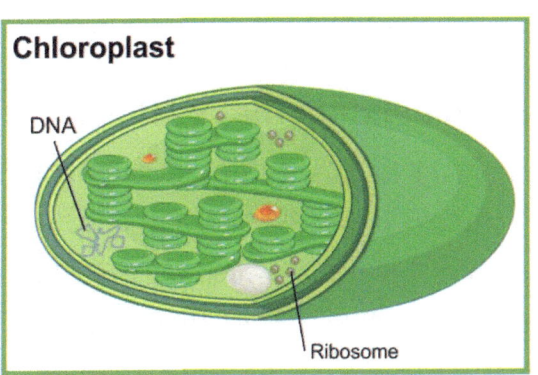

Endosymbiosis explains why mitochondria and chloroplasts have their own DNA; the ancestors of mitochondria and chloroplasts were free-living cells that had *their* own DNA!

------------------------------ **REVIEW** ------------------------------ U-9 L-10

I. In each blank, write the word that fits best. Choose from the words below.

   cells   DNA   link   archaeopteryx   birds   reptiles

The term _____ describes an organism (often a fossil) that has some traits common to a group of organisms older than its own group. The extinct _____ looked much like a bird, but it had traits much like those of _____. Other evidence of evolution is the fact that all of today's organisms are made of _____. In the cells of all organisms are molecules called _____ that direct the production of proteins.

II. Circle the word (between the brackets) that makes each statement correct.

   1. Fish and amphibians have matching bones in their [back / legs].

   2. During endosymbiosis, a tiny single-celled organism [was / was not] <u>digested</u> by a larger single-celled organism.

   3. Mitochondria and [chloroplasts / ribosomes] have their own DNA.

III. Amphibians, which evolved from fish, have jelly-like eggs much like fish's. Reptiles, which evolved from amphibians, have eggs with leathery coatings. Birds, which evolved from reptiles, have hard-shelled eggs. Do you see a pattern in the eggs? Describe it briefly.

261

Unit 9 -- Review What You Know ---

A. Unscramble the groups of letters to make science words. Write the words in the blanks.

1. BHDYIR / _____ = has one dominant and one recessive trait for a gene
2. LTFRAENAR / _____ = non-identical twins
3. SSIMOEI / _____ = division that forms sex cells
4. TIEVLOONU / _____ = change that results in new species
5. LOISFS / _____ = trace of a living thing from long ago
6. HIITENERD / _____ = passed from parents to offspring

B. <u>Write</u> the ending that best completes each statement.

1. A sperm cell has the same number of chromosomes as
   a. an egg cell   b. a body cell   c. a zygote   1. _____

2. One out of 4 pea-plant offspring are short when both
   parents are   a. pure tall   b. pure short   c. hybrid   2. _____

3. After a cell with 8 chromosomes divides by mitosis, each
   new cell will have [a. 4 b. 8 c. 16] chromosomes.   3. _____

4. The sex chromosomes of a human male are ......
   a. XX   b. YY   c. XY   4. _____

5. Blood does not clot normally in a person who has
   a. hemophilia   b. Huntington's disease
   c. sickle-cell disease   5. _____

6. Pink four-o-clock flowers are due to genes that are
   a. dominant   b. recessive   c. blending   6. _____

7. The number of chromosomes in a human sperm is
   a. 13   b. 23   c. 46   7. _____

8. A person with the genes **BB** for eye color is called
   a. pure blue   b. pure brown   c. hybrid   8. _____

9. Prints of ancient fern trees are found in ......
   a. ice   b. amber   c. coal   9. _____

10. The scientist who proposed the theory of
    natural selection is ......
    a. Mendel   b. McClintock   c. Darwin   10. _____

11. The scientist most responsible for the widespread
    acceptance of the concept of endosymbiosis is ....
    a. Margulis   b. Shubin   c. Darwin   11. _____

C. Apply What You Know

1. In humans, dark hair is dominant and light hair is recessive. Suppose a light-haired man marries a dark-haired woman. The man's body cells contain two genes for light hair, **dd**. The woman's body cells contain one gene for dark hair **Dd**. On the diagram below, fill in the boxes that show what genes for hair color will be in their sex cells. Then complete the Punnett square to show what hair-color genes their children can have.

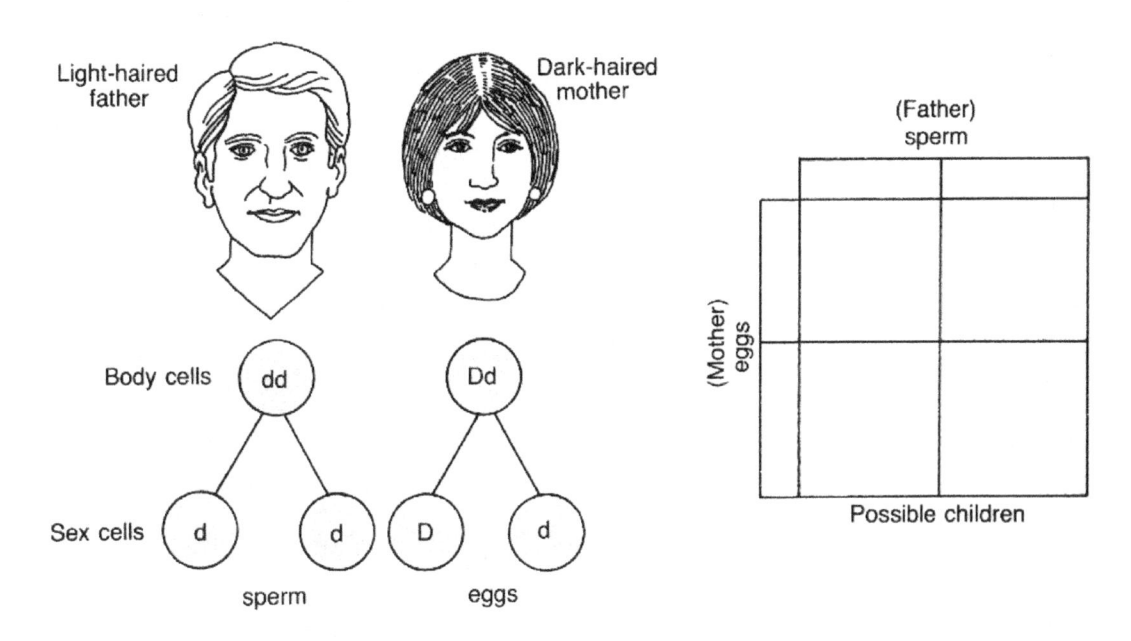

2. Circle the word (or words) [between the brackets] that makes each statement about the above couple and their children true:
    a. The mother is [hybrid / pure] for dark hair.
    b. The father is [hybrid / pure] for light hair.
    c. The chance that a child will have light hair is one in [two / four].
    d. A pair of parents with [**DD** and **dd** / **Dd** and **Dd**] could have a light-haired child.

3. A certain kind of chicken can have black, white, or gray feathers. Its color is determined by blending genes. A black chicken's body cells have two genes for black feathers, **BB**. A white chicken's body cells have two genes for white feathers, **WW**. A gray chicken's body cells have one gene for black and one gene for white, **BW**. Suppose that a black rooster mates with a white hen. On the diagram below, fill in the circles for the genes for color carried by their sex cells. Then complete the Punnett square, writing the genes for the sex cells and then the genes of their offspring.

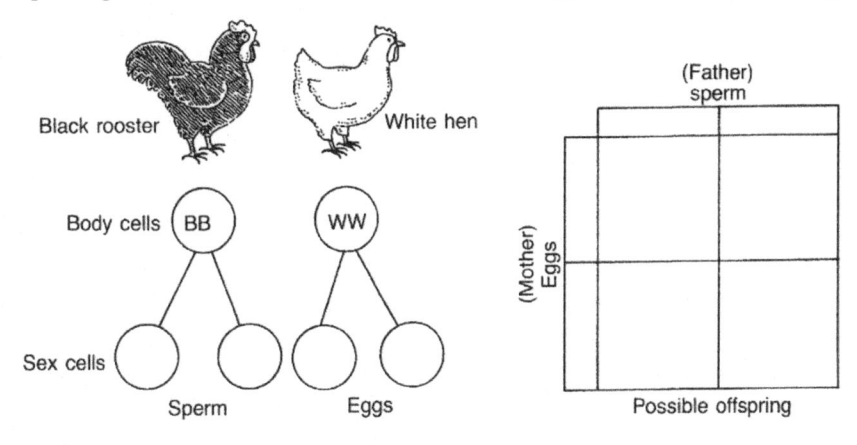

263

4. Suppose that a gray rooster is mated with a gray hen. Complete the boxes in the next diagram for the chickens' body cells and their sex cells. Then complete the Punnett square for their offspring.

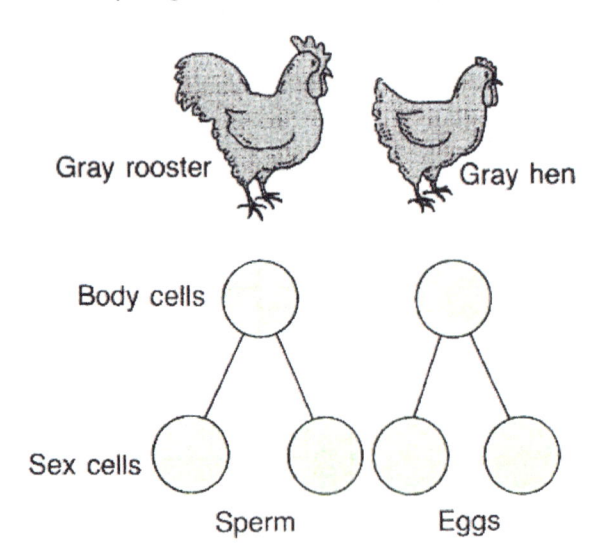

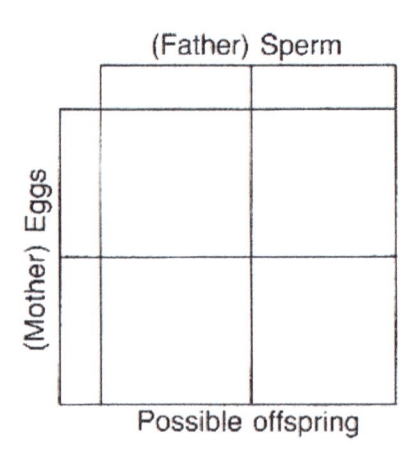

5. Circle the word (or words) [in brackets] that makes each statement true.

  a. The offspring of a black chicken and a white chicken are all [black / gray].

  b. Two black chickens [could / could not] have gray offspring.

  c. When two gray chickens mate, the chances are that [half / all] of their offspring will be gray.

  d. If a white chicken mates with a gray chicken, [none / half] of the offspring may be black.

D. Find Out More

1. In humans, color-blindness is caused by a defective gene attached to the **X** chromosome. Make a Punnett square, using the letters below. They stand for the possible combinations of the **X** and **Y** chromosomes for both normal and color-blind individuals:

  XX = normal female X^cX = carrier female X^cX^c = color-blind female
  XY = normal male X^cY = color-blind male

  Draw Punnett squares to show each of the following:

  1. The possible children of a carrier female and a normal male.

  2. The possible children of a color-blind male and a normal female.

2. Red hair seems to "skip a generation." What do you think that expression means?
_____
_____

SUMMING UP:
Review What You Have Learned So Far

A. Study the drawing of a flower, and note the five terms below. Place each of the five numbers from the image in the space above the matching term below:

___ ___ ___ ___ ___
stamen   sepal   petal   ovary   pistil

Now, for the statements below, circle the word that makes each statement true:

1. In an energy pyramid, a flower would be at the [top / bottom].
2. After pollination, a pollen tube grows into the [sepals / ovary].
3. When Mendel's pure-tall pea plants were crossed with his pure-short pea plants, the offspring were all [tall / short].
4. The part of a plant that carries on photosynthesis is the [root / leaf].

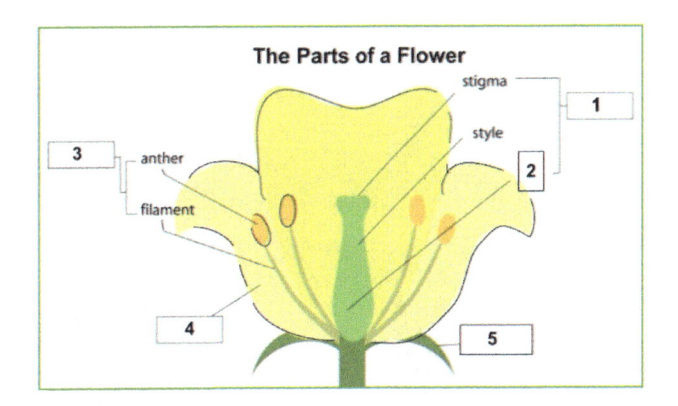

B. On the drawing below, write labels for each numbered part. Use the following labels:

kidneys	**spinal cord**	**cerebellum**	**adrenals**
biceps	**pancreas**	**lungs**	**salivary gland**

The number to the left of the statements below is the same as the number of the part. <u>Circle</u> the word (or words) [within the brackets] that makes each statement true.

1. The part of the brain that helps muscles work together is also the center for [sight / balance].
2. Messages go to and from the brain through a long nerve cord attached to the [cerebrum / medulla].
3. The contraction of the diaphragm and [rib muscles / trachea] causes air to rush into the body.
4. Millions of tiny filters remove the [carbon dioxide / urea] from the blood.
5. The digestive juice in the mouth contains an enzyme that starts to break down [protein / starch].
6. When the labeled upper-arm muscle [contracts / relaxes] the elbow bends.
7. When the body's "emergency" glands go into action, the heart [speeds up / slows down].
8. This same gland that makes a digestive juice also makes the hormone [thyroxine / insulin].

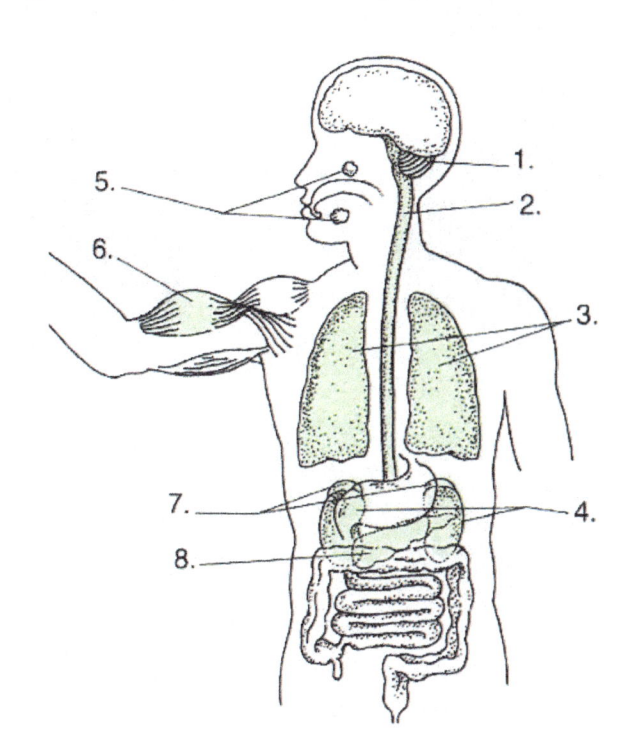

265

UNIT TEN

PROTECTING HEALTH

What are Infectious Diseases?   U-10 L-1

Exploring Science / Historical Steps

Hunting Down Lyme Disease. Old Lyme is a town in Connecticut. In 1975 its name was given to a new disease - Lyme disease. The first signs of the disease are skin rash, fever, chills, aches and pains. Many of its early victims were children. After a while, some of these children were back to normal. But some were left with arthritis, which causes pain in the joints.

At first, no one knew the cause of Lyme disease. Scientists usually have to find out what causes a disease before they are able to cure it. Like detectives, scientists who studied Lyme disease looked for clues.

One clue was that the victims all lived near each other. That meant that the disease might be "catching" (passed from person to person). Such diseases are usually caused by some kind of microbe (microscopic organism). Another clue was that the disease was found in a rural area. That could mean that some kind of insect or large organism might carry the microbe. Organisms that carry and spread the causes of diseases are called **vectors.**

In 1976, the vector was found. It was a tick. Ticks are related to spiders. Only a specific type of very tiny tick (the size of a sesame seed) carried the microbe that caused Lyme disease. Finally, in 1983, scientists identified the trouble-maker microbe. It was a specific bacterium shaped like a spiral.

In some people, common antibiotics won't kill this bacterium. Some victims have suffered for years! Fortunately, in 2020, scientists at Stanford University - **Jayakumar Rajadas** and **Venkata Raveendra Pothineni** found a new antibiotic against Lyme disease that works in everyone!

Then, in late 2021, **Erol Fikrig,** at Yale University, announced his team's discovery of an mRNA vaccine (see Lesson 2). It works not against the microbe, but against the tick! When fully developed, this vaccine should protect against several diseases carried by ticks.

A large portion of Lyme disease cases have happened in the north east and north central parts of the United States. However, it is wise to watch for very tiny ticks wherever you live.

➤ What are two possible ways to reduce the number of Lyme disease cases?

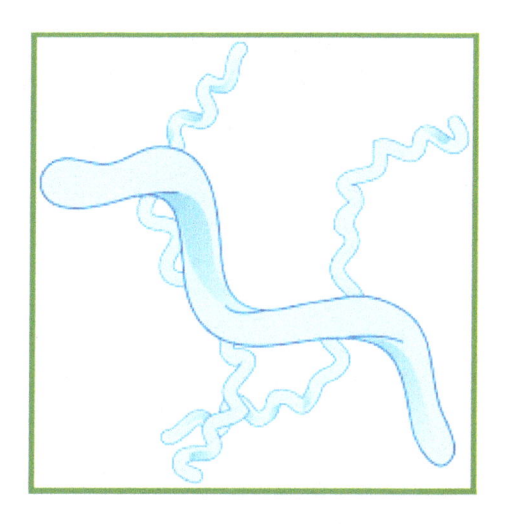

The bacteria that cause Lyme disease have a spiral shape. The bacteria are passed to humans by a tiny species of tiny ticks called the deer tick (Ixodes scapularis). Full grown deer ticks are 3 mm or smaller.

Infectious Diseases

Microbes that cause diseases, such as Lyme disease, are called **germs**. Diseases that are carried from person to person are called **infectious** (in-FEK-shus) **diseases**. In most infectious diseases (or **infections**) the cause is a microbe. A few infectious diseases are caused by larger organisms such as worms.

One group of microbes are the **bacteria**. Most bacteria are harmless, and some are very helpful. But some cause disease. Even though disease-causing bacteria are common, they only make us sick if they get past the body's defenses. When bacteria do make us sick, they make poisons (**toxins**) (TOK-sins). These toxins cause the **symptoms** (the signs) of the disease. Some common symptoms are pain and fever. Can you think of others? The table on the next page lists some common bacterial diseases.

Viruses also cause many diseases, including the common cold, COVID-19, AIDS, HPV, flus, and colds. The table on the next page lists many of the most common infectious diseases.

Some Organisms That Cause Infectious Diseases

A. Bacteria (grouped by shapes)

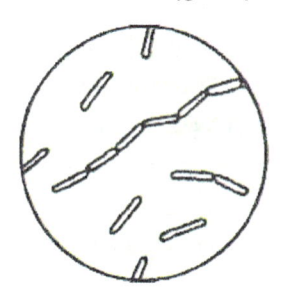

Bacilli (cause tetanus) 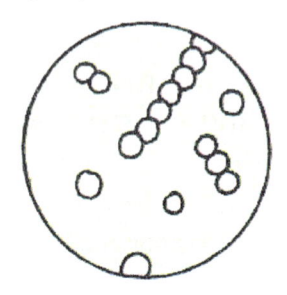   Cocci (cause strep throat) 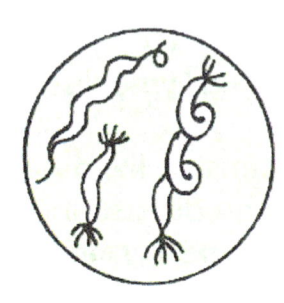   Spirilla (cause syphilis)

B. Viruses

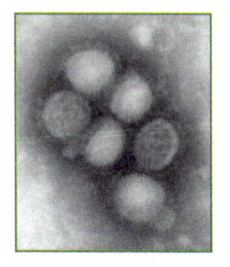

Influenza virus 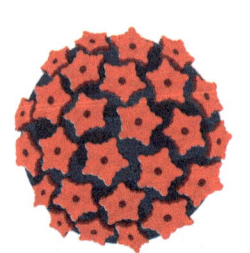   Virus of HPV 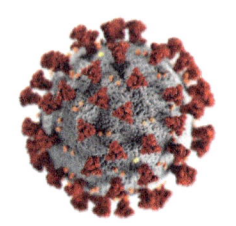   Virus of COVID-19

C. Protozoan

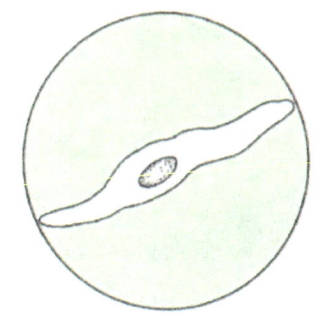

Malaria protozoan

D. Parasitic worms

Porkworm 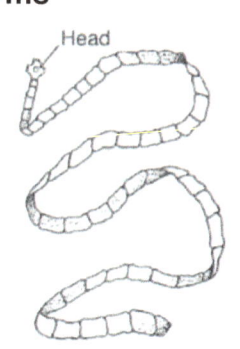   Tapeworm (Head)

Infectious Diseases

BACTERIAL DISEASES	How Spread	Description / Organs Affected
diphtheria	contact	deadly / respiratory
strep throat	contact	severe sore throat / heart
typhoid fever	food, water	deadly / digestive
pneumonia (some kinds)	air	can be deadly / lungs
tuberculosis	contact	can be deadly / respiratory
STIs (sexually transmitted) *gonorrhea / syphilis*	direct sexual contact	can be deadly / sex organs; brain and nervous system
botulism	damaged canned foods	deadly / nervous system
VIRUS DISEASES	**How Spread**	**Description / Organs**
COVID-19	air; contact	deadly / respiratory; others
colds	contact; air	fever; sneeze; cough; aches
influenza ("flu")	contact; air	respiratory / muscle aches
mumps	contact	swollen neck glands
measles	contact	rash; fever
polio	contact	nerve injury; paralysis
AIDS (Acquired Immune Deficiency Syndrome) (HIV)	direct sexual contact; bodily fluids	in time deadly if untreated; weakens immune system
hepatitis	contact	can be deadly; liver
rabies	bite of infected mammal	deadly; nerve; brain;
STIs - herpes, and HPV	direct sexual contact	HPV cancer trigger; sex organs
PROTOZOAN DISEASES	**How Spread**	**Description / Organs**
malaria	bite of infected mosquito	fever, chills; sometimes fatal
amebic dysentery	contaminated food, water	intestines; liver
FUNGUS DISEASES	**How Spread**	**Description / Organs**
ringworm	contact	itching; sores on scalp
athletes foot	contact	itching; sores on toes
WORM DISEASES	**How Spread**	**Description / Organs**
porkworm (trichina)	undercooked infected pork	pain; can be fatal; muscles
tapeworm	undercooked meat / fish	intestines

Protozoans and fungi can also cause infectious diseases. Another group of diseases are those caused by **parasitic worms**. The table lists some of these, too.

A disease that is "catching" is called a **contagious** (kun-TAY-jus) disease. Contagious diseases are usually passed from person to person by touching or other direct contact. Microbes may sometimes be passed by touching things that a sick person has used, such as an unwashed food dish. Some bacteria and viruses may be passed through the air, water, or food. Certain infectious diseases can only be "caught" by sexual contact. These are called **sexually transmitted infections** (STIs).

Some germs are passed to other people with the help of another organism - a vector. You might recall that Lyme disease bacteria are carried by ticks. Malaria protozoans are carried by mosquitoes. Rabies viruses are carried by mammals such as dogs or raccoons.

===

How A Virus Infects a Cell

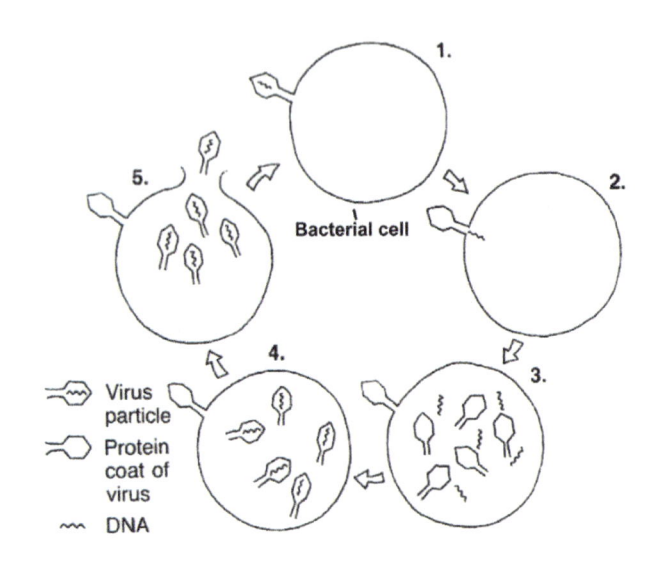

1. Virus attaches itself to the cell.
2. Virus DNA enters the cell.
3. Cell is "tricked" into making virus DNA and protein coats.
4. Virus parts are assembled.
5. Cell bursts open and releases many new viruses. Cell dies.

===

REVIEW — U-10 L-1

I. In each blank, write the word that fits best. Choose from the words below.

| parasitic | infections | toxins | botulism |
| STIs | symptoms | contagious | |

Infectious diseases are caused by microbes or _____ worms.

Infectious diseases may also be *called* _____. Bacteria produce

poisons, or _____ that can make us sick. Any disease that can be

"caught" is _____. A group of diseases that involve the sex organs

are called _____s.

II. To answer each of the following, use the table of infectious diseases. Circle the word or words (between the brackets) that makes each statement true.

A. Colds and flu are common [bacterial / virus] diseases.

B. Eating undercooked pork may result in infection by a [fungus / worm].

C. Gonorrhea is a disease passed to others by [sexual contact / mosquitoes].

D. Two kinds of germs that can cause pneumonia are viruses and [bacteria / worms].

E. Two diseases in which there is a rash and fever are [measles / mumps] and chicken pox.

F. Using food from a dented can is risky because the can may contain the toxin of [botulism / rabies].

III. People once thought that infectious diseases were caused by evil spirits. How did the invention of the microscope help change that belief?

_____

_____

How Does the Body Fight Infections? U-10 L-2

Exploring Science / Historical Steps

Battling Microbes. Your body is in a constant battle with microbes. Happily, you nearly always win! Why? The hero is your **immune system** - special cells and molecules that fight microbes. How does this system keep up with so many types of germs?

Remember natural selection? Our ancestors had to battle microbes, too. Their immune systems fought microbes. The humans who won the battle were able to live long enough to reproduce. To their offspring, they passed the genes that built good immune systems. Many generations later, most humans have strong defenses against germs.

But, for all of those years, the microbes were changing (evolving) too. Of course, there have always been many more microbes than humans. In addition, bacteria "grow up" and reproduce in much less time than we do. As for viruses, they don't even "grow up." Instead, they rapidly produce enormous numbers of copies of themselves within their victims' cells!

To make matters worse, a microbe that normally attacks only one type of animal may suddenly be able to attack a new animal - like us! How does this happen? A change (called a **mutation**) in a microbe's genes can allow the microbe to live in a new victim. For example, scientists think that the virus that causes **COVID-19** first reproduced only in bats; a gene mutated and allowed the virus to reproduce in human cells. Similarly, in the Fall of 2021, a large percentage of wild deer were found to be carrying this virus.

Over the centuries, some microbes have been able to defeat *most* humans' immune systems. Remember smallpox (from Unit 1, Lesson 3)? You may have also heard of the plague, or the 1918 flu. Millions of people died from these diseases!

There are even microbes around today that most people's immune systems can not defeat. One such microbe is **HIV** (shown above), the virus that causes AIDS.

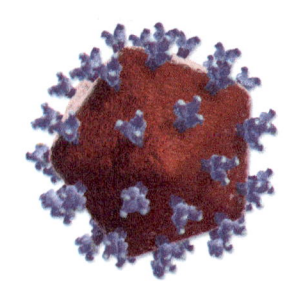

HIV - the virus that causes AIDS

Fortunately, humans have an additional powerful defense - science! But living things, even microbes, are complicated. It took years to control AIDS. Even today, the medicines that do so are too expensive for many people - so work continues.

For many months, the virus that causes COVID-19 (shown below) could not be stopped. It spread around the globe. A global disease is called a **pandemic**. The early 2020s will long be remembered as the years of the COVID-19 pandemic. Fortunately, science rescued us with effective vaccines (described in the section below).

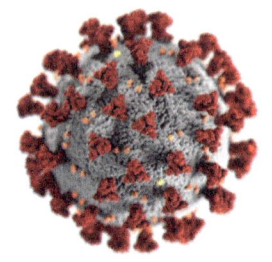

The virus that causes COVID-19

In a way, we have become spoiled by science. Many people expect scientists to have all of the answers, and to have them immediately! Some people even get upset if scientists change their minds as they uncover new information!

People who understand how science works realize that good scientists "listen to the data." This means that scientists admit when the results of an experiment prove that a hypothesis is wrong. Scientists learn, and they keep battling.

Body Defenses Against Infection

Bacteria and viruses that cause disease are common in our environment. Yet, most people stay free of infectious diseases. Why is that? The body has two main **lines of defense** against germs.

Our skin is part of our body's **first line of defense**. So are our breathing passages and our stomach. As long as our skin remains unbroken, it keeps germs out. Germs that are in our food are killed by the acid in our stomach.

Our inner nose, trachea and bronchi (the two tubes that branch from the trachea) are lined with **cilia**. These hairlike parts sweep out dust, which carries germs. The sticky mucus that these breathing parts produce also traps germs. A cough or a sneeze forces mucus and germs out of our body. Obviously, sneezes should always be "covered" so germs aren't shared!

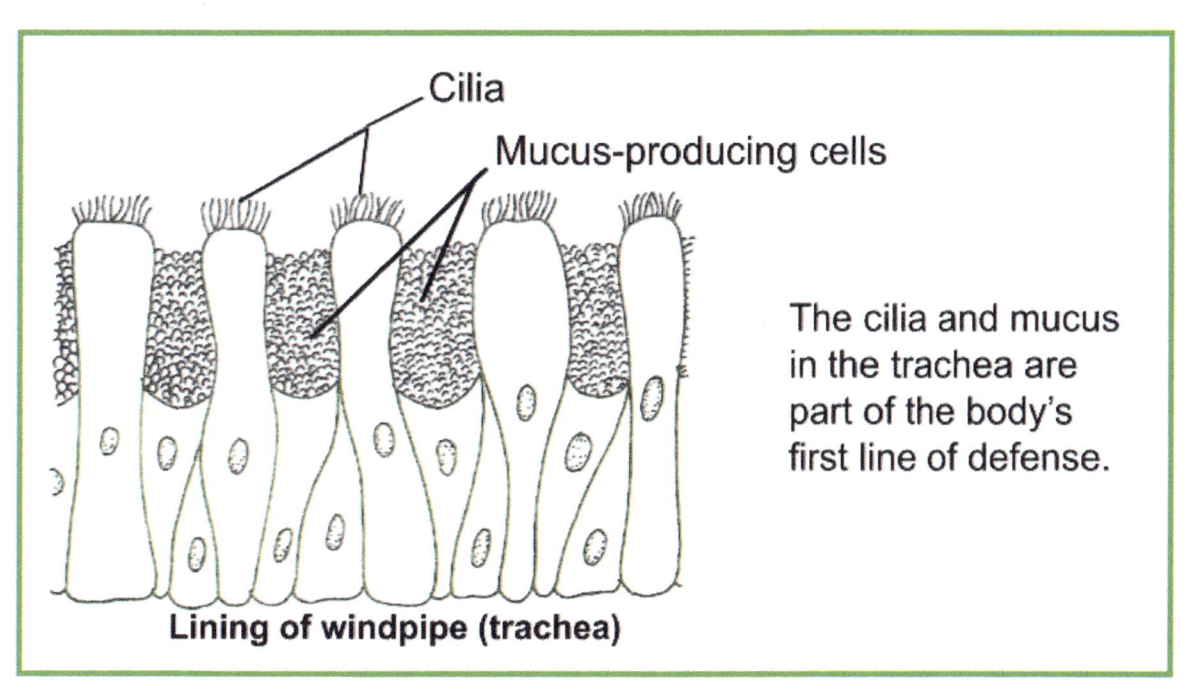

The cilia and mucus in the trachea are part of the body's first line of defense.

Lining of windpipe (trachea)

Sometimes disease germs get past the body's first line of defense. Then, the **second line of defense** takes over. This defense is in our blood. Here, there are several kinds of white blood cells.

Most white blood cells are made in the bone marrow. One type of white blood cell (shown below) eats bacteria and other germs.

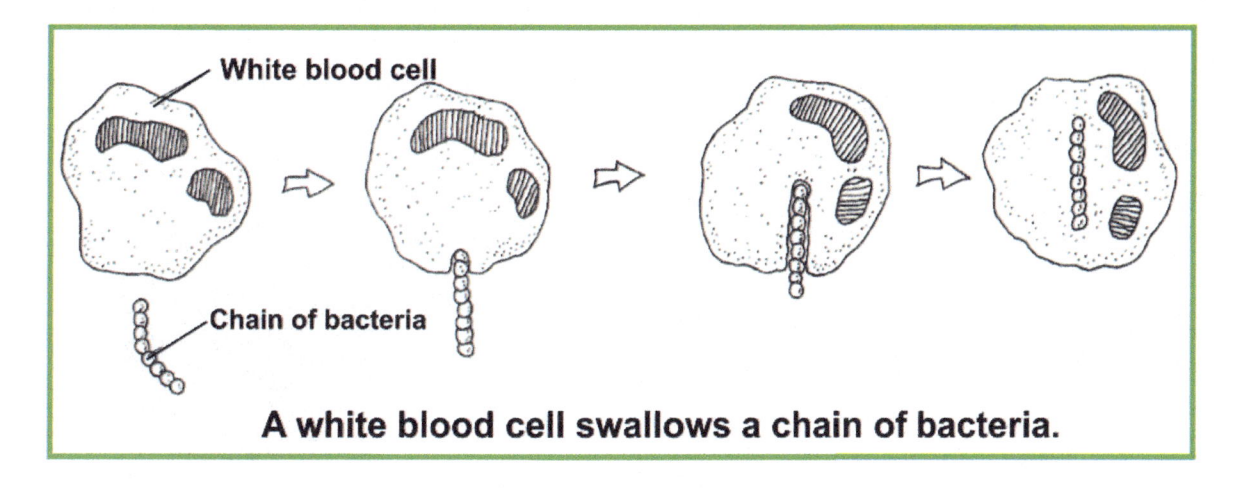

A white blood cell swallows a chain of bacteria.

273

Another type of white blood cell can make chemicals that attack germs. Still other white blood cells, called **T-cells**, help guide the bacteria-eating white blood cells to the invading germs. It is the T-cells that are attacked by the HIV virus - the cause of AIDS. As the HIV virus reproduces, many T-cells are killed. With too few T-cells, people with AIDS can't defeat the "normal" germs that healthy bodies can destroy. The "i" in AIDS stands for immune deficiency (which means a "too-weak" immune system).

Still another kind of white blood cell makes proteins called **antibodies**. For every type of microbe that invades the body, a special type of antibody is produced.

The first time that a germ gets into your body, it may make you sick. If all goes well, these white blood cells make antibodies for that kind of germ. The germs are gradually destroyed, and you recover. Plus, some of these antibodies stay in your blood. If the same kind of germ attacks again, the remaining antibodies attack, and the white blood cells remember how to make more antibodies rapidly. The invading germs are quickly killed. This time, you do not even get sick. You have become **immune** to that particular germ.

Actually, you can get **immunity** to some diseases in two ways. As just stated, one way is to have the disease. You may, for example, get sick with chickenpox, which is caused by a virus. After you get well, some of the antibodies against the chickenpox virus stay in your blood. These antibodies are ready to destroy any more chickenpox viruses that arrive.

A second way to become immune is by receiving a **vaccine** (vak-SEEN). Many vaccines are made from weakened or dead bacteria or viruses. You very likely received vaccines made this way - to protect you from measles, mumps, whooping cough, and polio. These shots are required before children enter school.

mRNA Vaccines

In 2020, scientists made a new kind of vaccine - called an **mRNA vaccine**. How do mRNA vaccines work?

First, recall the molecule called messenger RNA (mRNA) from Unit 9, Lesson 3 ("How Does DNA Make Proteins?"). Keep in mind that mRNA molecules tell our cells' ribosomes to build specific proteins.

Now, notice that the surface of the COVID-19 virus is covered with spikes. These are a type of protein.

Observe the diagram on the next page. While you read the following four steps, pay special attention to the underlined words in the reading. These same words appear - from left to right - in the large diagram.

1) The mRNA molecules in the vaccine contain the message to build spike proteins. These mRNA molecules enter the cell.

2) The mRNA (spike protein) molecules take their message to ribosomes in the cytoplasm.

3) The ribosomes make the amino acid chains that become millions of the spike proteins.

4) The immune cells in the blood "see" these new spike proteins, and build antibodies against them. Not only do the antibodies attack the spike proteins, the body now has the ability to rapidly make many more of these particular antibodies.

Later, if a vaccinated person inhales more of the COVID-19 virus particles, this person's immune cells immediately make huge numbers of antibodies. These antibodies attack and destroy the spike proteins on the virus. Destroying the spike proteins destroys the virus.

In 2021 and 2022, millions of people received mRNA vaccines. They worked beautifully!

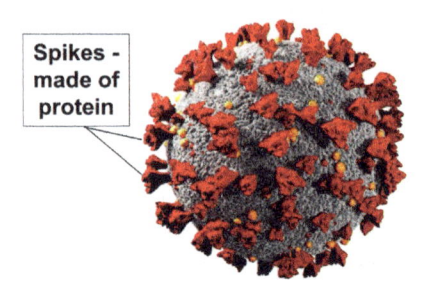

Spikes - made of protein

How the mRNA vaccine prepares the body to defeat the virus causing COVID-19

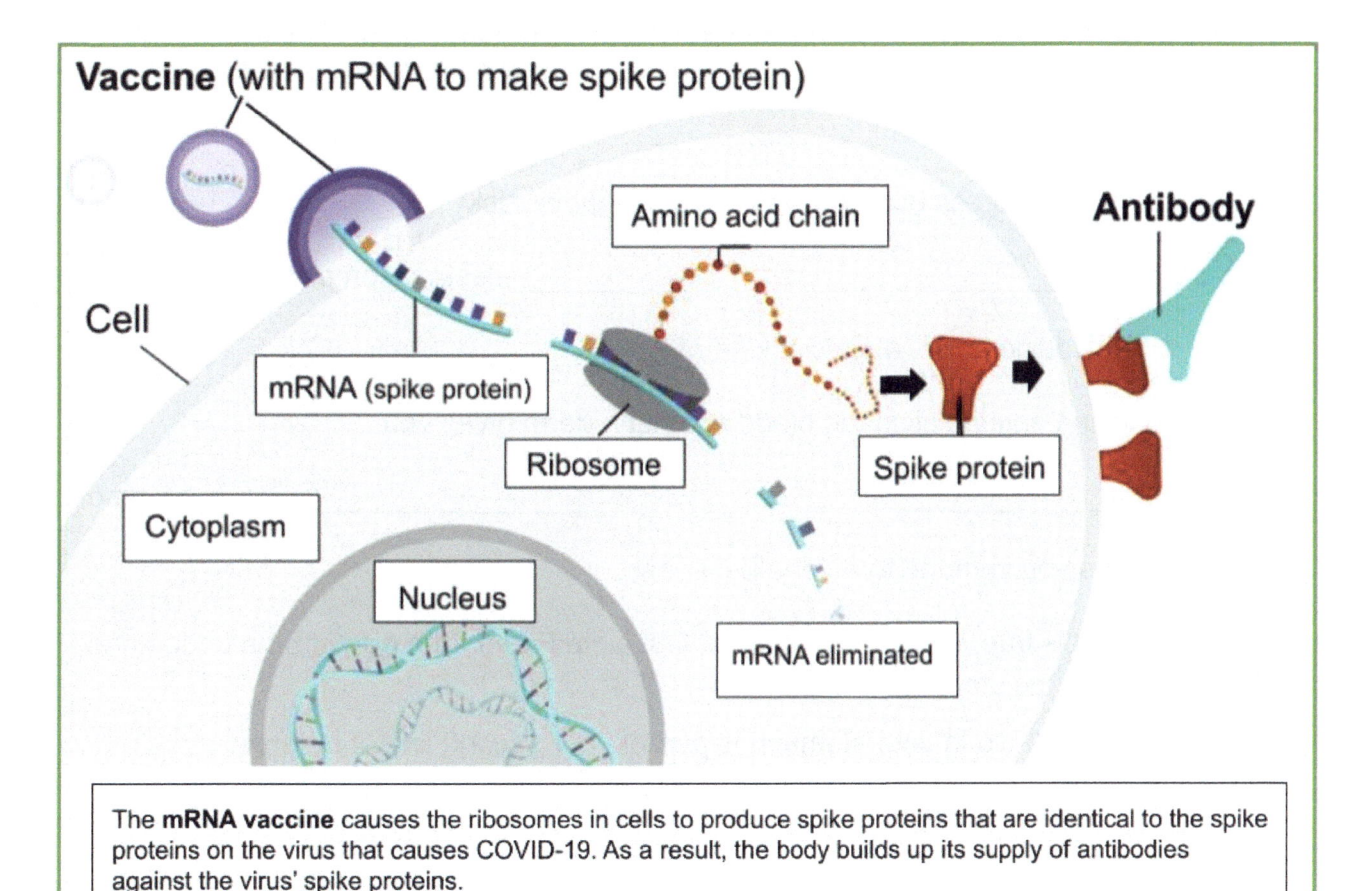

Vaccine (with mRNA to make spike protein)

Cell

mRNA (spike protein)

Ribosome

Amino acid chain

Cytoplasm

Nucleus

Spike protein

mRNA eliminated

Antibody

The **mRNA vaccine** causes the ribosomes in cells to produce spike proteins that are identical to the spike proteins on the virus that causes COVID-19. As a result, the body builds up its supply of antibodies against the virus' spike proteins.

➤ To Do Yourself — What Is One Way That the Body Traps Germs?

Make a model to show the action of mucus in your nose and throat passages.

You will need:

Cardboard, petroleum jelly, sand, talcum powder, hand lens.

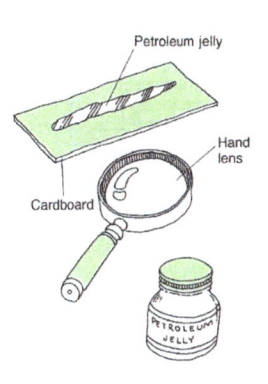

1. Cut a piece of cardboard 8 cm X 20 cm.
2. Spread a layer of petroleum jelly in the middle of the cardboard. Place some sand and powder about 5 cm in front of the jelly.
3. Close your eyes. Blow the sand and powder past the layer of jelly.
4. Clean up the excess sand and powder.

Questions

1. What does the petroleum jelly represent in the model? _____

2. What did you observe about the petroleum jelly? _____

3. What is one function of the mucus that lines your air passages? _____

REVIEW U-10 L-2

I. In each blank, write the word that fits best. Choose from the words below.

| vaccine | white blood cells | defense | cilia |
| antibody | mucus | bone marrow | immunity |

The body has two main lines of _____ against infectious diseases. Coughing forces out germs that are caught in sticky _____. Germ-eating _____ destroy some microbes. For every type of invading germ, a special kind of _____ is produced. Having antibodies in the blood against a germ gives you _____. A _____ is a kind of shot that makes you immune to a certain disease.

II. If the statement is true, write **T**. If it is false, first write **F** and then **correct** the underlined word(s) to make the statement true.

A. _____ The acid in your stomach is part of your <u>second</u> line of defense.

B. _____ Antibodies against measles can make you <u>immune</u> to measles.

C. _____ Cilia that help get rid of germs are found in the <u>bone marrow</u>.

D. _____ White blood cells that eat bacteria are part of your <u>first</u> line of defense.

III. Why is a person with AIDS at risk of getting many other diseases?

_____

_____

What is Heart Disease?

U-10 L-3

Exploring Science / Historical Steps

Heart, Balloons, and Straws. If you heard people mention hearts, balloons, and straws in the same sentence, you would probably think that they are talking about a party or a holiday celebration. In fact, these could be doctors discussing a treatment for people with a certain heart condition.

These patients have a buildup of **plaque** (PLAK) on their artery walls. Plaque is made of fats, protein and calcium. The thicker the layer of plaque, the less room there is for blood to flow through the artery. If plaque builds up in the arteries around the heart, it can cause severe heart problems.

In 1977, a German doctor - **Andreas Gruentzig** - first performed a now widely used procedure. A tiny balloon is inserted into an artery that leads to the heart. Then the balloon is pumped up. As it gets larger and larger, the balloon squeezes some of the plaque to the sides.

By the late 1980s doctors were also inserting a tiny straw-like tube, called a **stent**, into the space. With balloon surgery and stenting, the opening for blood to flow is much improved - and blood can once again feed the heart. In the United States, this procedure is performed about one million times each year!

➤ "Angio" means "blood vessel;" "plasty" means "forming." In a way, the balloon procedure is "forming a blood vessel." Guess what doctors call this procedure?

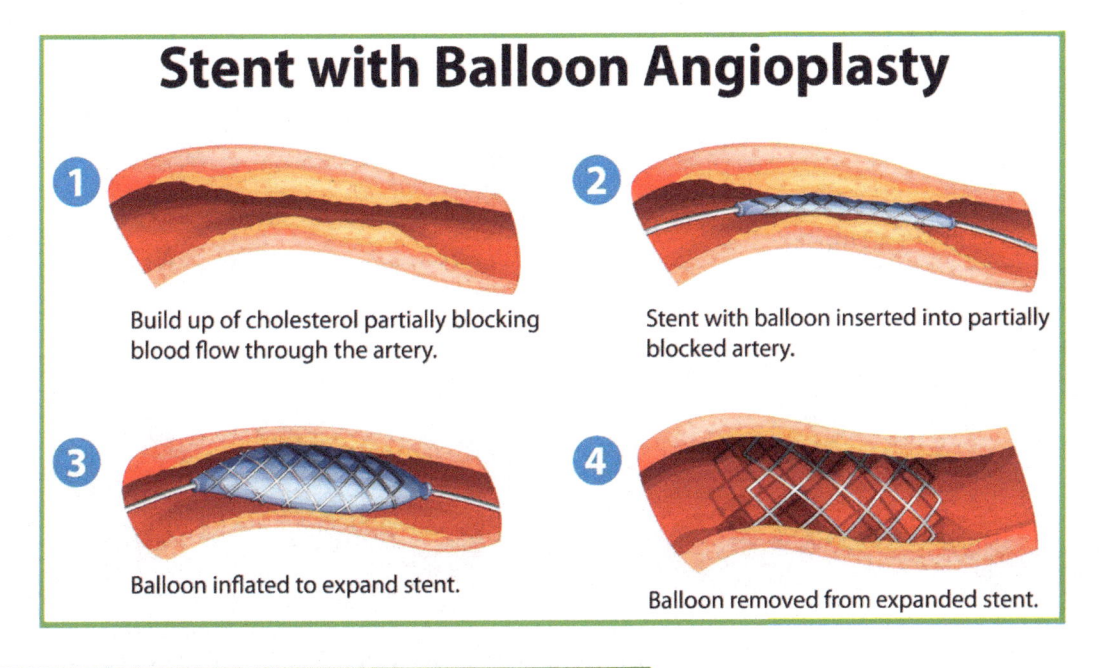

Stent with Balloon Angioplasty

1. Build up of cholesterol partially blocking blood flow through the artery.
2. Stent with balloon inserted into partially blocked artery.
3. Balloon inflated to expand stent.
4. Balloon removed from expanded stent.

Heart and Blood Vessel Disease

Did you know that more United States citizens die from heart diseases than from any other cause? Many people also die from blood vessel disease. Some people inherit the conditions that lead to heart or blood vessel disease. This is why doctors always ask about a patient's "family history" when it comes to diseases.

For some people, however, diet hastens the onset of such diseases. You just learned how surgeons deal with plaque that lines arteries. But what actually causes the plaque? It is often related to a diet that is too rich in fat, particularly **cholesterol**. Our body makes its own cholesterol. We do not need to eat it. Cholesterol is an animal product; plants do not contain cholesterol.

277

If there is too much cholesterol in the blood, some of it may stick to the inner walls of the arteries. As the cholesterol builds up, it causes plaque. The more plaque, the less blood that is able to get through the arteries. Do you see why nutritionists encourage people to eat lots of plants and only small amounts of meat?

In addition to eating well in order to reduce their chance of heart problems, many adults take medicines that lower the amount of cholesterol in their blood.

Observe the illustrations below. Remember that arteries carry blood to all parts of the body. Arteries called **coronary arteries** "feed" the heart's own muscle.

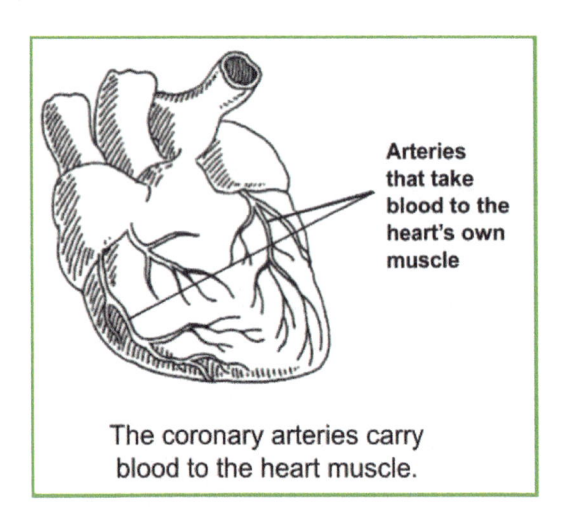

Arteries that take blood to the heart's own muscle

The coronary arteries carry blood to the heart muscle.

What if a tiny blood clot forms, but one of these arteries is clogged by fat? The small clot blocks the artery. No blood can pass through. The blood supply to a part of the heart muscle is shut off. Without oxygen and food, this piece of heart muscle dies. The result is a **heart attack**.

Often, the amount of heart muscle that is damaged is small. The patient is able to recover with medical care and proper exercise (to strengthen the rest of the heart). But sometimes, so much heart muscle is damaged that the heart attack victim dies. Everyone should know the warning signs of a heart attack. A person with any of these signs should call a doctor - or 911 - without delay.

Table 1 Warning Signs of Heart Attack

- Lasting "heavy" pain or discomfort in the center of the chest
- Pain that spreads to the arm, shoulder, neck, or jaw
- Sweating along with chest pain.
- Nausea and vomiting along with chest pain
- Shortness of breath

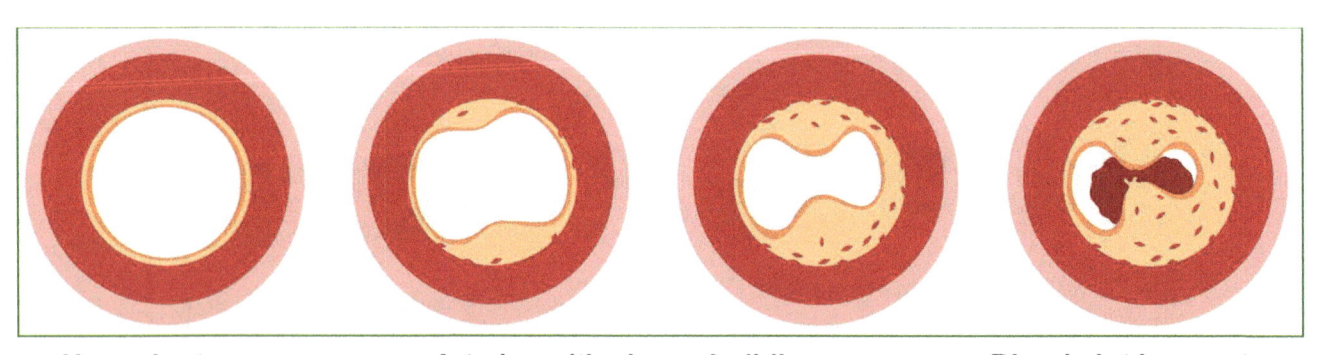

Normal artery   **Arteries with plaque building up**   **Blood clot in an artery clogged with plaque**

Do you know someone with **high blood pressure**? In order for their blood to pass through an artery, it pushes harder than normal on the artery walls. In other words, the pressure of the blood against the walls of the arteries is higher than normal.

Doctors aren't sure what causes most high blood pressure. A high salt diet, smoking, and stress seem to increase the chances of this disease.

High blood pressure is dangerous. It can lead to blood vessel disease. **Stroke** is a blood vessel disease that affects the brain.

A stroke may happen in two ways. One kind of stroke is caused by a blocked artery. A blood clot may block a narrowed artery in the brain. Another kind of stroke happens when an artery in the brain bursts. Often the reason it bursts is high blood pressure.

Either kind of stroke shuts off the blood that feeds part of the brain. What happens to the victim depends on the part of the brain that is affected. The stroke victim may have a loss of memory. They may be unable to speak, or to move some parts of the body. In all cases of stroke, some nerve cells die. If too many nerve cells die, the person dies. Table 2 lists the warning signs of stroke.

Table 2   Warning Signs of Stroke

- Sudden, temporary weakness or numbness of one side of the face, or one arm or one leg
- Temporary loss of speech, or trouble speaking or understanding speech
- Temporary dimness or loss of sight, especially in one eye
- A period of double vision
- Dizziness or unusual headaches
- Change in personality or mental ability

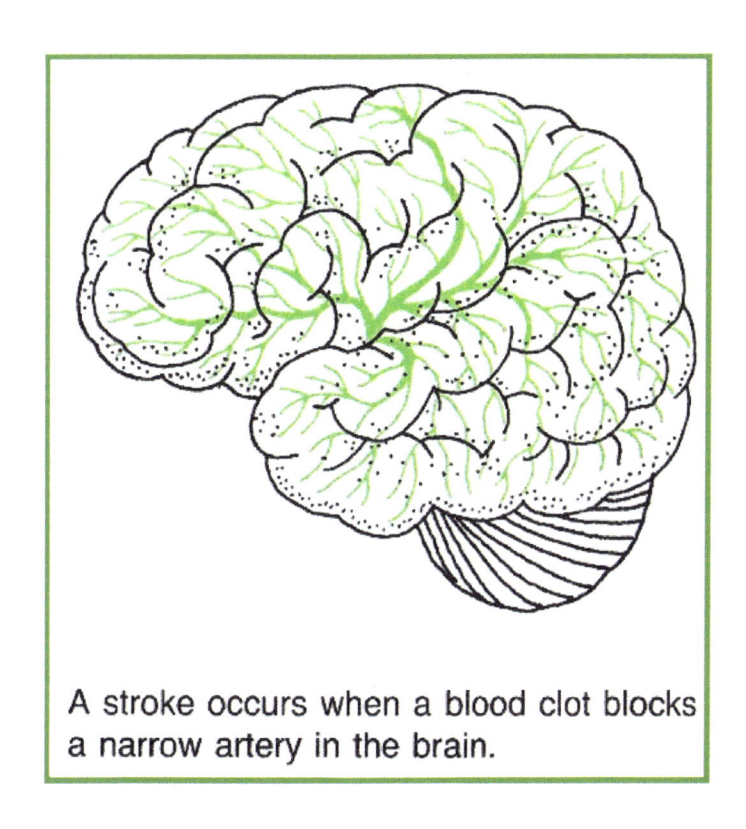

A stroke occurs when a blood clot blocks a narrow artery in the brain.

How To Prevent Blood Vessel Problems

Scientists have learned a great deal about how to prevent heart and blood vessel diseases. Their findings have led to the following advice:

➡ Eat less animal fat (which includes cholesterol). This means cutting down on meat (especially beef), and whole milk.

➡ Eat less salt. Over-salting is only a habit.

➡ If you smoke, stop. If you don't smoke, don't start.

➡ Exercise regularly. Jogging, swimming, and bicycle riding are among the types of activity that help keep the heart and blood vessels in shape. Caution: Before you start any exercise program, get your doctor's approval.

➡ Keep your weight normal. Extra weight stresses the heart.

> **To Do Yourself** How Can An Artery Become Blocked?

You will need:
  2 plastic straws, 2 medicine droppers,
  2 dishes, a pipe cleaner, lard, a partner

1. To make a model of a blocked artery, push each end of <u>one</u> straw into lard.
2. Use the pipe cleaner to push the lard into the straw. Do this until the straw is lined with lard.
3. Fill both droppers with water. Hold one straw over one dish, while your partner does the same with the other straw.
4. Both of you gently squeeze the water from the dropper into the top of the straw. Record what happens. Try it several times.

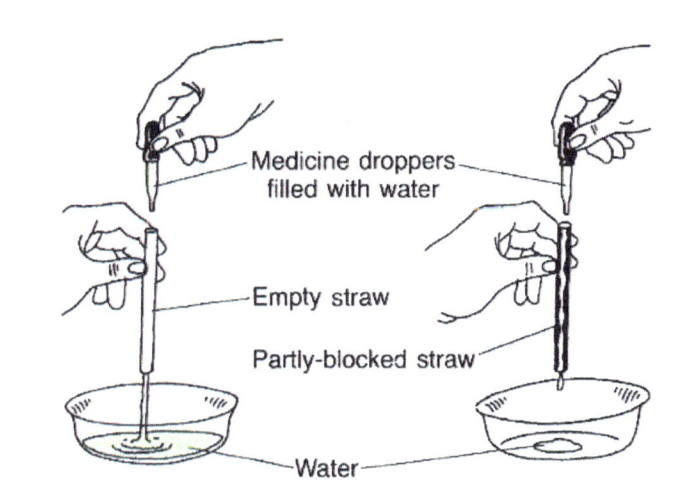

Questions

1. What does the straw represent in the body? _____

2. What does the lard represent in the body? _____

3. Which straw does the water pass through the fastest? The slowest? _____

REVIEW — U-10 L-3

I. In each blank, write the word that fits best. Choose from the words below.

  liver   cholesterol   arteries   heart attack
  veins   stroke   coronary   blood pressure

Two much _____ in the diet (and then in the blood) may lead

to fat-clogged _____ and make their walls become thicker. A

condition that may lead to heart or blood vessel disease is high

_____. If a _____

artery is blocked, some heart muscle dies. Either _____ or

_____ may result from a blood clot in an artery.

II. Circle the word (between the brackets) that makes each statement true.

  A. Pain in the chest is a sign of [stroke / heart attack].

  B. A broken blood vessel in the [brain / heart] can cause a stroke.

  C. A diet that is [high / low] in salt may help prevent high blood pressure.

  D. Loss of speech is one sign of a [heart attack / stroke].

  E. A diet that is [high / low] in fat may help prevent a heart attack.

What is Cancer? U-10 L-4

Exploring Science / Historical Steps

Eat Your Broccoli. For decades, parents told their children to finish everything on their plates. Today's parents are now careful not to push children to overeat. However, parents should still tell their children to eat their vegetables - especially foods like broccoli!

Why broccoli? This "miniature tree" vegetable, with its bright green stalk and darker green flowerets, is the source of a powerful chemical. Scientists have identified it as *sulforaphane*. This compound actually blocks the growth of certain kinds of tumors. Hence, it helps to prevent some cancers from forming.

In 1992, **Paul Talalay** and **Jed Fahey** of John Hopkins University published research that proved broccoli's importance. Sulforaphane seems to be able to jump-start the human body's production of certain kinds of enzymes. These enzymes help to break down and remove cancer-causing chemicals that have entered the body. Talahay and Fahey studied two groups of rats. One group received sulforaphane and a control group did not. The effects of the compound were dramatic. Many more rats in the group that did not get the compound ended up with tumors.

Sulforaphane is also in other vegetables from the broccoli family - cauliflower, brussel sprouts and cabbage. All have the mildly bitter flavor characteristic of this family of vegetables.

The good news is that sulforaphane is not destroyed by cooking or microwaving. So whether these vegetables are your favorites or not, it's smart to include them in your diet.

Talahay and Fahey followed this important work with many other studies. The term **chemoprotection** is now widely used to describe compounds in plants that help protect us from diseases. In fact, the name of the department at Johns Hopkins where these two men did their work is now known as the Cullman Chemoprotection Center.

➤ What other parts of the Healthy Eating Pyramid (from Unit 3) do you think might be part of an anti-cancer diet?

Broccoli contains sulforaphane that helps prevent cancerous tumors.

Cancer

How can what we eat help prevent cancer? Even for scientists, this is a hard question. Cancer is a puzzle. Scientists have found some of the pieces. One piece shows that there is sometimes a link between diet and cancer.

Just what is **cancer**? There are actually over 100 different diseases called cancer. All cancers are alike in one way. The growth and reproduction of cells gets out of control. Cancer is cell growth gone wild. Also, the new (cancerous) cells do not perform the jobs of the cells nearby. For example, cancerous liver cells do not act as normal liver cells.

Normally, cells divide and grow in an orderly way. As each living thing develops, new cells form until the body parts reach normal size. Then the division of cells - for growth - stops. After that, normal cells divide only to replace those that wear out or die.

But sometimes cells divide without stopping. In these cases a growth forms, called a **tumor** (TOO-mur). Most moles, warts, and other tumors stop growing. They do not get above a certain size. They do not spread to other parts of the body. Tumors of this kind are called **benign.**

Although most benign tumors are harmless, sometimes they must be removed or they will crowd the normal cells nearby.

Other kinds of tumors do not stop growing. Their cells may eventually spread to other parts of the body. This kind of tumor is called **malignant** (muh-LIG-nunt). Malignant tumors are also simply called cancers. A malignant tumor can be very dangerous. Why? As it grows and spreads, it crowds out and destroys healthy tissue. Unless the growth is stopped - or at least slowed down - the person who has cancer will die.

Cancer can occur anywhere in the body. **Leukemia** is cancer of the bone marrow. In this disease, too many white blood cells are made. Red blood cells are crowded out and their work is interfered with. In the same way, stomach cancer interferes with digestion.

The reason that cancers begin is not the same in every case. Too much or not enough of certain foods seems to be involved in some kinds of cancer.

In other cancers, chemicals called **carcinogens** (kar-SIN-uh-jens) are part of the cause. A few of the chemicals known to cause cancers in animals (like us) are listed in the table at the right. Smoke from tobacco is likely the top killer. Somehow, a carcinogen changes normal cells into cancer cells. **Radiation** (ray-dee-AY-shun) from X-rays, the sun, or nuclear sources can also start the growth of some cancers.*

When detected early, many cancers can be treated. And many can be cured. One method of treatment is to cut away the cancer, by **surgery**. If this is done before the cancer spreads, the patient may be cured. **Radiation** from X-rays or radioactive chemicals may also be used to kill cancer cells. These methods must be used carefully. The use of medicines - **chemotherapy** (keem-uh-THER-uh-pee) - is another treatment that doctors use. All treatments have the best chance of success when the cancer is detected early. Everyone should know the early warning signs of cancer (listed on the next page).

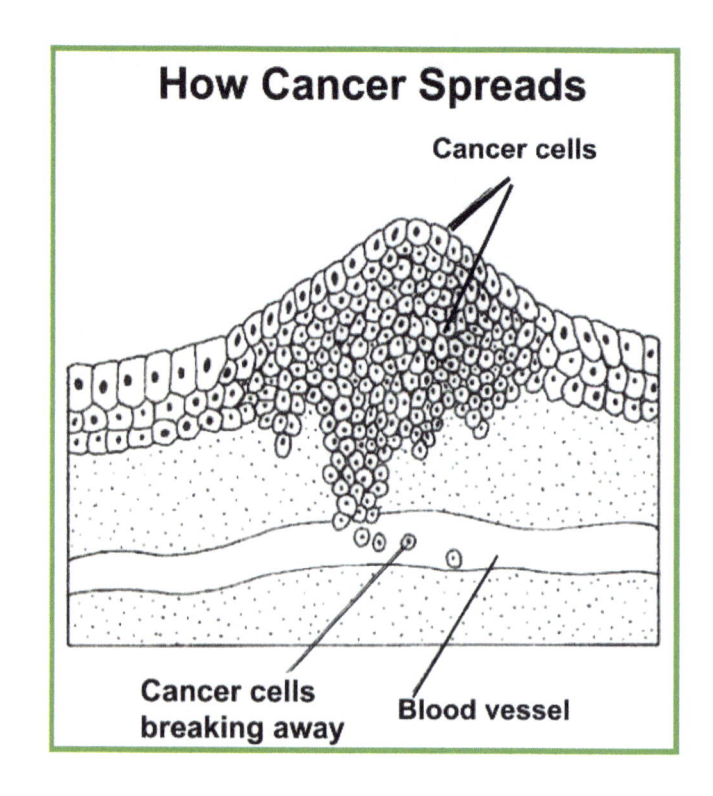

How Cancer Spreads

Cancer cells

Cancer cells breaking away

Blood vessel

--- Some Carcinogens ---

Chemical	Sources	Enters Body By
tobacco	-cigarettes -cigars -pipes	-smoking, -inhaling smoke from others
nitrites	-processed meats	-eating bacon, hot dogs, etc
asbestos	-insulation -heatproofing & fireproofing materials	-breathing near the sources
radon	-natural in some areas	-breathing when it becomes concentrated
-air pollution	diesel trucks and cars	-breathing when it becomes concentrated

Know Cancer's Warning Signals!

Change in bowel or bladder habits

A sore that does not heal

Unusual bleeding or discharge

Thickening or lump in breast or elsewhere

Indigestion or difficulty in swallowing

Obvious change in wart or mole

Nagging cough or hoarseness

If you think you have any of the early warning signs of cancer, it does not mean that you have cancer. It means that you should see your doctor.

[American Cancer Society]

➤ To Do Yourself   What Is In Burning Tobacco?

You will need:

Adult supervision, heat-proof test tube, test tube holder, burner, cotton ball, water, cigarette tobacco.

1. Place a pinch of tobacco into a test tube.
2. Plug the top with a wad of cotton.
3. Hold the tube as shown, over a flame.
4. Move the bottom of the tube over the flame for about 3 minutes. Observe the cotton in the test tube.
5. Allow the tube to cool. Remove the cotton and observe the brown tar on it.

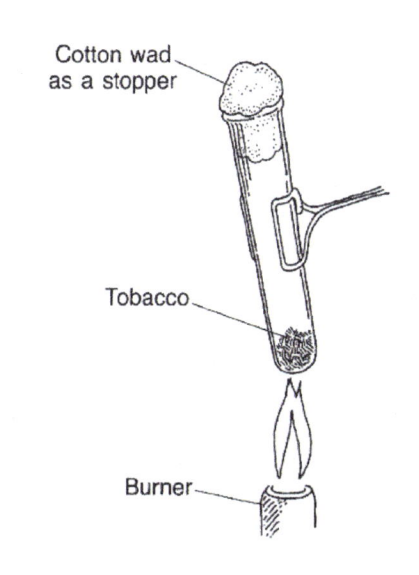

Questions

1. If the burning tobacco's smoke were breathed into the body, where would the tar collect? _____

2. What might happen when too much tar gets into the body? _____

REVIEW — U-10 L-4

I. In each blank, write the word that fits best. Choose from the words below.

division	surgery	chemotherapy	carcinogen
malignant	benign	tumor	radiation

A cancer-causing chemical is a _____. Three methods of treating cancer are _____, _____, and _____. An abnormal growth of cells is called a _____. Growths that are _____ are cancers.

II. Record **B** for benign or **M** for malignant to identify the kind of tumor described by each statement.

A. ____ Grows out of control

B. ____ Is usually harmless

C. ____ Destroys healthy tissue

D. ____ Does not spread to other parts of the body

E. ____ Can be very dangerous

F. ____ Leukemia is one example

G. ____ Includes most moles and warts

III. What two steps could people take to help *prevent* both heart disease and cancer?
Hint: One is something that you <u>should</u> do, and one is something that you should <u>never</u> do.

_____

_____

_____

_____

*
The work of the famous DNA research scientist, **Rosalind Franklin**, (mentioned in Unit 9, Lesson 2) involved a great deal of work with a special type of X-ray. It is very likely that these X-rays were the cause of her tragic death from cancer at age 37.

What is Drug Abuse? — U-10 L-5

Exploring Science / Historical Steps

Think Before You Drink. The huge Long Island Railroad diesel is on a late-night run. Engineer Thomas Cavanagh keeps his eyes on the track ahead. Suddenly, he sees a yellow van. The safety gate is down, but the van is trying to get around it. Cavanagh slams on the train's brakes, but it is already too late. The train crashes into the van, and nine teenagers are dead. Blood tests show that the teens had been drinking.

Tragic stories like this one started many people thinking. One was **Paul Anastas**, a teacher-coach from Massachusetts. Anastas lost two teen athletes in accidents in which the driver had been drinking. In 1983, Paul began an organization called **SADD** - Students Against Driving Drunk. In 1997 the name was changed to Students Against Destructive Decisions. SADD now has chapters in every state.

Members of SADD campaign against drinking and driving. They get the message out to students everywhere - it is very uncool to get in a car with someone who has been drinking, smoking pot or using any type of drug.

Since its founding, SADD has expanded its message. The organization now helps students learn not only about alcohol and drug abuse, but how to make better choices and enjoy a healthier lifestyle.

➤ Drunk drivers of all ages can have fatal accidents, but there are many more among younger drivers. Why do you think that this is the case?

➤ What do you think is the #1 cause of fatal auto accidents? Hint: The answer is *not* drunk driving! [Hint: See the caption below.

Driving safely requires full focus. Any type of distracted driving is extremely dangerous.

Drug Abuse

Alcohol, the chemical present in beer, gin, whiskey, wine - is a drug. In fact, it is a powerful drug. Among people who use alcohol, one in ten will move from "social drinking" to "problem drinking." This affects not only them, but their family, friends, employers - in fact, all of society.

Some who use alcohol say that it helps them feel less shy - and even "high." But alcohol use frequently leads to terrible decisions - like drinking and driving, or violent behavior. When used by pregnant women, alcohol often causes birth defects; this sad condition is called **fetal alcohol syndrome**.

Nicotine, the key drug of cigarettes, e-cigarettes and most **vaping** products, is also a powerful drug. Users very rapidly become strongly addicted. Over time, the odds are high that smokers will have serious health problems - particularly heart disease and cancer. Getting nicotine by **vaping** is not only addictive, it carries a high risk of lung damage in a very short time.

285

Sadly, using snuff or chewing tobacco often dooms the user to mouth or throat cancer later; plus, many who start with these products soon switch to cigarettes in order to get nicotine without staining their teeth and spitting in public.

So just what is a drug? Any substance, except food, that causes a change in the body is considered to be a **drug**. Used properly, many drugs are medicines. Some examples are aspirin and penicillin. These drugs may be ordered by doctors to help reduce pain or fight diseases. But even drugs that are medicines can be used in the wrong way. The improper use of a drug is known as **drug abuse**. Using a medicine in the wrong *amount* is also drug abuse; so is using a medicine for the wrong reason.

How does drug abuse get started? Sometimes, after an injury or a surgery, strong drugs are needed to reduce pain. A patient may have a hard time deciding when to stop taking a pain-relieving drug - and become addicted.

Some people (often young people) start using a drug to "feel "high." They may abuse drugs to escape problems or to be accepted by others.

Drug abusers often develop a need for drugs. This is **drug dependence** (di-PEN-duns). Drug dependent people are called **addicts**. For some drugs, an addict needs the drug for its effect on the mind. This is an emotional dependence. But for some drugs, like cocaine, the dependence is physical as well. The addict needs the drug for the effect on the body; they feel ill without it.

Different drugs have different effects on the brain. A **depressant** (di-PRES-unt) is a downer. A **stimulant** (STIM-yuh-lunt) is an upper. A **hallucinogen** (hu-LOO-suh-nuh-jen) is a vision-producer.

Inhalants (in-HAYL-unts) include sprays and glues that some people sniff to get high. The very first use of an inhalant can cause brain damage, and deaths are not rare. The table on the following page lists some drugs of each type.

---------- **REVIEW** ---------- <u>U-10 L-5</u>

I. In each blank, write the word that fits best. Choose from the words below.

| emotional | hallucinogens | aspirin | addict |
| physical | drug abuse | stimulants | depressants |

Any improper use of a drug is _____. A need for a drug's effect on the mind is a(n) _____ dependence. A person who needs a drug is an _____. "Uppers" are drugs that are _____. "Downers" are drugs that are _____.

"Vision-producers" are _____.

II. Use the table of addictive drugs to answer the following. Circle the word (between the brackets) that makes each statement true.

   A. A depressant is a drug that [speeds up / slows down] the nervous system.

   B. A drug user who sees something that does not exist might have been taking a drug called [angel dust / speed].

   C. Grass and pot are other names for [cocaine / marijuana].

   D. Alcohol is a [stimulant / depressant].

   E. Vaping products carry a serious risk of [lung / liver] damage.

III. What <u>brief</u> answer do you think most smokers will give if you ask them these questions?
   1) Did you expect to become "hooked" when you began?
   2) Do you think that you will eventually have heart disease or cancer due to smoking?

----- Addictive Drugs -----

Types	Some Effects	Examples and Common Names	Examples and Street Names	Dependence E = emotional P = physical
Depressants (downers)	-slow down nervous system; slow heart and breathing rates; make user sleepy; often cause depression -opioids: block physical pain; are highly addictive	-Alcohol (beer; wine; whisky; vodka) -Opioids (Oxycontin; Vicodin; Fentanyl; heroine; morphine; codeine) -Barbiturates (barbs; blues) -Methaqualones (Quaalude; ludes) -Tranquilizers (Valium; Halcion)	-alcohol (beer; wine; vodka; whiskey) -heroin (horse; duge; smack; junk) -codeine (schoolboy) -barbiturates (barbs;blues) -morphine (M)	-E & can be P -E & P -E & P -E & P
Stimulants (uppers)	-speed up nervous system; make user anxious, excited, restless, wakeful; causes emotional and physical dependence -vaping can cause lung damage, nicotine addiction	-codeine (coke; snow) -amphetamines (speed; pep pills) -cocaine (crack; rocks; snow) -caffeine (cola; tea; chocolate -nicotine (cigarettes; cigars) -e-cigarettes (vaping);	-benzedrine (bennies; cartwheels) -amphetamines (speed; pep pills) -cocaine (crack; rocks; coke; snow) -dexedrine (dexies) -nicotine (cigarette; cigar; fag) -e-cigarettes (vaping; vape pens; hookah)	-E -E -E -E -E and P -E and P
Hallucinogens	user sees, hears things that do not exist (hallucinates); user loses coordination, behaves strangely; may cause depression or panic*; causes emotional dependence	-marijuana (grass; pot; weed) -LSD (acid; sugar) -PCP (angel dust) -mescaline (cactus)	-marijuana (grass; pot; weed) -LSD (acid; sugar) -PCP (angel dust) -mescaline (cactus)	-E -E -E -E
Inhalants	high risk of instant death; damage to multiple organs; slow down bodily functions; initial feeling of stimulation; causes emotional dependence	-spray paint; lighter fluid -model airplane glue; -amyl nitrate (poppers) -aerosols	-all types = huffing; bagging	-E (but causes physical damage to brain and, too often, death)
anabolic steroids	damage to kidneys, liver, heart; acne; stunted growth; emotional changes (e.g. paranoid jealousy, aggression ("roid rage"))	-anabolic steroids (man-made chemicals similar to the male sex hormone) taken illegally to increase muscle growth	-arnolds; gym candy; juice	-E and P

* a *National Institutes of Drug Abuse* study - surveying 281,000 people, ages 18 -31, from 2009 to 2019 - found that of those who labeled themselves as severely depressed, about a third who were *not* marijuana users said that they considered suicide, while more than half of those who used marijuana daily said that they considered suicide. [From an article in the Courier-Journal, 7/6/2021, p. 9B]

Unit 10 -- Review What You Know --

A. Use the clues below to complete the crossword.

Across
1. Shots to prevent diseases
4. Covid is one
5. "Bug" that carries Lyme microbe
9. Can carry rabies germs
10. Can be passed to other people

Down
2. Cancer causing chemical
3. Shuts off blood supply to part of brain
6. Drug dependent person
7. Cause ringworm and athlete's foot
8. Improper use of drug

B. Write the word (or words) that best completes each statement.

1. A vaccine can give immunity to
   a. heart attack   b. cancer   c. polio   1. _____

2. High blood pressure may lead to
   a. pneumonia   b. stroke   c. STIs   2. _____

3. Most warts and moles are
   a. malignant tumors   b. cancers   c. benign tumors   3. _____

4. Alcohol is a drug that is
   a. a depressant   b. a stimulant   c. an upper   4. _____

5. Smoking is unhealthy because it helps cause
   a. only cancer   b. only heart disease
   c. cancer and heart disease   5. _____

C. Apply What You Know

Observe the illustrations on the next page. Choose from the <u>terms</u> below to label each illustration. Then, use the letter that is next to each label to match one of the following statements. Put the <u>letter</u> (or <u>letters</u>) in the space after the number.

porkworm   **covid virus**   **trachea cell**
Lyme disease bacteria   **white blood cell**

1. _____ May be present in uncooked meat
2. _____ Its cilia sweep out germs
3. _____ Carried by a tick
4. _____ Can "swallow" germs
5. _____ , _____ , _____ Causes disease
   (This answer requires three letters).

A. _____   B. _____

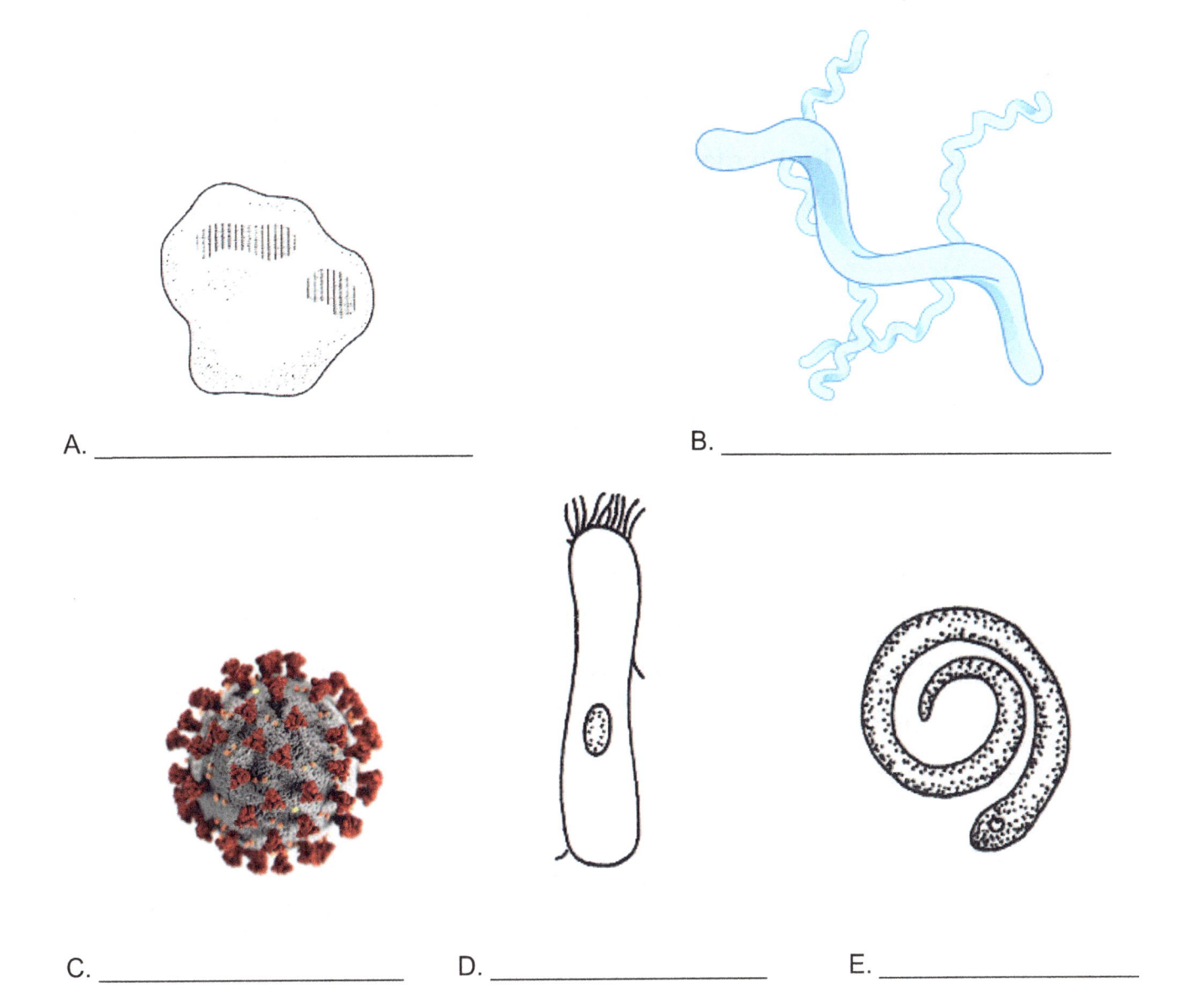

C. _____   D. _____   E. _____

D. Find out more.

1. Complete some research and compare recent and past deaths (for example in 1900) from the following causes:

    A) infections   **B)** heart disease   **C)** cancer

    Try to locate graphs showing the rates of these diseases. If you find only charts of numbers, create your own graphs. What do you think accounts for the changes in the numbers over the decades?

2. With parent supervision, collect labels from some over-the-counter medicines, vitamins, etc. Make a small chart of the items' names, the amount that is taken, and whether they warn against overuse.

3. Do you like to solve problems? Have you found your study of life science enjoyable? You ought to consider a career involving cell biology. From medical advances to discoveries in evolution, careers involving DNA and RNA are "exploding"!

    For a special challenge now - research the term **epigenetics**. Darwin would be shocked to learn that natural selection is actually not the only means by which species change! Ask your teacher if you may share what you've learned with the class.

289

SUMMING UP:
Review What You Have Learned So Far

A. Study the drawing. In the statements below, circle the word (or words) [between the brackets] that makes each statement about the scene true.

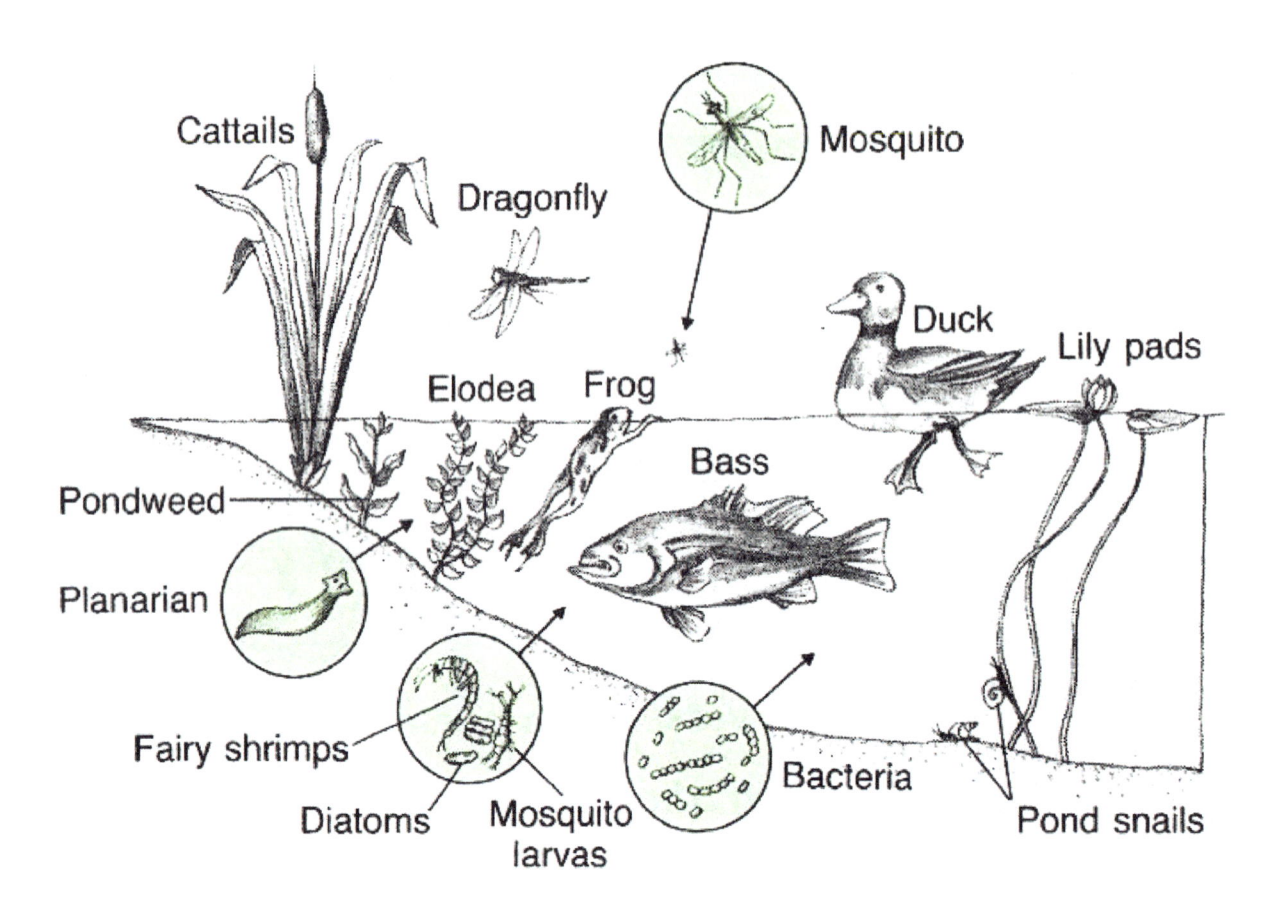

1. The air, water, and soil are all parts of the pond [community / ecosystem].

2. All of the lily pads in the pond make up a [population / food web].

3. Photosynthesis is carried out by the [bacteria / cattails].

4. The [mosquito larva / frog] is at a young stage of its life cycle.

5. The [bass / duck] is warm-blooded.

6. The [diatom / elodea] is classified in the protist kingdom.

7. The dragonfly is [a vertebrate / an invertebrate].

8. All [shrimp / planarians] are able to reproduce by regeneration.

9. [A duck / A bass] lays the most eggs.

10. Because the snail eats dead organisms, it is a [predator / scavenger].

B. Column A lists diseases or disorders. In column B are some ways to prevent or treat diseases or disorders. In each blank, write the letter (or letters) of each item from column B that matches an item in column A.

A	B
1. _____ Night blindness	a. Eat foods rich in calcium.
2. _____ Cancer	b. Reduce animal fats in the diet.
3. _____ Sickle-cell disease	c. Eat foods rich in vitamin A.
4. _____ Heart attack	d. Avoid carcinogens.
5. _____ Diabetes	e. Exercise regularly.
6. _____ Polio	f. Visit a genetic counselor
7. _____ Weak bones	g. Take insulin
8. _____ Hemophilia	h. Have shots of a vaccine

C. The statements below refer to the human body. If the statement is true, write T in the blank. If the statement is false, write F, and then correct the underlined word or words to make the statement true.

1. _____ The pituitary gland produces growth hormone.

2. _____ Most digestion takes place in the stomach.

3. _____ The blood platelets help in the clotting process.

4. _____ A ball-and-socket joint is found in the knee.

5. _____ The organs that filter urea from the blood are the ureters.

6. _____ Helpful bacteria live in the large intestine.

7. _____ The testes belong to both the reproductive and endocrine systems.

8. _____ A human embryo develops inside of the womb or ovary.

9. _____ White blood cells help to fight infections.

10. _____ Hormones are produced by the nervous system.

11. _____ Insulin is produced in the spleen.

12. _____ Clotting in the blood vessels can cause a stroke.

13. _____ Goiter is a disease that affects the thyroid gland.

14. _____ Blood platelets are produced in the pancreas.

15. _____ The largest bone in the human leg is the ulna.

Careers in Life Science

Health Care - On the Scene and Behind the Scenes. Does it surprise you to know that there are over 200 different jobs in health care? Did you know that over a third of doctors are female, and nearly a tenth of nurses are male? Maybe you would enjoy a career in health care!

Of course, there are doctors, nurses, and others who have direct contact with sick people. Without these "on-the-scene" workers, we would not have hospitals. But hospitals - and other health-care settings such as doctor's offices and clinics - also have many "behind-the-scenes" workers. These workers may never see the people they serve, but their work is very important.

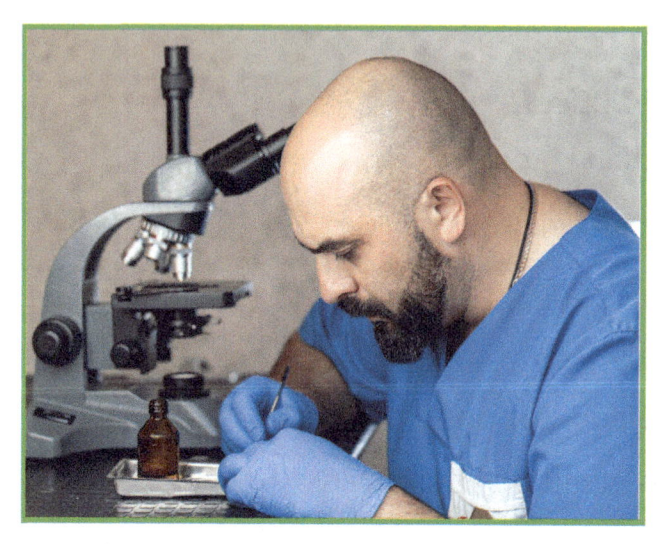

A nurse helps keep a young patient calm.

Nurse Aides and Nurses. Working as a hospital volunteer may help you discover if you would enjoy a career as a nurse. There are various levels of nursing-related work - with the responsibilities and the pay increasing as you are better trained. A four to eight weeks course after high school can train you to be a certified nursing assistant (C.N.A.). A year of training after high school would prepare you to become a licensed practical nurse (L.P.N.). To become a registered nurse (R.N.), you need two or more years of study after high school. Some nurses go on to earn a bachelor's degree (a four year college degree).

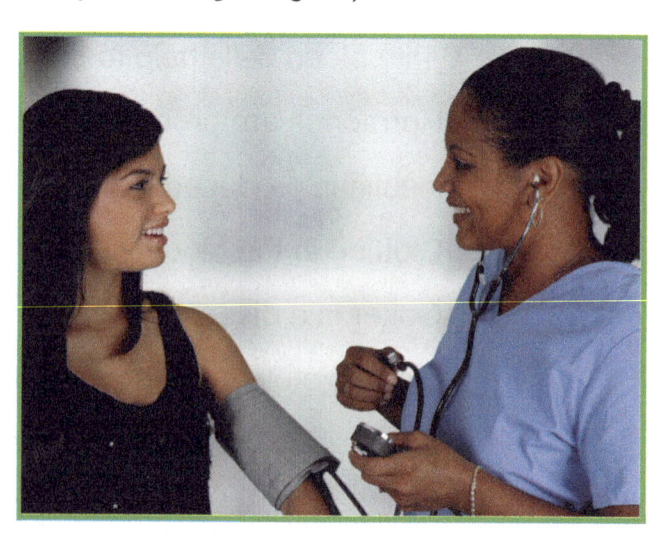

A medical laboratory assistant helps determine the cause of a disease.

Medical Laboratory Assistants and Technicians. Do you enjoy working with microscopes and test tubes? The "hands-on" part of studying life science is also part of working in a medical laboratory. Some high schools offer courses that prepare you for a job as a laboratory assistant. Or you may qualify as an assistant after a year of on-the-job training. Two or more years of further study are needed to become a medical technician.

A nurse checks a patient's blood pressure.

Glossary

Acquired Immune Deficiency Syndrome (AIDS): the disease (caused by the HIV virus) that attacks the immune system

addict: a person who becomes dependent upon a drug

air sacs: (alveoli): tiny bag-like structures in lungs where oxygen moves from the air into the blood, and carbon dioxide moves from the blood into the air

algae: simple, plant-like protists that produce food by photosynthesis

amino acids: molecules that are the building blocks of proteins

amphibians: the class of cold-blooded vertebrates, most of which spend part of their life cycle in water and part on land, and include frogs and salamanders

anabolic steroids: manufactured chemicals that imitate the male sex hormone

anther: pollen-producing structure located at the top of the stamen

antibodies: germ fighting molecules made by white blood cells

antibiotic: medicine that kills bacteria or slows their growth

anus: the opening through which solid wastes (that remain after digestion) pass out of the body

aorta: the largest artery in the body

appendix: a small, thin sac attached to the large intestine

archaea: 1. one of the three domain levels of classification 2. single-celled organisms without a nucleus, but with chemical traits different enough from bacteria to be in a different domain

arteries: blood vessels that carry blood away from the heart

asexual reproduction: the production of offspring from only one parent

ATP: battery-like molecules that are able to rapidly receive and give up energy

atrium (pl: atria): an upper chamber of the heart that receives blood from the veins and sends it to a ventricle

axon: the long narrow extension of a nerve cell that carries a message away from the main cell body

Bacteria: (singulular: **bacterium**): 1. one of the three domain levels of classification 2. simple, single-celled organisms without a nucleus; some are helpful and some cause diseases

base: half of a "rung" in a DNA ladder

base pair: two bases that join to form a rung in a DNA ladder

behavior: all of the actions of a living thing

benign tumor: a usually harmless growth of cells that does not spread

bias: an opinion based on previous experiences that must not be allowed to interfere with scientific thinking

biceps: the muscle on the front of the upper arm that contracts to bend the elbow

bile: a juice, made by the liver, that breaks fats into tiny droplets

biodiversity: the number of different living things in an area

biology: the study of living things

biomagnification: the accumulation of a substances in organisms from the bottom to the top of a food chain

biome: a large area of land with the same climate and a specific kind of climax community

biosphere: the part of the earth where life exists

bladder: an elastic sac that stores urine until it passes through the urethra and outside of the body

blending: a combination of genes in which neither is dominant or recessive - instead both show in the offspring

brain: the center of control in the nervous system, made up of the cerebrum, cerebellum, and medulla

bronchus (pl. **bronchi**): the tube through which air passes from the trachea (windpipe) into the lung

budding: asexual reproduction that produces two cells of different sizes

bulb: a short, thick underground stem with modified leaves that store food; can asexually reproduce the plant

Calorie: a measure of food energy

cambium: growth tissue in the stem of a plant; produces xylem and phloem cells

cancer: a malignant tumor or other growth of cells that spreads and destroys healthy tissue

capillaries: microscopic blood vessels that connect arteries with veins

carbohydrates: a group of nutrients that includes starches and sugars

carbon dioxide: a gas given off during cell respiration and used during photosynthesis

carcinogens: chemicals that cause cancer

cardiac muscle: involuntary muscle tissue of the heart

carnivore: a consumer that eats meat

cartilage: tough but flexible tissue that is found in parts of the skeleton, such as the outer ear and the center of the nose

cell: the tiny unit of which living things are made

cell membrane: the outer "skin" of a cell

cell respiration: the "burning" of food in which sugar combines with oxygen to produce carbon dioxide and water, with the release of energy (that is captured in ATP)

cell wall: the stiff covering (outside the membrane) of cells of plants and some microbes (e.g. bacteria; algae)

cellulose: a nonliving woody material in the cell walls of plants; fiber

cerebellum: the part of the brain that controls balance and the working of muscles

cerebrum: the part of the brain that has control centers of thought, memory, the senses, and voluntary muscles

chemoprotection: the use of molecules in medicines or foods to prevent disease

chlorophyll: the green substance that cells use to make food during photosynthesis

chloroplasts: the green cell parts that contain chlorophyll

cholesterol: a substance produced by the body and contained in animal-based foods - over time it can contribute to the narrowing of the inside of arteries

chromosomes: thread-like parts in the cell nucleus; made up of genes (which are made of DNA)

cilia: hair-like parts of cells (in breathing passages); sweep out dirt and germs

circulatory system: the transport system that moves blood through the body in a circular pattern

class: a classification level made up of similar orders of organisms

climate: the pattern of weather in a place over a long period of time

climax community: a stable community at the end of a succession

clot: a net of threads and trapped blood cells that help stop bleeding from a cut

cocoon: the silky shell, spun by the larva of some insects, which protects the insect during the pupa stage

cold-blooded: having a body temperature that changes when the outside temperature changes

community: all of the populations that live together in a certain place and interact with one another

compound: a kind of matter made up of two or more different elements joined together

conditioned response: a learned response in which one stimulus takes the place of another

conifer: 1. the class (a classification level) of cone-bearing plants 2. an individual cone-bearing plant

coniferous: referring to conifers (cone-bearing plants)

consumer: an organism that feeds on other living things

contagious disease: diseases that can be passed to other organisms by contact or other means

control: the group in a controlled experiment that does not receive a change

controlled experiment: compares a group that does, with a group that does not, receive a change

covid: the common name for the virus that caused the pandemic of 2020

CRISPR: a method for removing or adding specific base pairs of DNA

cutting: a cut off stem or leaf that can be used to asexually reproduce some plants

cyanobacteria: photosynthetic bacteria that produced the earth's first oxygen atmosphere; were long incorrectly called blue green algae

cycle: a series of steps that lead a material or living thing back to where it started

cytoplasm: the material in a cell that is inside of the cell membrane and outside of the nucleus

D

Deciduous: trees that shed their leaves in the cool/cold seasons

decomposer: a consumer that obtains energy by breaking down wastes or dead organisms

deficiency disease: a disease caused by a lack of a vitamin or mineral

depressants: drugs that slow down the heart rate, breathing system, and nervous system

diabetes - type 1: inherited disease in which the pancreas is unable to produce insulin, without which body cells are unable to take in sugar from the blood

diabetes - type 2: disease in which blood sugar is not well controlled, often associated with obesity and beginning as an adult

diaphragm: the sheet-like muscle below the lungs that is used in breathing

digestion: the process of breaking down foods into useful forms

digestive system: group of organs that break down (digest) food into molecules that the body can use

DNA: the large molecule (shaped like a double helix) that makes up genes (which make up chromosomes)

domain: the top classification level - made up of kingdoms; all organisms are classified into one of three domains

dominant gene: a gene that always shows its trait in the organism

double helix: a twisted ladder shape; often used to describe the structure of DNA

drug: any substance, other than food, that causes a change in the body

drug abuse: the improper use of drugs

drug dependence: the need for a drug

E

Eardrum: a thin sheet-like tissue between the outer ear and the middle ear; it vibrates when struck by sound waves

eating disorder: mental conditions wherein a person loses control of their food intake and has a confused body image; common types include anorexia and bulimia

ecologist: a biologist who studies ecology

ecology: the study of the interactions between living things and the environment

ecosystem: a living community and its nonliving environment

egg: a female reproductive cell of a plant or an animal

element: one of the basic kinds of matter that combine to form molecules and compounds

embryo: an early stage in the development of a plant or animal when the organism is still attached to its parent

endangered species: a living thing in danger of becoming extinct

endocrine system: a group of glands that produce chemical messengers called hormones that move directly into capillaries

endorphins: natural brain chemicals that reduce pain and produce positive feelings

environment: everything that is around a living thing

enzymes: proteins that help to digest foods

epigenetics: the study of how genes switch on and off, and the influence of the environment on these actions; a means of evolution in addition to natural selection

esophagus: the tube that carries food from the mouth to the stomach

ethology: the study of animal behavior

eukarya: one of the three domains in which scientists group all organisms; cells of eukarya have a nucleus

evaporate: to go into the air as a gas (vapor)

evolution: the process of change through time that results in new types of living things

excretion: getting rid of wastes

exhale: to move air out of the body

extinct: a species that has died out

Family: a classification level made up a group of genera

fats: a group of nutrients that supply fuel for energy, cushion organs, and are used to make some cell structures (e.g. cell membranes)

fatty acids: substances formed as a result of the digestion of fats

feces: the solid waste that forms in the large intestine

fertilization: the joining of the nuclei of an egg and a sperm

fetus: a human embryo after two months of development

fiber: a complex carbohydrate that the human body cannot digest, but that is needed for healthy digestion; cellulose

fission: a type of asexual reproduction by splitting; produces two cells of equal size

food chain: a way of showing the order in which food energy passes from one organism to another

food pyramid: a guide that can be used to help people choose foods necessary for a healthy diet

food web: a way of showing how two or more food chains are linked together

formula: a way to write the symbols for a compound

fossils: traces or remains of living things from long ago

fraternal twins: non-identical twins resulting from two eggs and two sperm

fruit: a ripened ovary (fleshy or dry) containing the seeds of a plan

fungi (sing: **fungus**): non-photosynthetic decomposers with some traits that appear plant-like

Gallbladder: the organ under the liver that stores bile

gastric juice: the digestive juice made by, and released into, the stomach

gender: the identity of a person in regards to traits considered socially to be male or female

gene: a portion of DNA (within a chromosome) that controls the inheritance of a trait

genetic code: the sequence bases on the mRNA that signals specific amino acids to be added during the production of a protein; each amino acid is signaled by a set of three bases

genetic disorder: disease that a person can inherit from his or her parents

genetic engineering: methods that alter genes

genus (pl: **genera**): the classification level made up of a group of species that are alike in many ways and are closely related

germs: microbes that cause infectious diseases

global warming: the increase in earth's temperature; usually in reference to the increase caused by human activities

glycerol: a substance formed as a result of the digestion of fats

grafting: joining a cut-off stem of one plant to the stem of another plant that is rooted

Habitat: the place where an organism lives

hallucinogens: drugs that cause people to see, hear, smell, or taste things that do not exist, and that often cause people to act in strange ways

hemoglobin: the iron-containing protein that allows red blood cells to carry oxygen

hemophilia: the genetic disorder in which blood does not clot normally

herbivore: a consumer that eats plants

HIV (human immunodeficiency virus): the virus that causes AIDS

hormones: chemical messengers that control activities in the body

hotspots: areas of the earth containing exceptional biodiversity

hybrid: an organism that has two unlike genes for a trait

hydrate: to increase the amount of water

hypothesis: a guess or possible answer to a scientific question

Identical twins: twins resulting from one sperm and one egg (and therefore containing all of the same genes)

immune: having resistance, or immunity, to a germ

immunity: the body's ability to resist a disease

incubate: to keep eggs warm while the embryos in them are developing

infectious diseases: diseases that are able to be passed to others

infer: making a prediction or conclusion based on observation and reasoning

inhalants: substances that give off vapors or fumes that can be inhaled

inhale: to take air into the body

inherited: passed from parents to offspring

insulin: the hormone that controls the use of sugar in the body

invasive species: non-native organisms whose populations are causing problems in their new territory

invertebrates: animals without backbones

in vitro: produced / taking place in the laboratory (not in an organism)

in vivo: produced / taking place in an organism

involuntary: a response that is not under an organism's control

Joint: the location where bones meet

Kidneys: organs that filter wastes from the blood, and make urine

kingdom: one of the six classification levels - made up of phyla

Large intestine: the part of the digestive system where solid wastes are formed as water is removed from food

larva: the second stage in the life cycle of an insect that has complete metamorphosis

larynx: the voice box; box-like structure at the top of the trachea and containing the vocal cords

lens: **1.** the part of the eye that focuses light on the retina; **2.** the curved piece of clear material in a microscope that magnifies and focuses light

lichens: a "dual" organism made of a fungus and a photosynthetic microbe (algae or cyanobacteria) - working together - that often serves as the pioneer organism

life functions: all of the essential activities that organisms carry out in order to stay alive (e.g. moving, growing, using food, reproducing)

ligaments: bands of tissue that tie bones together at joints

limiting factor: anything in the environment that puts limits on where an organism is able to survive and maintain its population

Malignant tumor: a cancerous growth

mammal: a warm-blooded animal with body hair and mammary glands

mammary glands: the glands in mammals that produce milk for feeding their young

marrow: the center of long bones, where blood cells are made

medulla: the part of the brain that controls breathing, heartbeat, and many involuntary actions

meiosis: the division of cell nuclei in order to form sex cells

menstruation: the release of the lining of the uterus; also called a period

messenger RNA: the phase of RNA that carries the message for the creation of a protein - from the nucleus to a ribosome

metamorphosis: the change in body form during an animal's life cycle

methane: a gas that contributes to global warming. One source is cow burping.

microbe: an organism so small that it can be seen only with a microscope

minerals: chemical elements that help the body work properly and build bones, teeth, and blood

mitochondria: structures in eukaryotic cells that produce energy via cell respiration, and trap it in ATP molecules

mitosis: the process in which a body cell's nucleus divides to form two nuclei (in order to then form two daughter cells)

molecule: the smallest portion of a substance (made of more than one atom) that has the properties of that substance

molt: to shed an outside skeleton during growth

motor nerve: a nerve that carries messages from the spinal cord and brain to other parts of the body

mRNA: (**messenger RNA**) name given to a segment of RNA involved in the production of a protein; [see also RNA]

mucus: the sticky substance produced in the trachea (windpipe) and bronchi

mutation: a change in a gene that produces a change in the offspring

Native plant: plants naturally growing in a region

natural selection: the most widely accepted process by which the evolution of new species occurs, based on random mutations and competition among members of the same species

nerve: a bundle of nerve cells (fibers) that carry messages

nervous system: the group of organs whose job it is to control and carry messages throughout the body

neuron: another name for a nerve cell

niche: the way that an organism fits into its habitat; the role it plays

nitrates: compounds (vital to a plant's building of proteins) that contain nitrogen, oxygen and other elements

nitrogen: an element (with the symbol **N**) that is vital to the production of proteins

nonbinary: not claiming to be one or the other, as in male or female

nonvascular plants: plants that lack tubes to carry water

nucleus: a cell's center of control

nutrients: materials, in foods, that are needed by the body

nutrition: the study of foods and how the body uses them

nymph: the stage between egg and adult in the life cycle of an insect with incomplete metamorphosis

Obesity: the condition of being heavier than the weight considered to be healthy for a specific height and body frame

offspring: new organisms that result from reproduction

omnivore: a consumer that eats both plants and animals

optic nerve: the large nerve that carries messages from the eye to the brain

order: the classification level consisting of groups of similar families

organ: a group of tissues that work together to do a special job

organism: a living thing

ovary: the egg-producing reproductive organ in plants and female animals

oxygen: the element in the air that organisms use during cell respiration (to get energy from food)

Paleontology: the study of fossils to learn about the past

pancreas: the gland that produces digestive enzymes as well as the hormone insulin

pandemic: the situation in which an infectious disease spreads worldwide or at least into several nations

parasites: organisms that obtain their nutrients from other living things in which or on which they live

parasitic worms: worms that cause infectious diseases

pattern recognition: a means of problem solving in which observations outside of the norm are sought

penicillin: the drug, made from the mold *penicillium*, used to treat bacterial diseases

penicillium: the mold that produces penicillin

penis: the male sexual organ involved in intercourse

peristalsis: wavelike contraction and relaxation of muscles; common in the organs through which nutrients move during digestion

petals: the outer parts of a flower (usually colored) that help to protect the inner parts, and that attract pollinators

petri dish: two-part flat, circular, clear dish with the upper portion being slightly larger and fitting over the lower portion

phloem: the tissue in plants that moves food through the plant

photosynthesis: the process by which photosynthesizing cells, in the presence of light, put together carbon dioxide and water to make sugar and oxygen

phylum (pl: **phyla**): the classification level made up of a group of similar classes

pioneer plants: the first plants to grow in a specific habitat

pistil: the female reproductive part of the flower (contains the ovary)

pith: soft tissue in a plant stem that stores food

pituitary gland: the "master gland" that makes growth hormone, and other hormones that control other glands

placenta: the organ within the uterus that brings food to the embryo in many pregnant mammals

plaque: the material (made of fat, protein and calcium) that builds up on the inside of artery walls

plasma: the liquid part of the blood

platelets: cells in the blood that help form clots

polio: a viral disease that causes paralysis; vaccination has eliminated this disease from the U.S., but not globally

pollen: a powder-like material made of individual pollen grains (which contain plant's sperm cells)

pollination: the transfer of pollen from a stamen to a pistil

pollute: to add something harmful to the environment

population: all of the members of one species that live in a area

pore: a tiny opening of a sweat gland in the skin

predator: an animal that eats other animals

prey: an animal that a predator eats

primate: the group of mammals with enlarged brains, grasping hands, complex social groups, and a reliance on sight; includes monkeys, lemurs, apes, humans

producer: a green plant (or other photosynthesizing organism) that makes food

proteins: **1.** nitrogen-containing molecules, (made of linked amino acids) that form many of the basic cell structures as well as enzymes **2.** built in ribosomes according to DNA's "instructions"

protist: a single-celled microbe that has a true nucleus

protozoans: single-celled animal-like protists that eat other organisms; some live in colonies

pulse: the expansion of arteries due to the heart's action of pushing blood with each contraction

Punnett square: a chart used to predict the heredity of a trait

pupa: the non-eating stage of an insect that undergoes complete metamorphosis

pupil: the part of the eye through which light enters

pure: having two "like" genes for a trait

Recessive gene: a gene whose trait is hidden when a dominant gene for the same trait is present

reflex: an involuntary action in response to a stimulus

regeneration: the growing back of lost parts

reproduce: to produce more on one's own kind

reptiles: the class of cold-blooded vertebrates that includes turtles, snakes, lizards, alligators, and crocodiles

respiration: breathing

respiratory system: the group of organs involved in breathing

response: a reaction to a stimulus

retina: the thin tissue that lines the inside of the eye, receives light, and sends messages to the optic nerve

ribosomes: structures in eukaryotic cells that follow instructions from mRNA and align amino acids to build proteins

RNA: a molecule, similar to one side of a portion of DNA, that carries the code from DNA to a ribosome so that a protein may be produced

Saliva: the liquid in the mouth produced by the salivary glands, and containing an enzyme that helps digest starch

scavengers: animals that eat dead animals, speeding decomposition

scrotum: in male mammals, the external sac of skin that contains the testes

seed: in flowering plants, the embryo and its food, surrounded by a protective coat

seed pod or **seedpod**: the fruit (ripened ovary) containing seeds in some flowering plants (particularly beans and peas); also called the husk

semicircular canals: the three fluid filled, tubular structures of the middle ear that sense position

sensory nerve: a nerve that carries messages from the sense organs (including the skin) to the spinal cord and brain

sepals: the leaf-like parts of flowers that protect the flower bud before it opens

sex chromosomes: the **X** and **Y** chromosomes - that determine the gender of a baby

sexual reproduction: reproduction from two parents

sexually transmitted infections (STIs): diseases passed by sexual contact

sickle-cell anemia: a genetic disorder in which red blood cells have the shape of a sickle; this reduces the ability of blood to carry oxygen

skeletal muscle: voluntary muscle, attached to the skeleton (as in the arms and legs)

small intestine: the part of the digestive system where most digestion and absorption takes place

smooth muscle: involuntary muscle, as in the stomach and intestines

species: a group of organisms that have all of the same structures, and are able to reproduce and have offspring that are also able to reproduce

sperm: the male reproductive cell of a plant or an animal

sphincter: donut-shaped muscle

spinal cord: the nerve cord that relays messages between the brain and other parts of the body

spores: cells involved in the asexual reproduction of fungi, nonvascular plants, ferns, and conifers

stamen: the male reproductive part of a flower, the top of which is the pollen-producing anther

stem cells: cells that have the ability to mature into a variety of cell types

stent: a tiny straw-like device that is inserted into a blood vessel to keep it open

stigma: sticky top of the pistil on which pollen lands during pollination

stimulants: drugs that speed up the heart rate, breathing rate, and nervous system

stimulus (pl. **stimuli**): a message received from the environment, or any change in the environment to which an organism responds

stroke: loss of brain cells due to the blocking of an artery in the brain or to the rupture of a blood vessel in the brain

succession: a series of changes in a living community

sweat glands: tiny parts in the skin that form sweat

symptom: a sign, such as pain or fever, of a disease

system: a group of organs that work together to do a special job

Tendon
Tendon: a strong but somewhat flexible band of tissue that connects a muscle to a bone

testis (pl. **testes**): the male reproductive organ in an animal; produces sperm

theory: an explanation of a scientific idea that has been thoroughly tested and is widely accepted

thyroid gland: a gland in the neck that produces the hormone thyroxin

thyroxin: the hormone (made by the thyroid gland) that controls the rate at which the body produces energy

tissue: a group of similar cells that do a special job

toxin: a poison that is produced by some bacteria

trachea (windpipe): the tube through which air moves to and from the lungs

traits: characteristics, which may be inherited, that identify organisms as individuals

triceps: the muscle on the back of the upper arm that contracts to straighten the elbow

tuber: a fleshy underground stem that can reproduce the entire plant

Umbilical cord
Umbilical cord: the cord containing blood vessels that attaches a mammal embryo to the placenta

urea: the waste formed when amino acids are broken down; found in urine and sweat

ureter: the tube that carries urine from the kidney to the bladder

urethra: the tube through which urine (and in males, sperm) leaves the body

urine: the liquid waste produced while the kidneys clean the blood

uterus (also called the womb): organ in which mammal embryos develop

Vaccine
Vaccine: a man-made material that stimulates the immune system to build antibodies against a germ

vacuoles: bubble-like cell parts that store water, food, or wastes

vagina (birth canal): **1.** in human females, the flexible, tubular opening involved in sexual intercourse; **2.** the passage for the baby during birth

valve: a door-like structure in the heart that prevents blood from flowing backwards

vascular plants: plants that have tubes (veins) for carrying water, minerals, and foods

vector: an organism that transmits a disease-causing organism (usually a microbe) to other organisms

vegetative propagation: asexual reproduction of plants from growing parts

veins: **1.** tubes that carry liquid in plants **2.** blood vessels in humans that move blood to the heart

ventricle: a chamber of the heart that sends blood out of the heart and into the arteries

verify: proving the outcome of an experiment by repeating it and obtaining the same results

vertebra (pl. **vertebrae**): one of the small bones that make up the backbone of an animal

vertebrates: animals that have a backbone

villi (sing: **villus**): tiny finger-like parts in the walls of the small intestine that absorb digested foods into the blood.

viruses: particles smaller than bacteria that cause many diseases

vitamins: nutrients needed by the body to help it work properly and to prevent diseases; vitamins are not a source of fuel

voice box: see larynx

voluntary: an action that is under the control of the animal

Warm-blooded: having a body temperature that stays more or less the same no matter the outside temperature

windpipe (trachea): the tube through which air moves to and from the lungs

Xylem: plant tissue that moves water and minerals up from the roots, through the stem, to the leaves

Yolk: the stored food supply for some animal embryos

Zygote: the cell that results from the fertilization of an egg by a sperm

---

Photo Credits

Photos used in this book are from the public domain or via permission.
With the exception of those listed below, all photos were purchased from **Depositphotos** and **Dreamstime**.

-Text cover: (Chimpanzee mother touching young) Depositphotos / @ GUDKOVANDREY

-Unit 1: **5**, (Pasteur) Engraving by T. Johnson after painting by L Bonnat, Prints and Photographs Division, Library of Congress, LC-USZ62-80072;
9, (Jenner) Prints and Photographs Division, Library of Congress, LC-USZ62-9465;
30, (Rachel Carson), Courtesy Linda Lear Center for Special Collections and Archives, Connecticut College; (Book cover - Jane Goodall), Courtesy JGI Photo Library, Jane Goodall Institute;
31, (International recycling symbol) recycling.com

-Unit 2: 39, (Leeuwenhoek) Prints and Photographs Division, Library of Congress, LC-USZ62-11252;
40, (Cork image) NIH, From NIH/NLofMed/Digital Collections; NLM Image ID: A012510;
46, (Linnaeus portrait) {{PD-1996}} Johan Henrik Scheffel, Public domain, Wikimedia Commons;
52, (Burbank) Prints and Photographs Division, Library of Congress, LC-USZ62-125645

-Unit 4: **100**, (Alexis St. Martin) Harris & Ewing Collection, Prints and Photographs Division, Library of Congress,, LC-DIG-hec-38711; **112,** (William Harvey), Courtesy of National Library of Medicine, Order Number: A033168

-Unit 7: 172, (Fleming) Courtesy of National LIbrary of Medicine, Bethesda, Md, NLM ID 101415056

-Unit 9: **234**: (McClintock at cabinet) Courtesy of National LIbrary of Medicine, NLM Unique ID:101584613X258; (McClintock holding award) Courtesy of National LIbrary of Medicine / Reproduction Number LC-DIG-ppmsca-12441; **249**, (Dr. Ferguson) Fair use image, Blackpost.org, 12/26/2019, author: Secret Charles-Ford

-Unit 10: **267**, (Spirilli) Clipart; (tick) CDC, Public Health Image Library, Jim Gathany;
268, (COVID-19) CDC, Public Health Image Library, Alissa Eckert, MSMI, Dan Higgins, MAMS;
272, (HIV) Wikipedia/ Bruce Blaus;
285 (Beer bottles/wrecked car) QuoteInspector.com

Primary Index

Note: People's names are in bold print.
At the end of this primary index are additional indexes.

A

abdomen, 126
absorb, 105
absorption, 104
acne, 128, 140
adapted, 254
addicts, 286
adrenal glands 160-61
adrenaline, 160
AIDS, 268-9
air pollution, 282
alcohol, 285-7
algae, 20, 40, 50, 68
alligator, 202
alpha waves, 155
alveoli (air sacs), 122-3
alveolus (air sac), 122-3
amber, 257
ameba 40, 44, 171
American Chestnut tree, 176, 182
amino acids, 65, 80, 102, 229-30, 232, 274-5
amphibians, 36, 200
Anastas, Paul, 285
anemia, 85
anemone, 55, 196
angioplasty, 287
animal cell, 38, 42, 43
animal kingdom, 54
anoles, 174
antacid, 105
anther, 181
antibiotic, 172
antibodies, 229, 274-5
antioxidants, 91
anus, 99, 105
aorta, 108, 113
appendectomy, 104
appendicitis, 104
appendix, 99, 104
archaea, 50
archaeopteryx, 259
armadillo, 57
arteries, 108-9, 112, 277-8
arthritis, 133, 267
arthropods, 55
artificial heart, 114
artificial kidney, 126
asexual reproduction, 170, 172
   -in plants, 174
   -in animals, 174-5
Asian bush honeysuckle. 29, 180
Asian carp, 29
athletic trainers, 119
ATP, 71-2
atria, 111
atrium, 110, 113
axon, 149-50

B

bacteria, 50, 105, 268
Bakker, Dr. Bob, 256
balanced diet, 90-1
balanced ecosystem, 31
ball-and-socket joint, 132
basking, 200
bears, grizzly, 87
Beaumont, William, 100
behavior, 143
benign tumor, 282
beta waves, 185
bias, 10, 112, 244
biceps, 135
bile, 102
binary fission, 170
biodiversity, 12, 24-5
biofeedback, 155
biologists, 5
biology, 5
biomagnification, 20, 30
biome, 25
biosphere, 5
birds, 56, 204
birth canal, 207-8
Biuret solution, 81
bladder, 127, 208
blending traits, 204
blind spot, 166
blindness, 8
blinking, 166
Bloch, Dr. C.E., 82
blood, 106-7
blood cells, 44, 106-7, 273
blood clot, 278-9
blood pressure (high), 278-9
blood transfusions 106
blood vessel disease, 277, 279
blood vessels, 104, 108
bloodstream, 108
blubber, 79
blue babies, 115
bodily wastes, 126
body cells, 219-20, 225
body temperatures, 73, 128, 155
bone cell, 44
bone marrow, 282
bones, 130-33
brain, 149-50
brain map, 153
breathing, 125
brine shrimp eggs, 2
bromothymol blue, 124
bronchi, 121-22
bronchus, 121
Brown, Louise, 207
bud, -yeast, 171
      -plant, 177
budding (yeast), 171
bulb, 177
bumblebees, 180
Burbank, Luther, 52

C

calcium, 83, 140
calories, 87
cambium, 75
cancer, 51, 123, 281-3, 285-6
capillaries, 108-10, 112-13, 128

carbohydrates, 78-9, 102
   -simple, 79   -complex 79
carbon, 66
carbon dioxide, 14, 69, 110, 124
carbs (carbohydrates), 79
carcinogens, 282
cardiac, 136
cardiac muscle, 136
carnivore, 18, 58
carrier, 250

Carson, Rachel, 30
cartilage, 121, 132
cartilage rings, 122-3
cast, 258
caterpillar, 213
cell, 40, 65
cell membrane, 40
cell respiration, 71-2, 121, 125
cell wall,
 -in plants, 42
 -in algae, 42
 -in bacteria, 42, 172
cellulose, 42, 79
centipede, 213
central nervous system, 149
centrifuge, 106
cerebellum, 150, 154
cerebrum, 150, 153
Charpentier, Emmanuelle, 249
chemical digestion, 100
chemical messenger, 159
chemoprotection, 281
chemotherapy, 282
chicken egg, 205-6
chicken pox, 274
chlorine, 66
chlorophyll, 42, 69
chloroplasts, 42, 69, 260
cholesterol, 80, 91, 277-8
chordates, 55
chromosomes, 219-21
cilia, 121, 273
circulate, 108, 112
circulatory system, 108-9
class, (classification), 48
climate, 25
climate change, 31
climax community, 22
climax plants, 22
clinical social worker, 167
clone, 175-6
clot, 278-9
clownfish, 196
cnidaria, 55
cocaine, 286
cocoon, 211
cold-blooded animals, 205
community, 12
compare, 9
complete metamorphosis, 210-11
complex carbohydrates, 79
complex organisms, 45

compound, 66
compound microscope, 40
conclusion, 6
conditioned response, 155
conditioning, 155
cone (male and female), 187
coniferous forest, 25-6
conifers, 53
consumers,
 -first level, 16
 -second level, 16
 -third level, 16
contagious diseases, 270
contract, 110, 135
control, 9, 281
control group, 10. 281
controlled experiment, 10
cornea, 146
coronary arteries, 278
covering cells, 44
COVID-19, 268-9, 272, 274
CPR, 124
Crick, Francis, 223
CRISPR, 249-50
cross pollination, 182
cultivator, 223, 226
cutting, 178
cyanobacteria, 50
cycle, 14
cytoplasm, 40

D

Darwin, Charles, 254
DDT, 30
decay bacteria, 18
deciduous forest, 25-6
decomposers, 18
deficiency disease, 83
dehydrated, 83
dementia, 149
depressants, 286-7
dermatologist, 128
desert, 25-6
diabetes, 90, 160
dialysis, 126
diaphragm, 125
diarrhea, 105
diatoms, 50
digestion, 98
digestive system, 97-8
dinosaurs, 256
DNA, 42, 48, 50-1, 220, 224, 228-9, 231, 260
DNA fingerprinting, 238
Dolly, cloned sheep, 175

domain, 48, 50
domain archaea, 50
domain bacteria, 50
domain eukarya, 50
dominant genes, 238
dominant plants, 22
dominant traits, 235
double helix, 223
Doudna, Jennifer, 249
downers, 286
Down's syndrome, 252
Drew, Dr. Charles, 106
Drosophila melanogaster (fruit fly), 237
drug, 286
drug abuse, 286
drug dependence, 286
dry fruit, 185
duckbill platypus, 207
Dudrick, Dr. Stanley, 102

E

eagle, bald, 30
ear, 147
eardrum, 147
eating disorder, 90
echinoderms, 55
ecologists, 5
ecology, 5
ecosystems, 12
egg, 181
egg layers (mammals), 57
egg tube, 197, 205, 207
electron microscope, 39
elements, 65
elodea, 42
embryo, 108, 169, 184-5, 194
emerald ash borer, 210
emperor penguins, 204
endangered species, 31
Endangered Species Act, 30, 87
endocrine glands, 160
endocrine system, 160
endorphins, 159
endosymbiosis, 261
energy, 71-2
energy pyramid, 20
environment, 2
enzymes, 98, 102
epiglottis, 97
equation, 70-2
erection, 207
esophagus, 98, 100

ethology, 30, 143
eukarya, 50
European starling, 29
evolution, 254, 257
excretion, 127
exhale, 123, 124-5
exoskeleton, 210
experiment, 6
external fertilization, 198, 200
extinct, 31, 256
Exxon Valdez, 47
eye (human), 146
eye (potato), 177

F

fad diet, 81, 90
Fahey, Jed, 281
Fallopian tube, 207, 209
family (classification), 48
fats, 66, 78-80
fatty acids, 102
feces, 105
Felis, 46, 47
Felis domesticus,
Ferguson, Dr. Angelia, 249
ferns, 53
fertilization, 193, 226
   -external, 198
   -internal, 198
   -in flowers, 181-2
fertilize, 193
fertilized egg, 192
fetal alcohol syndrome, 285
fetus, 209
fiber, 78
Fikrig, Dr. Erol, 267
filament, 181
fingerprints, 238
first line of defense, 273
fish, 56, 196-99
fission, 170
fixed joints, 132-3
flatworms, 55
flavors (sensing), 147
Fleming, Sir Alexander, 172
flowering plants, 53
flowers, fertilization in, 181-2, 184
flu, 268-9
fly, 4, 5
flying mammals, 57
food chain, 16
food desert, 90

food makers, 16
food web, 18
forests, 22, 24-6
formula, 66
fossil fuels, 49
fossils, 256
Franklin, Rosalind, 223, 284
fraternal twins, 246
frogs, 200-1, 203
fruit, 184, 186
fruit flies, 237
fuels, 71, 78, 81
function, 223
fungi, 14, 18, 51

G

gallbladder, 102
gap (between neurons), 149-50
gardeners, 63
gases, 65, 66
gastric juice, 100
gender, 244
gender identity, 244
gene editing, 249
genera, 48
genes, 219-21
genetics, 219, 221
genetic disorders, 250
genetic engineering, 228
genetically modified, 176
genus, 48
geo, 144
geotropism, 144
germinate, 185
germs, 268, 275
GH (growth hormone), 160, 228
gills, 197
gizzard, 101
glands,
   -digestive, 100, 102
   -endocrine, 160
global warming, 31, 49, 68, 204
glycerol, 102
GMO, 176, 182
goiter, 85, 86
Goodall, Jane, 30
grafting, 52, 178
grains, 91
grasslands, 25, 27
grizzly bear, 87

growth hormone (GH), 160, 228
grub, 210
Gruentzig, Dr. Andreas, 277

H

habitat, 24, 28
hallucinogens, 286-7
hamstrings, 137
hare, snowshoe, 28
Harvey, William, 112
Healthy Eating Pyramid, 91, 281
hearing, 147
heart, 109-10
heart attack, 278
heart disease, 114
heat transplant, 114
Heimlich, Dr. Henry, 97
Heimlich "Hug," 97
hemoglobin, 106
hemophilia, 251-2
herbivores, 18
herpetologists, 200
herps, 200
high blood pressure, 278
hinge joint, 132
hip replacement, 134
HIV (virus), 51, 269, 272-3
Homo neanderthalensis, 130
Homo sapiens, 130
honeysuckle bush, 29, 180
hoofed mammals, 58
Hooke, Robert, 40
hormones, 159, 176
horse,
   -evolution of, 257
   -zebra birth, 169
horsefly, 14
hotspots, 24
HPV virus, 51, 268-9
human genome project, 227
human growth hormone, 228
human reproductive system, 208
Huntington's disease, 250, 251
hybrid, 235, 239-40
hydra, 1
hydrated, 83
hydrogen, 66
hypothesis, 6

I

ice ages, 31
identical twins, 246
infectious, 268
inferring, 9
inhale, 121
immune, 274
immune system, 272
immunity, 274
imprinting, 143
in vitro, 207
in vitro fertilization, 207
in vivo, 205
inborn behavior, 156
incomplete metamorphosis, 211-12
incubation, 206
Indian corn, 234
inhalants, 286-7
inhale, 121, 125
inherited traits, 219
inner ear, 147
insulin, 160
intercourse, 208
internal fertilization, 198
internet, 6
intestines, See large intestine; small intestine
invasive species, 29
invertebrates, 54
involuntary action, 150
involuntary muscle, 136
involuntary response, 143
iodine, 84
iris, 146
iron, 66, 83-4, 91
IVF, 207

J

Jackson, Bo, 134
Jeffreys, Sir Alec, 238
Jenner, Edward, 9
joint-legged animals (arthropods), 55
joints, 132
jumping beans, 210
jumping genes, 210
Just, Dr. Ernest, 193

K

kelp, 31, 50, 68
kidneys, 127
 -artificial, 126
kingdom, 48

L

labor, 205
landscape architects, 63
landscape maintenance contractors, 63
large intestine, 98, 102, 105
larvae, 210-11
larynx, 122-3
laughter, 159
learned behavior, 155
leaves, 75
Leeuwenhoek, Anton van, 39
lens (eye), 146
lens (magnifying), 40
leukemia, 282
lichens, 23, 26
life cycle, 198
life functions, 1
ligaments, 132
limiting factor, 28
lines of defense, 273
Linnaeus, Carolus, 46
liver, 99, 102
living things, 1-5
lizards, 174
Lorenz, Konrad, 143
Loricifera, 54
lung capacity, 122
lungs, 113
Lyme disease, 267
lynx, 28

M

maggots, 4
malignant tumor, 282
Malpighi, Marcello, 112
mammals, 55-9,
 -reproduction, 207-9
mammary glands, 207
mammoths,
mandible, 131
Margulis, Lynn, 260
marine, 50
marrow, 108, 131, 273
marsupial, 57
matter, 65
McClintock, Barbara, 234
McCollum, Dr. E.V., 82
measurement, 6
meat, 79, 80
mechanical digestion, 100

medical laboratory assistant, 292
medulla, 150, 154
meiosis, 225
memory, long term, 92
memory, short term, 92
Mendel, Gregor, 235
menstruation, 208
mental health technician, 167
mercury, 20
messenger RNA, 228, 230-32
metamorphosis, 200, 210
methane gas, 31, 65
Mexican arrow plant, 202
microbes, 40, 272
microscope, 40
microsphere, 65
middle ear, 147
milk,
 -in diet, 88, 91
 -mammals, 207
Miller, Stanley, 65
Minamata bay, 20
minerals, 83
mitochondria, 41, 71-2, 260
mitosis, 224-26
model building, 233
mold, 51, 172-3
molecules, 66, 98
mollusk (soft bodied), 55
molt, 54, 211
monarch butterfly, 211
Morgan, Thomas Hunt, 237
mosquitoes, 270
moss, 52
motor nerves, 150
Mount Everest, 121
Mount St. Helens, 22
mouth, 98, 100
MRI (Magnetic Resonance Imaging), 244
mRNA, 228, 274-5
mucus, 121, 273
Mullis, Kary, 238
muscle cells, 44
muscles, 134-6
mushroom, 14, 51
mutation, 254, 272

N

native plant, 180
native species, 29
natural selection, 254, 272

Neanderthals, 130
nerve cell, 44, 45, 149
nerve ending, 146
nerve fibers, 150
nerves, 44, 45
nervous system, 45, 149
neurologist, 92
neuron, 44, 149
neurosurgeon, 153
niche, 28
nicotine, 285, 287
nitrate, 14-6
nitrogen, 14-6, 18
nitrogen cycle, 14-5
nonbinary, 244
nonvascular plants, 52
nonwoody stems, 75
nose, 147
nucleus, 40
nurse aides, 292
nurses, 292
nutrients, 78
nutrition, 91
nutritionists, 91
nymph, 211-12

O

obesity, 90
objectively, 244
observation, 6
odor, 147
offspring, 170
oil spill, 49
oils (vegetable), 80
omnivore, 18
On the Origin of Species, 255
optic nerve, 146
order (classification), 48
organisms, 1
organ system, 44
organs, 44
outer ear, 147
ovary
   -in animals, 207-8
   -in flowers, 181-2
   -human, 160-1, 193
overdose, 83
oxygen, 3, 66, 68, 74, 110
oxygen/carbon-dioxide cycle 14

P

pacemaker, 114
pain, 151
paleontologist, 256
paleontology, 256
pancreas, 99, 102, 160-1
pandemic, 272
paramecium, 50
parasite, 50, 76
Pasteur, Louis, 5
patterns (recognizing), 9
Pauling, Linus, 233
Pavlov, Ivan, 155-6
Pence, Valerie, 176
Penfield, Dr. Wilder, 153
penguins, 204
penicillin, 172, 286
Penicillium, 172
penis, 207
peppered moths, 253
period (menstrual), 208
periodic table, 65
peristalsis, 100
petals, 181
petri dish, 172
phloem, 75
photosynthesis, 70, 72
phototropism, 145
phylum, 48, 52
physical therapist, 119
pioneer community, 22
pioneer plants, 22
pistil, 181
pitcher plant, 16
pith, 75
pituitary gland, 160-1
pivot joint, 132
Pizzo, Chris, 121
placenta, 209
plague, 272
planarian, 174
plant cell, 38, 42, 43
plant kingdom, 48, 52
plant tissue culture, 176
plants, 48, 74-6, 176, 180, 184
plaque, 277-8
plasma, 106
platelets, 107
platypus, 57
pollen, 180-1
pollen tube, 182
pollination, 182
pollution, 31
population, 12
pore, 128
porifera, 55
porkworm, 268-9

Pothineni, Venkata Raveendra, 267
pouched mammals, 87
predator, 16
pregnancy, 209
pregnant, 269
prey, 16
Priestly, Joseph, 74
prints, 257
primates, 58
producers, 16
proteins, 15, 16, 65, 66, 78, 80, 81, 91, 100, 102, 228-9, 231-2
protists, 50
protozoa, 40, 50,
pseudopodia, 40
psychiatric nurse practitioner, 167
psychiatric social worker, 167
psychiatric technician, 167
psychiatrist, 149, 167
psychologist, 167
psychology, 167
pulse, 110
Punnett square, 239
pupa, 211
pupil, 143, 146
pure (trait), 235

Q

"quad" muscle, 137

R

rabies, 269-70
radiation, 282
radiation therapy, 282
rainforest, 24-6
Rajadas, Jayakumar, 267
Raynaud's disease, 155
reasoning, 156-7
recessive genes, 238
recessive trait, 235
rectum, 98, 105
red blood cells (RBCs), 106
red clover, 180-1
red meat, 91
Redi, Francesco, 4
reflex, 151
regeneration, 174
reproduction, 3
   -in higher animals, 193-209
   -in humans, 207-9
   -in plants, 176-87

reproduction
 -in simple organisms, 170-1, 172-3
reptile, 56, 200-1
respiration, 121, 125
respiratory system, 121
response, 2, 143, 151
retina, 146
ribosomes, 42, 230-2, 274-5
ribs, 125
RNA, 51, 228, 230-1
robots, 1
rodents, 57
root cells, 76
root hairs, 76
roots, 76
roundworms, 55
Royal Society, 10

S

Sabin, Dr. Florence, 108
SADD (Students Against Drunk Driving), 285
saliva, 100-1
salivary glands, 99
salts, 66
scab, 107
scavengers, 18
scientific name, 46, 48
scientific thinking, 5-6, 9-10
scion, 178
scrotum, 207-8
sea anemone, 55, 196
sea slug 175
sea mammals, 58
sea urchin, 193-4
Sedlacek, Dr. Keith, 155
seed, 80, 282, 184-5
seed coat, 184-5
seed pod or seedpod, 184
seeing, 146
segmented worms, 55
self-pollination, 182
semicircular canals, 148
sense organs, 146
sensory nerves, 150
sepals, 181
Servheen, Chris, 87
sex cells, 219-20, 225
sex chromosomes, 244
sexual reproduction, 170
sexually transmitted infections, 269-70
sickle cell disease, 249-50
Silent Spring, **30**

silks (corn), 183
Simard, Suzanne, 14
simple carbohydrates, 79, 102
skeletal muscle, 135
skeleton, 131
skin cell, 41, 44
skinks, 174
skull, 131
sleep, 90, 92
sliding joint, 133
sloth, 57
small intestine, 98, 100, 102
smallpox virus, 9, 272
smell, sense of, 147
smoking, 278, 282, 285
smooth muscle, 136
snail, 55,195
snakes, 56, 175, 200-1,
snowshoe hare, 28
social worker, 167
sodium, 66, 83, 84
sodium chloride, 66
solid fats, 80
space colony, 14
specialized, 44
species, 12, 48
sperm, 181
sperm tube, 187, 208
sphincter, 105, 146
spike protein, 274-5
spinal cord, 131,149-50, 152
spiny anteater, 207
spirogyra, 50
sponges, 55
spontaneous generation, 4
spores, 172
St. Martin, Alexis, 100
stamens, 181
starch, 79, 102
starfish, 174
stem cells, 175, 249
stems, 75
stent, 277
STI (sexually transmitted infections), **U10L1**
stigma, 181-2
stimulants, 286-7
stimulus, 143, 151
stock, 52, 178
stomach, 98, 104-5
stretch, 136
stroke, 278-9
structures, 48, 223

Students Against Drunk Driving, 285
succession, 22
sulforaphane, 281
surgery (for cancer), 282
sweat, 128
sweat glands, 128
symbiotic relationship, 196
symbiosis, 196, 261
symptoms, 83, 268
systems (within an organism), 45

T

T cells, 274
tadpoles, 200-1, 203
Talalay, Paul, 281
taproot, 76
tassels (corn), 183
taste, 147
temperature, body, 73, 128, 155
tendons, 135
tentacles, 196
test-tube babies, 207
testes (pl. testis), 160-1, 193, 207-8
theories, 6
thyroid gland, 160-1
thyroxin, 160
ticks, 267
tidal peritoneal dialysis, 126
Tiktaalik, 259
tissue, 44
tissue culture, 176
tobacco, 282-3, 285
tongue, 147
Tontlewicz, Jimmy, 124
toothless mammals, 57
touch, sense of, 146
toxins, 268
trachea, 121-2
traits, 219
transfusion, 106
transgenic, 176
transport system, 108
trial-and-error learning, 156-7
triceps, 135
tropical rainforest, 24-5, 68
tropism, 144
trunk-nosed mammals, 58
Tschanz, John, 126
Tschanz, Ken, 126
tuber, 177,179

tuberculosis, 108, 269
tumor, 281
tundra, 24-5
turning joint, 132
twins, 219, 246
type 2 diabetes, 90

U

umbilical cord, 209
upper, 286
urea, 127
ureter, 127
urethra, 127, 207-8
Urey, Harold, 65
urinary bladder, 127
urine, 127
uterus, 207-8

V

vaccination, 10
vaccines, 10, 51, 274-5
vacuoles, 42
VAD, 114
vagina, 207-8
valve, 108, 110, 113-4
vaping, 285
vascular plants, 52
vector, 267, 270
vegetables, 91

vegetative propagation, 176
vegetarian, 80
veins, 52, 108-9, 112-3
ventricle, 110-11, 113
ventricular assist device (VAD), 114
Venus fly trap, 16
verifying, 10
vertebrae, 54-6, 131
vertebrates, 54
veterinarian, 217
Victoria Gray, 249
villi, 104
virus, 51, 268
  -replication, 270
vitamin A, 82
vitamin C, 83
vitamin D, 82, 83
vitamins, 83-4, 105
vocal cords, 123
voice box, 123
voluntary action, 152
voluntary muscle, 135
voluntary response, 143

W

warm-blooded animals, 265
wastes, bodily, 126
water, 83

water cycle, 14
Watson, James, 223
white blood cells, (WBCs), 106-8, 273
wildlife (in cities), 12
Wilkins, Maurice, 223
windpipe (trachea), 121
wing pads, 211-12
womb, 207-8
woody stems, 75
woolly mammoth, 31, 257

X

X chromosome, 244-7
Xenoturbella, 54
xylem, 75

Y

Y chromosome, 244-7
yeast, 51, 170-1
yolk, 193, 198, 205
yolk sac, 198

Z

zebra, 169
zoo veterinarians, 217
zoos, 217
zoo keepers, 217
zygote, 193-4, 198, 208

Index 2) Terms not included in the reading, but listed on tables or on a small number of illustrations

adenine, 230
amebic dysentery, 269
amphetamines, 287
anabolic steroids, 287
asbestos, 282
ascorbic acid, 84
athletes foot, 269
bacilli, 268
bacteria diseases, 268-9
barbiturates, 287
beriberi, 85
botulism, 269
caffeine, 287
carpels, 131
clavicle, 131
cocaine, 287
cocci, 208
codeine, 287
colds, 269
cystic fibrosis, 252
cytosine, 230

deer tick, 267
diphtheria, 269
Down's syndrome, 252
e-cigarettes, 287
femur, 131
fentanyl, 287
fibula, 131
fungus diseases, 269
genetic code, 230
gonorrhea, 269
guanine, 230
hepatitis, 269
herpes, 269
humerus, 131
influenza, 268, 269
LSD, 287
malaria, 268
malaria protozoan, 268
mandible, 131
marijuana, 287
measles, 269

mescaline, 287
metacarpals, 131
metatarsals, 131
methaqualone, 287
mumps, 269
muscular dystrophy, 252
niacin, 84-5
night blindness, 84, 85
nitrites, 282
opioids, 287
parasitic worms, 270
PCP, 287
patella, 131
pellagra, 85
pelvis, 131
periodic table, 65
phalanges, 131
PKU disease, 252
pneumonia, 269
polio, 269
potassium, 83

porkworm, 268-9
protozoan diseases, 268-9
radius, 131
radon, 282
riboflavin, 84
rickets, 85
ringworm, 269
scapula, 131
spinal column, 131
spirilla, 268
sternum, 131

strep throat, 269
syphilis, 269
tapeworm, 268, 269
tarsals, 131
Tay-Sach's disease, 252
thiamin, 84
thymine, 230
tibia, 131
tobacco, 282, 283, 285
tranquilizers, 287
trichina (porkworm), 269

Turner's syndrome, 252
typhoid fever, 269
ulna, 131
uracil, 233
virus diseases, 268-9
vitamin B, 84
vitamin E, 84
vitamin K, 84
worm diseases, 268-9

Index 3) Entries mentioned only as potential research topics

Alzeimer's disease, 149
blood types, 118
botanist, 190
complete and incomplete
   protein, 94

E.O. Wilson Biodiversity
   Foundation, 12
epigenetics, 289
Half-Earth Project, 12, 24
mycorrhizae, 14

Parkinson's disease, 149
Shubin, Neil, 259
Tiktaalik, 259
transposon, 234
Wilson, E.O., 12

Index 4) Terms from an optional section of Unit 9, Lesson 3 on DNA's production of proteins

amylase, 229
base, 229
base pairs, 227, 229, 230
complementary bases, 230
genetic code, 230
messenger RNA
single-stranded, 230
sugar-phosphate, 23

Made in the USA
Columbia, SC
22 April 2022